Deutsche Gesellschaft für Kristallographie

25. Jahrestagung

27. bis 30. März 2017

Karlsruhe

Abstracts

Table of content

Plenary Talks

PL1
Atomic-scale Analysis of Energy Materials using Advanced Atom Probe Tomography

D. Raabe[1]
[1]Max-Planck-Institut für Eisenforschung GmbH, Düsseldorf, Germany

Recent progress in developing and using correlative methods for the joint analysis of complex materials using by Atom Probe Tomography (LEAP 3000, LEAP 5000) in conjunction with Electron Microscopy (SEM, TEM, STEM, Cs corrected Titan Themis) and Field Ion Microscopy is presented. Measurements are conducted on the same atom probe sample tips and in some cases atomic resolution is reached.

Examples from functional and structural energy-related materials are presented including segregation effects in multicrystalline silicon solar cells and their relation to cell efficiency, superalloys for advanced turbines, high strength steels and hydrides.

PL2
MicroED opens a new era for biological structure determination

T. Gonen[1]
[1]Janelia Research Campus, Howard Hughes Medical Institute, Laboratory of Molecular Electron Microscopy, Ashburn, United States

My laboratory studies the structures of membrane proteins that are important in maintaining homeostasis in the brain. Understanding structure (and hence function) requires scientists to build an atomic resolution map of every atom in the protein of interest, that is, an atomic structural model of the protein of interest captured in various functional states. In 2013 we unveiled the method MicroED, electron diffraction of microscopic crystals, and demonstrated that it is feasible to determine high-resolution protein structures by electron crystallography of three-dimensional crystals in an electron cryo-microscope (CryoEM). The CryoEM is used in diffraction mode for structural analysis of proteins of interest using vanishingly small crystals. The crystals are often a billion times smaller in volume than what is normally used for other structural biology methods like x-ray crystallography. In this seminar I will describe the basics of this method, from concept to data collection, analysis and structure determination, and illustrate how samples that were previously unattainable can now be studied by MicroED. I will conclude by highlighting how this new method is helping us understand major brain diseases like Parkinsons disease.

PL3
Magnetism in quasicrystals and their periodic approximants

A. Kreyssig[1]
[1]Ames Laboratory and Dept. of Physics and Astronomy, Iowa State University, Ames, United States

Quasicrystals consist in well-ordered atomic structures with point symmetries that are inconsistent with translational order. Since their discovery by Dan Shechtman in 1982, one key question is still unanswered, if antiferromagnetic order can develop in these compounds despite the geometric frustration caused by 3fold and 5fold rotational symmetries, and the lack of periodicity. Several theoretical studies have admitted the possibility of ordering of localized magnetic moments on quasilattices. Nevertheless, no quasi-antiferromagnets have been discovered to date. The recently discovered i-*R*-Cd quasicrystals (*R* = rare-earth elements) [1] and their periodic *R*Cd6 approximants with similar local structural motives offer a set of model systems for attaining a deeper understanding of how the magnetism evolves from a conventional lattice in the approximant phase to an aperiodic quasicrystal given the chemical simplicity associated with binary compounds and the well-defined local magnetic moments of rare-earth elements.

After a brief introduction, the talk will present our magnetization and scattering studies demonstrating complex long-range antiferromagnetic order in the periodic *R*Cd6 approximants [2, 3] and spin-glass like behavior in i-*R*-Cd quasicrystals [4, 5], and will discuss these results in context to earlier observations in more complex magnetic aperiodic materials.

References:
[1] A. I. Goldman *et al.*, Nature Mat. **12**, 717 (2013). [2] M. G. Kim *et al.*, Phys. Rev. B **85**, 134442 (2012). [3] A. Kreyssig *et al.*, Phil. Mag. Lett. **93**, 512 (2013). [4] T. Kong *et al.*, Phys. Rev. B **90**, 014424 (2014). [5] T. Yamada *et al.*, Phys. Rev. B **94**, 060103(R) (2016).

PL4
A Century after the Braggs – On Precision and Accuracy of Single Crystal X-Ray Results

K. Woźniak[1]
[1]University of Warsaw, Department of Chemistry, Warszawa, Poland

Although everything seems to be already well known in the field of routine structural single crystal X-ray analysis and more than 1.1 mln organic, inorganic and macromolecular structures have been solved and refined so far, even commonly used approaches in X-ray diffraction and models of electron density applied should be critically re-evaluated from time to time. It is incredible that the Independent Atom Model (IAM) of electron density, effectively introduced a century ago, is still the most common model of electron density used in structural analysis. One would even say that its success has dominated the whole field of X-ray diffraction for the past century and for years now plays, in my opinion, quite a negative role. When IAM was introduced, Max von Laue, the Braggs, and their colleagues, were using home-made pieces of equipment which could have hardly supplied qualitative information on diffraction spots. In consequence, the errors associated with the model of electron density used were overshadowed by far larger diffraction hardware errors. However, within the past century, there has been an overwhelming progress in design and production of X-ray hardware which is made for needs of both small laboratories and large scale facilities. This progress in sophisticated X-ray hardware should also accelerate progress in the quality and complexity of models of electron density used to interpret experimental results. The use of the 100 years old IAM effectively proves that even with the most modern scientific tools, one can step backward and do *ca.* 100 years old crystallography.

In my lecture, I will discuss precision and accuracy of single crystal X-ray results obtained for multiple measurements of single crystals of oxalic acid as a function of resolution of X-ray data and quality of electron density model applied (IAM, multipole model (MM), Hirshfeld Atom Refinement (HAR) and Transferable Aspherical Atom Model of electron density (TAAM)). I will present a detailed comparison of structural, thermal and electronic parameters obtained for the same multiple diffraction data sets collected for single crystals of oxalic acid when different models of electron density are refined against collected intensities of reflections complemented by comparison to single crystal neutron diffraction and theoretical results[1]. Some practical suggestions will be presented how to estimate and improve the quality of single crystal X-ray diffraction structural results. Among others with the newer models, one can obtain more precise and accurate information on positions of H-atoms [2] or energy of intermolecular interactions in crystals.

References

[1] R. Kamiński, S. Domagała, K. N. Jarzembska, A. A. Hoser, W. F. Sanjuan-Szklarz, M. J. Gutmann, A. Makal, M. Malińska, J. M. Bąk, K. Woźniak, *Statistical analysis of multipole-model-derived structural parameters and multipole-model-derived charge-density properties from high-resolution X-ray diffraction experiments,* Acta Crystallographica, A70 (2014) 72-91.

[2]M. Woinska, S. Grabowsky, P. M. Dominiak, K. Woźniak, D. Jayatilaka, *Hydrogen atoms can be located accurately and precisely by routine X-ray crystallography,* Science Adv., 2 No. 5 (2016) e1600192.

Acknowledgments

Polish National Science Centre MAESTRO grant decision number DEC2012/04/A/ST5/00609 is acknowledged.

Talks

MS01 - New approaches in structural biology

MS01-01
Serial Synchrotron Crystallography using a High Viscosity Extrusion injector device on beamline MASSIF-3 at the ESRF

D. von Stetten[1], M. Nanao[1], U. Zander[1,2], M. Weik[3], G. Schiro[3], J. P. Colletier[3], N. Coquelle[3,4], E. de la Mora[3], J. Woodhouse[3], R. B. Doak[5], B. Shoeman[5], M. Gruenbein[5], M. Kloos[5], P. Theveneau[1], F. Dobias[1], S. Aumonier[1], A. Royant[1,3], I. Schlichting[5], G. Leonard[1], C. Mueller-Dieckmann[1]
[1]European Synchrotron Radiation Facility, Structural Biology, Grenoble, France
[2]EMBL, Grenoble, France
[3]Institut de Biologie Structurale, Grenoble, France
[4]ILL, Grenoble, France
[5]Max Planck Institute for Medical Research, Heidelberg, Germany

Serial Femtosecond Crystallography (SFX) on XFEL sources has recently provided a lot of new results by collecting diffraction images from many protein microcrystals, using free-standing liquid jets, streams of a viscous medium, or solid mounts. Here we present the experimental setup and X-ray diffraction data recorded from protein microcrystals embedded in a grease matrix, using a High Viscosity Extrusion (HVE) injector [Botha et al., Acta Cryst D71, 387-397 (2015)] at the MASSIF-3 beamline at the ESRF synchrotron. Up to a million still images were recorded within half an hour using an Eiger 4M detector, so that even a moderate hit rate of a few percent easily yielded tens of thousands of diffraction patterns that could be successfully indexed and which were of sufficient quality for structure solution.

Synchrotron sources provide a pseudo-continuous X-ray beam which favors this kind of experiment, as no synchronization is needed to ensure that the X-ray beam hits the crystals embedded in the supporting medium. As a proof of concept, microcrystals of insulin were used in an injector experiment and gave diffraction data allowing the structure to be solved using molecular replacement. Furthermore, experimental phasing of gadolinium-derivatised lysozyme crystals was possible with only a few thousand diffraction still images.

While femtosecond resolution in time-resolved experiments certainly requires an XFEL source, many other experiments can and should be done at synchrotron sites. Also, the availability of such a serial crystallography setup on a more readily available synchrotron beamline allows the preparation and optimization of conditions for XFEL experiments.

MS01-02
Analysing structure and function of TRAP transporters by combining X-ray crystallography and pulsed EPR.

J. Glaenzer[1], M. F. Peter[1], G. H. Thomas[2], G. Hagelueken[1]
[1]University of Bonn, Bonn, Germany
[2]University of York, York, United Kingdom

The tripartite ATP-independent periplasmic (TRAP) transporters are a widespread class of membrane transporters in bacteria and archaea. As indicated by their name, TRAP transporters are independent of ATP hydrolysis and some representatives have been shown to rely on a Na+ gradient and membrane potential to power the transport mechanism. This is considered a reason why TRAPs are especially widespread in marine microorganisms. Molecules known to be transported by TRAP transporters range from small organic acids including C4-dicarboxylates, larger sugar acids like N-acetylneuraminic acid (Neu5Ac) to amino acids. Most TRAP transporter substrates contain a carboxylic acid group, which is specifically recognised by the SBP of the transporter.

A typical TRAP transporter consist of three structural domains: a high affinity substrate binding protein (SBP) and two trans-membrane domains (TMDs) with four and twelve predicted trans-membrane helices. So far, structural information is only available for the SBP domain, but biochemical experiments suggest a complex interplay between the three domains during transport events. Using a combination of X-ray crystallography and EPR, we have analysed the dynamics of substrate scavenging by the SBP and could show that the SBP only exists in two defined states in solution. Further, we have mutated functionally important residues of the SBP and ivestigated their impact on the dynamics of the SBP. Finally, a high-resolution crystal structure of the SBP in complex with a short peptide provides important new insight into the transport mechanism.

MS01-03
MeshAndCollect and more: Synchrotron Serial Crystallography at the ESRF

U. Zander[1]
[1]EMBL, Grenoble, France

Recent advances in Macromolecular Crystallography (MX) have given rise to serial crystallography methods on both XFELs and synchrotrons. By making use of highly brilliant microfocus synchrotron beamlines and fast readout direct photon counting detectors, the feasibility of collecting high-resolution data from many micrometer-sized crystals, either by injecting them into the beam or by mounting them statically on a sample holder has been successfully applied. However, processing data that has been collected in such a way is still a challenge. With MeshAndCollect we have developed a synchrotron serial crystallography method that allows protein crystallographers to collect almost automatically diffraction data [1]: a large number of microcrystals are mounted on a sample holder that is scanned with a low-intensity X-ray beam. The collected images are inspected automatically for protein diffraction patterns and diffracting positions are ranked. Subsequently, data wedges over a small rotation range (e.g. 10°) are collected at positions starting with those showing strongest diffraction. These wedges are integrated using standard software tools such as XDS or DIALS. Two complementary methods have been deployed to select partial datasets to be best merged: The first, named "ccCluster", uses hierarchical cluster analysis and is based on isomorphism represented in the correlation coefficient between datasets. The second one performs combination of datasets by genetic algorithm selection ("CODGAS" [2]) and optimizes data quality based on data metrics such as R-values and $\langle I/\sigma(I)\rangle$. The MeshAndCollect method has been successfully used on a diverse range of systems with crystal sizes down to 5 μm, including several test cases for SAD phasing and is deployed on all ESRF MX beamlines.

References:
[1] MeshAndCollect: an automated multi-crystal data-collection workflow for synchrotron macromolecular crystallography beamlines. U. Zander, G. Bourenkov, A.N. Popov, D. de Sanctis, O. Svensson, A.A. McCarthy, E. Round, V. Gordeliy, .C Mueller-Dieckmann and G. A. Leonard
Acta Cryst. (2015). D71, 2328-2343
[2] Merging of synchrotron serial crystallographic data by a genetic algorithm. U. Zander, M. Cianci, N. Foos, C. S. Silva, L. Mazzei, C. Zubieta, A. de Maria and M. H. Nanao
Acta Cryst. (2016). D72, 1026-1035

Fig.1 The MeshAndCollect workflow.

Fig.2 Schematic representation of the data merging progress by a genetic algorithm.

Figure 1

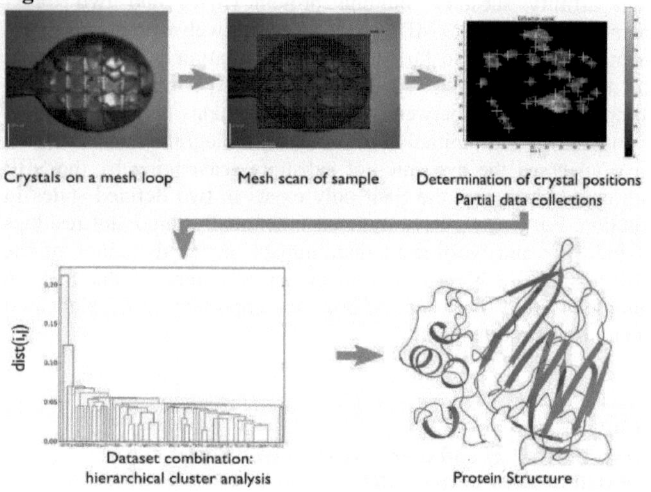

Figure 2

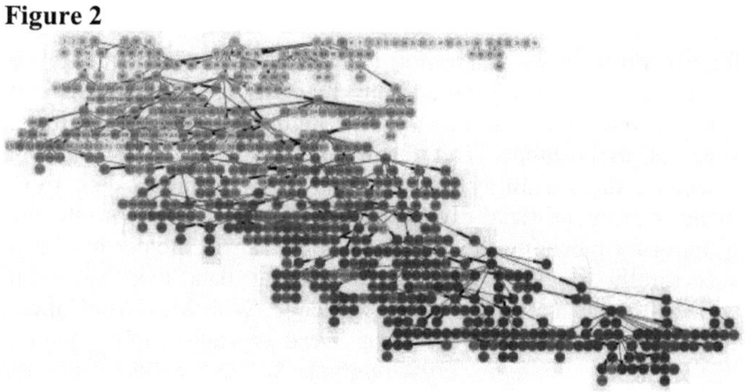

MS01-04

AUSPEX: a diagnostic tool for graphical X-ray data analysis

A. Thorn[1], J. Parkhurst[2], P. Emsley[3], R. A. Nicholls[3], M. Vollmar[2], G. Evans[2], G. Murshudov[3]
[1]Universität Hamburg, Centre for Ultrafast Imaging, Hamburg, Germany
[2]Diamond Light Source, Didcot, Oxfordshire, United Kingdom
[3]MRC Laboratory of Molecular Biology, Cambridge, United Kingdom

Contamination of macromolecular diffraction data recorded at cryogenic temperatures by diffraction from ice (*ice rings*) is a much bigger and more widespread problem than commonly recognized. Ice rings are notoriously difficult to account for in integration, and can significantly affect the quality of structural models. In extreme cases, it can even prevent structure solution. Furthermore, contamination with ice diffraction is hard to recognize from images produced by modern pixel detectors, and even harder to identify in already processed data.

In order to address this problem, a new software tool, AUSPEX, has been developed. AUSPEX facilitates ice ring detection, allowing for visual inspection of the observed intensity (or amplitude) distribution and automatic ice ring detection. AUSPEX can also be used to investigate the structure and distribution of errors within crystallographic data sets; this reveals a number of other pathologies in data processing and conversion.

Figure 1 *Plot of intensity vs resolution as generated by AUSPEX. Ice rings are clearly visble as a rise in the average intensitiy at a certain resolution. Grey bars show typical ranges for hexagonal ice diffraction.*

Figure 2 *Ice rings are a much more widespread problem than commonly recognized. Using plots as the one shown above, we found that 85 out of 400 data sets selected randomly from the PDB were contaminated with ice diffraction.*

Figure 1

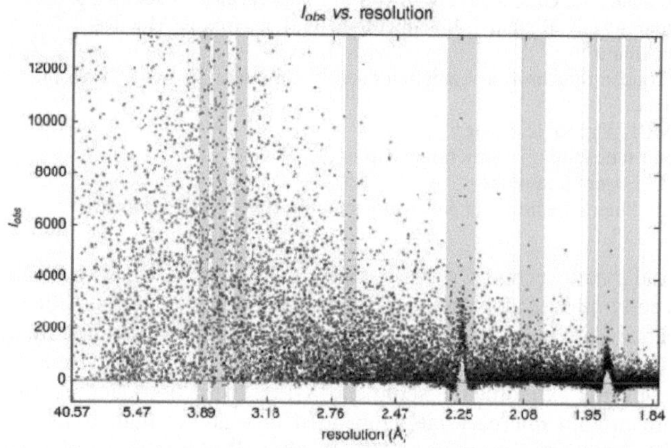

Figure 2

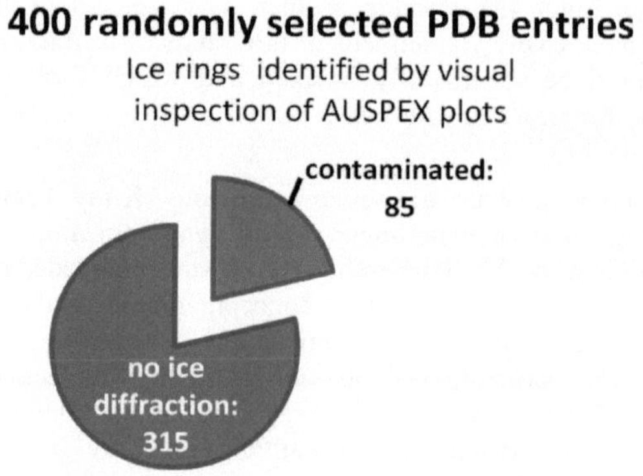

MS01-05

P11 at PETRA III: A Versatile Beamline for High-Throughput and Serial Crystallography

A. Burkhardt[1], S. Panneerselvam[1], O. Lorbeer[1], T. Pakendorf[1], B. Reime[1], J. Meyer[1], P. Fischer[1], N. Stuebe[1], P. Roedig[1], M. Warmer[1], A. Meents[1]
[1]DESY, Photon Science, Hamburg, Germany

P11 at PETRA III is dedicated to structural investigations of biological samples at different length scales. The beamline provides two state-of-the-art endstations: an X-ray microscope which is currently under construction [1] and a crystallography experiment which is operated between 5.5 and 30 keV [2]. In future, a third experimental endstation for serial crystallography and pump-probe experiments will be available.

Basis of beamline design was to make full use of the excellent source properties of PETRA III and to deliver most of the photons from the source into a very small focal spot at the sample position. This makes P11 ideally suited for structural investigations of microcrystals and serial crystallography experiments using liquid delivery systems [3] or Silicon microchips [4, 5].

The P11 optics concept involves the generation of a secondary source at 65.5 m using three dynamically bendable KB mirrors located in the P11 optics hutch. A second KB system which is installed further downstream can be used for refocusing the X-ray beam down to 4 × 9 μm² (v × h) with full flux from the source (2 × 10^{13} ph/s at 12 keV). Smaller beam sizes down to 1 × 1 μm² with more than 2 × 10^{11} ph/s in the focus can be obtained by slitting down the secondary source.

In addition, the crystallography endstation at P11 is equipped with a fast automatic sample changer and a large capacity storage Dewar which offers excellent conditions for high-throughput crystallography and fast crystal screening. Together with the high photon flux provided and the Pilatus 6M detector in place, a full data set at P11 can be typically collected in less than 2 min.

References:
[1] A. Meents *et al.*, Proc. SPIE 8851, 88510K (2014); [2] A. Burkhardt *et al.*, EPJ Plus 131:56 (2016); [3] F. Stellato *et al.*, IUCrJ 1, 204 (2014); [4] P. Roedig *et al.*, Sci. Rep. 5, 10451 (2015); [5] P. Roedig *et al.*, J. Appl. Cryst. 49, 968 (2016).

MS01-06
High-resolution 3D printed injectors for time resolved serial crystallography

M. Heymann[1]
[1]Max-Planck-Institute of Biochemistry, Martinsried, Germany

Reliable sample delivery is essential to biological imaging using X-ray Free Electron Lasers (XFELs). Continuous injection using the Gas Dynamic Virtual Nozzle (GDVN) has proven valuable, particularly for time-resolved studies. However, many important aspects of GDVN functionality have yet to be thoroughly understood and/or refined due to fabrication limitations. We report the application of 2-photon polymerization as a form of high-resolution 3D printing to fabricate high-fidelity GDVNs with submicron resolution. This technique allows rapid prototyping of a wide range of different types of nozzles from standard CAD drawings and optimization of crucial dimensions for optimal performance.

We prototyped a range of sample injectors, including stable sub-micron diameter jets operating at below 2 ul/min flow rates, sheet-jets, multiple flow focusing and mixing jets, as well as tailored lipidic cubic phase injectors. These sample injectors where validated in house and during multiple beamtimes at LCLS-CXI hutch at Stanford Linear Accelerator.

MS02 - Structural chemistry under non-ambient conditions

MS02-01
Novel nitrogen-rich iron nitrides synthesized at high-pressure high-temperature conditions

M. Bykov[1], E. Bykova[1,2], G. Aprilis[1,3], K. Glazyrin[2], I. Chuvashova[1,3], E. Koemets[1], L. Dubrovinsky[1]
[1]University of Bayreuth, Bayerisches Geoinstitut, Bayreuth, Germany
[2]DESY, FS-PE, Hamburg, Germany
[3]University of Bayreuth, Laboratory of Crystallography, Hamburg, Germany

Transition metal nitrides have many important applications in modern technology and science. In particular, iron nitrides occupy an important role in the development of strong permanent magnets [1]. Furthermore, the significant amount of nitrogen has entered the Earth"s iron-containing core during the accretion [2]. Therefore, studying of Fe-N alloys may give a clue to the processes in the deep Earth interiors [3]. Iron nitrides have a very complex phase diagram, and can exist in many different phases such as α''-$Fe_{16}N_2$, γ'-Fe_4N, ε-Fe_3N, ζ-Fe_2N, and FeN [4]. Among them, the most nitrogen-rich compound FeN has only been synthesized in the form of thin films with ZnS and NaCl structure types. Therefore, a number properties of FeN (e.g., the bulk modulus) remain unverified since single crystals are not available. To our knowledge no experimental studies have so far dealt with the iron nitrides with N:Fe ratio > 1. However, several theoretical calculations have predicted spinel-type Fe_3N_4 and iron pernitride FeN_2 [5]. Application of high pressure is known to be able to stabilize phases with unusual compositions and stoichiometries due to the significant role of the *PV* term in the expansion of the Gibbs energy.

We have studied the chemical reactions between Fe and N_2 in a diamond anvil cell in the pressure range up to 135 GPa and temperatures up to ~2700 K. For this study we have used the newly-developed double-sided laser-heating system, which allows simultaneous laser-heating of the sample and single-crystal X-ray diffraction data collection. By using this system installed at the beamline P02.2 at DESY we have synthesized four iron nitrides, which were never described before. Namely these are FeN_4, FeN_2, Fe_3N_2 and B8-type FeN. We also have solved and refined crystal structures, that they possess at high pressures and high temperatures. We have shown that increasing pressure leads to the formation of iron nitrides with higher nitrogen content. In this contribution the details of the experiments, crystal structures and relations between them will be discussed.

References:
[1] Y. Jiang et al., Sci. Rep. **6**, 25436 (2016)
[2] B. Johnson et al., Earth-Science Rev. **148**, 150 (2015)
[3] S. Minobe et al., Geophys. Res. Lett. **42**, 5206 (2015)
[4] W. Y. Ching et al., Appl. Phys. Lett. **80**, 2904 (2002)
[5] M. Wessel et al., Chem. - A Eur. J. **17**, 2598 (2011)

MS02-02
Synthesis of binary hafnium oxides at high pressures and high temperatures

L. Bayarjargal[1], W. Morgenroth[1], N. Schrodt[1], B. Winkler[1], V. Milman[2], C. Stanek[3], B. P. Uberuaga[3]
[1]Goethe-Universität, Institut für Geowissenschaften, Abt. Kristallographie, Frankfurt am Main, Germany
[2]Dassault Systemes, BIOVIA, Cambridge, United Kingdom
[3]Los Alamos National Laboratory, Materials Science and Technology Division, Los Alamos, United States

The system Hf-O has been extensively studied, as hafnia, HfO_2, has outstanding physical properties which have led to its wide-spread application [1]. The useful properties of hafnia have prompted the search for further compounds beyond the known suboxides and HfO_2 in this binary system, especially as theoretical studies [2-3] imply that further stoichiometric binary compounds may be obtained at high pressures. In addition to hafnia, another suboxide (Hf_6O) have been synthesized in a previous study [4]. Up to 20 at% oxygen can be dissolved in metallic hafnium where, at low temperatures, the oxygen atoms are ordered over the octahedral sites in the hcp structure of α-Hf [4]. Theoretical studies [5, 6] conclude that Hf_6O, Hf_3O and Hf_2O are in fact stoichiometric phases, which are stable below 300-700 K and at ambient pressure.

The aim of this study is to synthesize new HfxOy compounds from Hf and HfO_2 at high pressures and temperatures. We loaded diamond anvil cells with KCl, Hf and HfO2 and performed powder diffraction experiments at the Extreme Conditions Beamline P02.2 at PETRAIII. The sample was heated using a laser up to 3000 K between 12 and 34 GPa. After laser heating experiments two new binary phases have been synthesized additionally to all known high pressure phases of Hf and HfO2. Density functional theory calculations have been employed to explore the new structures. The first new phase has a chemical composition of Hf_8O_7 and a tetragonal structure [7]. This structure is based on a fcc Hf packing with oxygen atoms occupying octahedral interstitial positions. The second phase has a composition close to Hf_6O, where oxygen atoms occupy octahedral interstitial sites in a hcp Hf packing.

Authors acknowledge financial support from DFG, BMBF (05K13RF1) and from Los Alamos National Laboratory, USA. Parts of the research were carried out at the light source PETRA III at DESY.

References:
[1] J. Wang et al., J. Mat. Sci. 27 (1992) 5397-5430 and
[2] K.-H. Xue et al., Phys. Rev. Lett. 110 (6) (2013) 065502.
[3] J. Zhang et al. Phys. Rev. B 92 (2015) 184104.
[4] M. Hirabayashi et al. J. Phys. Soc. Jpn. 35 (2) (1973) 473-481.
[5] A. V. Ruban et al., Physical Review B 82 (2010) 134110.
[6] B. P. Burton and A. van de Walle, Calphad 37 (2012) 151-157.
[7] L. Bayarjargal et al., High Pressure Research (2017) accepted.

MS02-03
High-pressure synthesis and crystal structure of a new orthorhombic modification of Ir_4B_5

P. Benedikt[1], N. Christopher[2], W. Björn[2], H. Hubert[1]
[1]University Innsbruck, Inst. of General, Inorganic and Theoretical Chemistry, Innsbruck, Austria
[2]University of Frankfurt, Inst. für Geowissenschaften, Abteilung für Kristallographie, Frankfurt, Germany

Over the last few years, the class of transition-metal borides attracted major interest within the scientific community due to some outstanding physical properties such as a high hardness, extremely low compressibility and superconductivity [1, 2]. Up to now, most syntheses were carried out under ambient conditions and only very few transition metal borides were synthesized under high-pressure conditions [1, 3].

One focus lies on the synthesis and characterisation of new iridium borides under high-pressure/high-temperature conditions. The syntheses were carried out with a high-pressure device consisting of a hydraulic 1000 t press and a Walker-type module with eight tungsten carbide cubes.

The structure determination revealed that we were able to synthesize a new iridium-boride β-Ir4B5, which crystallizes in the orthorhombic space group *Pnma* (No. 62) with the parameters a = 1077.2(2), b = 284.4(1), and c = 605.2(2) pm, with $R1$ = 0.0286, $wR2$ = 0.0642(all data) and Z = 2. The structure can be compared to the closely related ambient pressure phase α-Ir4B5 (formerly IrB1.35). Both structures can be described as a stacking of alternating puckered boron and iridium layers in the c direction but the stacking sequence itself varies. Whereas in α-Ir4B5 the stacking is without any displacements (ABAB), β-Ir4B5 is built up by an alternating stacking sequence ABA"B" of boron and iridium layers [4]. The new phase β-Ir4B5 crystallizes in a new structure type that contains, amongst other building units, unique boron ribbons (Figure 1).

In summary, we were successful in synthesizing a new modification of Ir4B5 crystallizing in a new structure type in the orthorhombic space group *Pnma*. It furthermore represents the first 5d transition metal boride synthesized under high-pressure conditions.

The FWF and the DFG are gratefully acknowledged for funding in the framework of an ERA-chemistry project (I 1636-N19 and WI-1232-401-1).

Figure 1: Crystal structure of β-Ir4B5

References:
[1] H.-Y. Chung, M.B. Weinberger, J.B. Levine, A. Kavner, J.-M. Yang, S.H. Tolbert, R.B. Kaner, *Science* **2006**, 316, 436-439.
[2] J. Nagamatsu, N. Nakagawa, T. Muranaka, Y. Zenitani, J. Akimitsu, *Nature* **2001**, 410, 63-64.
[3] E. Bykova, H. Gou, M. Bykov, M. Hanfland, L. Dubrovinsky, N. Dubrovinskaia, *J. Solid State Chem.* **2015**, 230, 102-109.
[4] T. Lundstrom, L.-E. Tergenius, *Acta Chem. Scand.* **1973**, 27, 3705-3711.

Figure 1

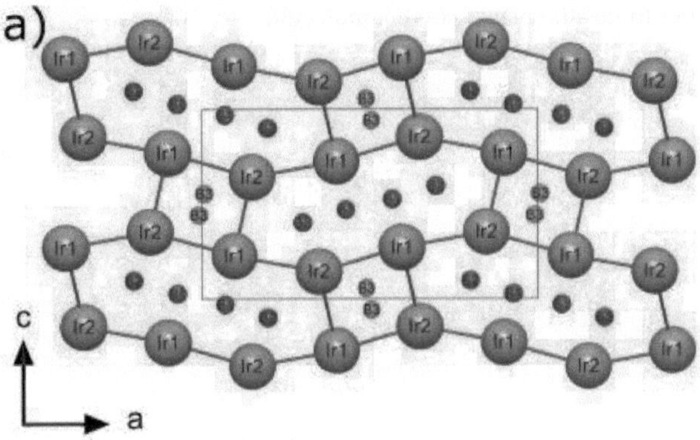

MS02-04
The photo-induced and temperature reversible single-crystal-to-single-crystal transformation of a two-dimensional polymer

G. Hofer[1,2], F. Grieder[1,2], A. D. Schlüter[2], T. Weber[1]
[1]ETH Zurich, Department of Materials, X-ray Platform, Zürich, Switzerland
[2]ETH Zurich, Department of Materials, Polymer Chemistry, Zürich, Switzerland

Recently, polymer chemists further developed the well-known concept of single-crystalline photo-dimerization [1] to create a two-dimensional (2D) network in the single-crystalline state: a 2D polymer [2,3,4,5]. To achieve this, the monomer is crystallized in a layered arrangement. Irradiation of the obtained crystals with light causes the monomer to form covalent bonds between adjacent monomer molecules, resulting in a macro molecular 2D-polymer single crystal. Heat treatment reverses the bond formation and the original monomer crystal is recovered. The polymerization and depolymerization can be paused any time by removing the crystals from the triggering source, allowing for detailed ex-situ structural investigations for any achieved conversion. This provides highly valuable insights into the structural changes associated with the conversion from monomer to polymer crystal and back.

The focus of our study is on the 2D polymer developed by Kory et al [3]. We address models obtained for monomer, intermediate and polymer structures of this isosymmetrical, but non-isostructural phase transformation. In-house and synchrotron X-ray diffraction was employed to study the reaction kinetics and to follow structural changes of the polymerization and depolymerization reaction in detail. Structure refinements revealed the space group to be $R3$ for all crystals. Upon polymerization, the lattice parameters at 100 K significantly increase from a = 19.3280(1) Å, c = 28.858(1) Å to a = 19.5350(3) Å, c = 29.0439(9) Å. The unit-cell parameters show a hysteresis-like behavior upon polymerization and depolymerization. The aim of this study is to understand the structural changes during the forth and back reaction to get insight into fundamental properties of 2D polymerization. Of additional interest is the polar solvent molecule, which partially changes orientation during polymerization and therefore the crystal structure"s dielectric momentum.

References:
[1] Wegner, G. (1971). *Makromol. Chem.* **145**, 85–94.
[2] Payamyar, P., et al. (2016). *Chem Commun.* **52**, 18–34.
[3] Kissel, P., et al. (2014). *Nat. Chem.* **6**, 774–778.
[4] Kory, M. J., et al. (2014). *Nat. Chem.* **6**, 779–784.
[5] Lange, R. Z., et al. (2017). Submitted to *J. Am. Chem. Soc.*

Figure 1: Monomer and polymer single layer. Red, blue, green: monomer molecule, black: solvent molecule.

Figure 1

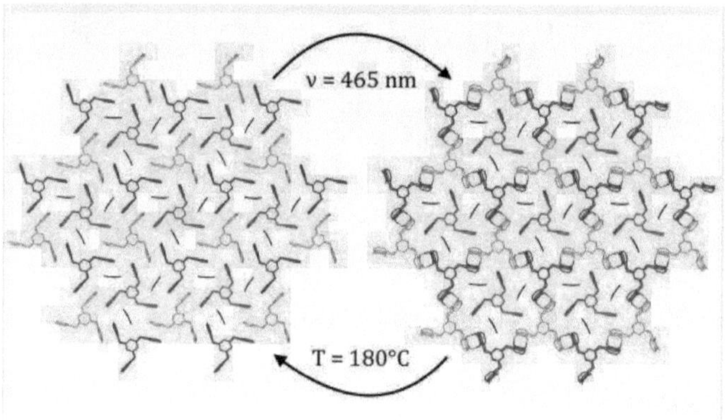

MS02-05
Diffusion Pathways in Ion Conductors: Making the Most of Neutron-Diffraction Data

D. Wiedemann[1]
[1]Technische Universität Berlin, Institut für Chemie, Berlin, Germany

Sensors, electrolytes, energy storage systems: the fields of application for solid-state ion conductors are diverse. Scientific, as well as economic, interest in them is still on the rise. High-temperature neutron diffraction gives access to averaged positions of the atomic nuclei and their precise displacements, even when dealing with notoriously elusive chemical species like lithium ions. Notably, the high cost and ethical implications of such experiments, which are predominantly conducted at nuclear research reactors, demand the best possible evaluation (*cf.* Fig. 1).

If they are of adequate quality, data acquired from single-crystal or powder diffraction permit modeling of anharmonic thermally activated displacement. This enables the visualization of diffusion pathways and the determination of associated migration barriers *via* evaluation of probability-density function (PDF, see Fig. 2) and effective one-particle potential (OPP).[1]

But what about low-quality datasets or model failure? In these cases, an examination of the scattering-length density reconstructed *via* maximum-entropy methods (MEM) may at least yield semiquantitative results.[2] As a supplement, heuristic means, or with difficult data, topological analyses are helpful. Within them, the framework of static species is searched for voids that are accessible for mobile ions. Representations of the procrystal-void surface provide quick insight into possible pathways, whereas Voronoi–Dirichlet partitioning (VDP) allows classifying them according to their suitability for certain ions.[3]

References:
[1] D. Wiedemann, M. M. Islam, S. Nakhal, A. Senyshyn, T. Bredow, M. Lerch, *J. Phys. Chem. C* **2015**, *119*, 11370–11381; D. Wiedemann, S. Nakhal, J. Rahn, E. Witt, M. M. Islam, S. Zander, P. Heitjans, H. Schmidt, T. Bredow, M. Wilkening, M. Lerch, *Chem. Mater.* **2016**, *28*, 915–924.
[2] D. Wiedemann, S. Indris, M. Meven, B. Pedersen, H. Boysen, R. Uecker, P. Heitjans, M. Lerch, *Z. Kristallogr. – Cryst. Mater.* **2016**, *231*, 189–193.
[3] S. Nakhal, D. Wiedemann, B. Stanje, O. Dolotko, M. Wilkening, M. Lerch, *J. Solid State Chem.* **2016**, *238*, 60–67.

Fig. 1. Methods for evaluation of neutron-diffraction data with different quality demands.

Fig. 2. Example for PDF analysis: optimal Li^+ diffusion path in γ-$LiAlO_2$ at 1202 °C (light gray: PDF isosurface; Max: bottleneck).

Figure 1

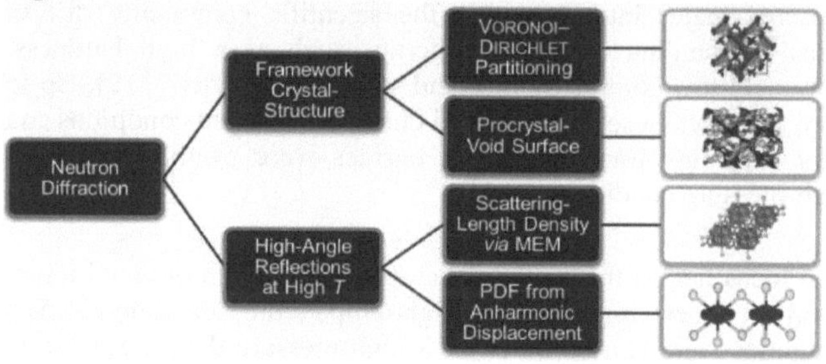

Figure 2

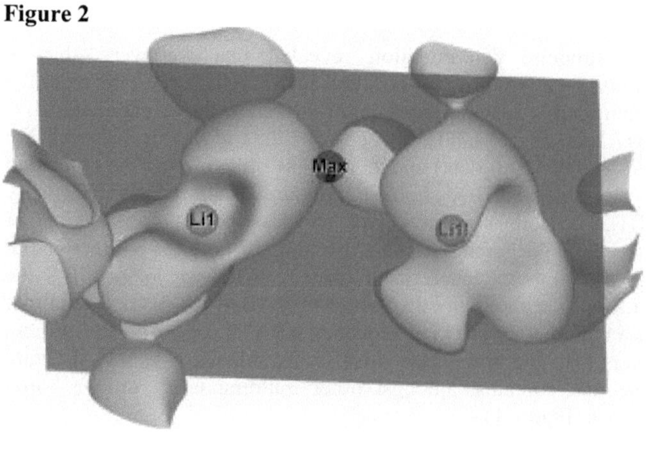

MS02-06
Thermoelectric properties and atom mobility in argyrodite-type $Cu_8GeSe_4Te_2$

S. Schwarzmüller[1], D. Souchay[1], D. Günther[1], A. Gocke[1], I. Dovgaliuk[2], O. Oeckler[1]
[1]Leipzig University, Faculty of Chemistry and Mineralogy, IMKM, Leipzig, Germany
[2]ESRF, The European Synchrotron, Grenoble, France

Some complex chalcogenides such as Cu_8GeSe_6 are good ionic conductors. Mobile atoms are also favorable for thermoelectric materials; however sulfides and selenides often exhibit low electrical conductivity and phase transitions (at 50 °C for Cu8GeSe6).[1] These may lead to material degradation and thus limit the perspective for potential applications. The present study aims at stabilizing cubic structures of Cu8GeSe6 and suppressing phase transitions by partial substitution of S or Se by Te.

Starting from the elements millimeter-sized single crystals of $Cu_8GeSe_4Te_2$ were obtained in sealed silica glass tubes. Diffraction data were measured at the Swiss-Norwegian beamline (SNBL) at the ESRF (Grenoble) under argon atmosphere up to 550 °C. Physical properties were determined using polycrystalline samples.

$Cu_8GeSe_4Te_2$ crystalizes in a cubic argyrodite-type structure (space group *F*-43*m*, *a* = 10.43 Å) derived from that of Cu_8GeSe_6 (space group *P*6_3*cm*) by partially substituting Se atoms with Te. Similar to substitutions with halogen atoms,[2] the introduction of a second chalcogen stabilizes the cubic argyrodite type. Single crystal structure refinements show rather large prolate displacement ellipsoids for Cu atoms already at room temperature compared to Cu_8GeSe_6, indicating that $Cu_8GeSe_4Te_2$ is a potential Cu ion conductor. Cu positions exhibit severe disorder that changes with temperature, when ionic conductivity is enhanced. Whereas Cu_8GeSe_6 features low thermal and electrical conductivity (0.3 W m^{-1} K^{-1} and 0.03 S cm^{-1}, respectively, from 50°C to 400 °C) with a Seebeck coefficient between 65 and 180 μV K^{-1}, $Cu_8GeSe_4Te_2$ is a promising thermoelectric material with higher electrical conductivity (220 – 70 S cm^{-1}) and better Seebeck coefficient (100 – 210 μV K^{-1} from 50 to 400 °C). The selenide telluride reaches ZT values up to 0.45 at 400°C where it outperforms phonon-liquid electron-crystal (PLEC) materials such as Cu_7PSe_6.[3] Thus, $Cu_8GeSe_4Te_2$ is an intriguing thermoelectric material, which is stable for applications up to 550 °C.

References:

[1] M. Onoda, M. Ishii, P. Pattison, K. Shibata, A. Yamamoto, G. Chapuis, *J. Solid. State. Chem.* **1999**, *146*, 355.
[2] T. Nilges, A. Pfitzner, *Z. Kristallogr.* **2005**, *220*, 281.
[3] K. S. Weldert, W. G. Zeier, T. W. Day, M. Panthöfer, G. J. Snyder, W. Tremel, *JACS* **2014**, *136*, 12035.

MS03 - Materials Science

MS03-01
Structure – property relationship in spincrossover nanocomposites

D. Schaniel[1], P. Durand[1], S. Pillet[1], B. Fournier[1], K. Rassoul[1], E. E. Bendeif[1]
[1]Universite de Lorraine, CRM2, Vandoeuvre les Nancy, France

Recently we reported on the elaboration and characterisation of a spin-crossover nanocomposite exhibiting an extremely wide magnetic hysteresis loop around room temperature [1]. These nanocomposites were obtained by impregnating silica host matrices with the spin-crossover compound $[Fe(Htrz)_2(trz)]BF_4$. The host matrices offer well defined and monodisperse pore sizes of 6 and 20 nm leading to the growth of $[Fe(Htrz)_2(trz)]BF_4$ nanoparticles of corresponding diameter inside the porous silica network. The magnetic properties such as the transition temperature, the hysteresis width, the abruptness of the spin transition and the residual high-spin or low spin fractions depend on the particle size and differ from the bulk reference material.

In order to elucidate in more detail the interaction between molecular guest and inorganic matrix arising from the temperature-induced structural changes in the guest molecules we followed the thermal spin transition in $[Fe(Htrz)_2(trz)]BF_4$ nanocomposites using total X-ray scattering. Modelling the pair distribution function (PDF) allowed to identify the contribution of the high- and low-spin phases as a function of temperature using a methodology based on the differential PDF analysis coupled to a singular value decomposition procedure. In this manner, a one-to-one correspondence between the structural analysis and the magnetic properties could be established.

References:
[1] Durand, P; Pillet, S.; Bendeif E.-E.; Carteret, C.; Bouazaoui M.; El Hamzaoui, H.; Capoen, B.; Salmon, L.; Hébert, S.; Ghanbaja, J.; Aranda, L.; Schaniel, D. J. Mater. Chem. C 2013, 1, 1933

MS03-02
Characterization of domain wall dynamics in lead zirconate titanate using X-ray photon correlation spectroscopy

S. Gorfman[1], A. Bokov[2], A. Davtyan[3], Z. G. Ye[2], A. Zozulya[4], U. Pietsch[3], M. Sprung[4], C. Gutt[3]
[1]Institute of Crystallography / University of Freiburg, Freiburg, Germany
[2]Simon Fraser University, Department of Chemistry and 4D labs, Burnaby, Canada
[3]University of Siegen, Physics, Siegen, Germany
[4]Deutsches Elektronen-Synchrotron DESY, Hamburg, Germany

Complex domain patterns in ferroelectrics and multi-ferroics may be easily rearranged by external influence of thermal, electrical or mechanical nature. Such rearrangements often greatly contribute to the remarkable physical properties of ferroic materials – either of pyroelectric, dielectric, piezoelectric or elastic ones. Direct observation of complex domain patterns is often challenging. The existing techniques (e.g. optical birefringence, electron / confocal Raman or piezoresponse force microscopies) do not bring any unambiguous conclusions about the physical laws governing domain wall motion.

We present an X-ray Photon Correlation Spectroscopy (XPCS) [1] experiment for the investigation of temperature-driven dynamics of notoriously complex domain microstructures in a single crystal of PbZr0.54Ti0.46O3 (PZT) close to the morphotropic phase boundary. XPCS - the recent coherent synchrotron diffraction technique - offers an opportunity to measure the length-scale specific and the motion specific information: it involves a collection of speckles - extremely detailed two-dimensional patterns of scattering of a coherent synchrotron X-ray beam at a sub-micrometre domain patterns. It is assumed that the correlations between different speckle patterns reflects the correlation between the corresponding real space domain patterns.

We collected the set of ~11 000 speckle patterns in the diffraction geometry and at the P10@PETRAIII beamline, as the temperature of the crystal was gradually cooling down from 650 K and 350 K, passing through the phases of tetragonal (>600 K) and monoclinic (< 550 K) symmetry [2]. We introduce original statistical analysis of the correlation coefficients, which shows that domain wall motions in the monoclinic phase can be described as a superposition of overlapping modes, where each mode develops as the series of countable domain wall movements (jerks).

References:
[1] O. G. Shpyrko. X-ray photon correlation spectroscopy. J. Synchrotron Rad, 21, 1057 (2014)
[2] A. A. Bokov, X. Long, and Z.-G. Ye Optically isotropic and monoclinic ferroelectric phases in PbZr/TiO3(PZT) single crystals near morphotropic phase boundary. Phys. Rev. B 81, 172103 (2010).

Figure 1

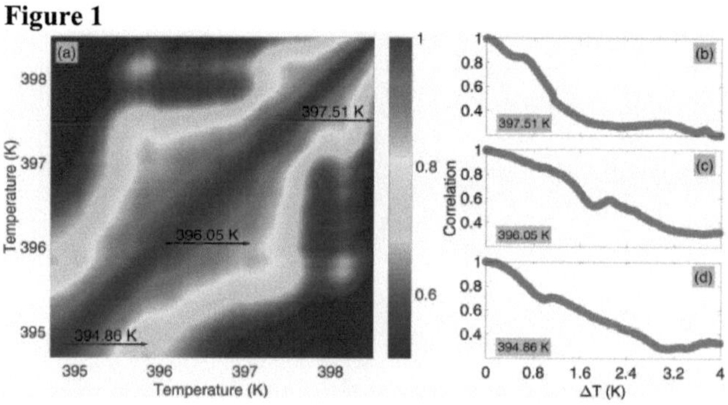

MS03-03
Correlated Three-Dimensional Imaging of Dislocations: Insight into the Onset of Thermal Slip in Semiconductor Wafers

D. Hänschke[1], A. Danilewsky[2], E. Hamann[1], S. Bode[3], L. Helfen[1,4], T. Baumbach[1,3]
[1]Karlsruhe Institute of Technology (KIT), Institute for Photon Science and Synchrotron Radiation (IPS), Eggenstein-Leopoldshafen, Germany
[2]University Freiburg, Crystallography, Institute for Geosciences, Freiburg, Germany
[3]Karlsruhe Institute of Technology (KIT), Laboratory for Applications of Synchrotron Radiation (LAS), Karlsruhe, Germany
[4]European Synchrotron Radiation Facility (ESRF), Grenoble, France

During thermal processing of semiconductor wafers any microscopic flaw at the surface may trigger nucleation and expansion of dislocations. These can rapidly evolve into so-called slip bands, extending considerably into initially defect-free substrate areas and entailing large-scale warpage and atomic steps at the surface. An understanding of the origin of slip bands and the associated plastic deformation, as well as predicting the evolution with suitable theoretical models is both of scientific interest and industrial concern.

We report on a novel correlative approach for non-destructive three-dimensional (3D) imaging of dislocation networks in almost perfect crystals even with considerable lateral extension and its application to the investigation of the early slip band formation in Si wafers.

Three methods are combined: X-ray diffraction laminography (XDL), X-ray white-beam topography (XWBT), and circular differential interference contrast microscopy (CDIC). The XDL scan geometry is particularly well suited to image the 3D defect arrangement in laterally extended specimen like wafer with a few microns spatial resolution. Based on this information, we may identify individual defects in an XWBT dataset by means of forward simulation. This enables the determination of Burgers vectors orientations even for more complex defect arrangements. Finally, optical microscopy links each dislocation to surface steps created by moving dislocation end points during defect propagation.

We show the results for the characterization of dislocation networks induced into Si wafers after controlled mechanical damaging and subsequent thermal annealing. In the context of surficial features, the 3D paths of dislocations inside complex arrangements and the associated Burgers vector distribution are determined, precisely. We relate our observations to theoretical predictions of the driving forces and identify the mobilization and multiplication of pre-existing dislocations as the dominating mechanism during thermally induced plastic deformation. The undisturbed long-range emission of dislocations from regenerative sources is discovered.

Concluding, we demonstrate the capabilities of the proposed approach and provide new insight into the early slip band formation in thermally treated Si wafers.

MS03-04
Assessment of solid electrolytes by Voronoi-Dirichlet partitioning — Crystallographic, energetic, and economic parameters for Na and Al conductors

F. Meutzner[1], W. Münchgesang[1], T. Nestler[1], M. Zschornak[1], R. Schmid[1], A. A. Kabanov[2], M. Ureña de Vivanco[1], A. P. Shevchenko[2], T. Leisegang[1,2], V. A. Blatov[2], D. C. Meyer[1]
[1]TU Bergakademie Freiberg, Institut für Experimentelle Physik, Freiberg, Germany
[2]Samara National Research University, Samara, Russian Federation

The possibility to store electrical energy enables not only mobile electronics and electric cars, but will be of paramount importance for the advancing penetration of renewables in the energy grid. Due to the capricious nature of sustainable energy carriers, in times of high production, energy will be stored for times of low or no production in order to balance out offer and demand. Especially Na-based chemistries have shown high promise and are increasingly researched. Other elements could be imagined as well, e.g. Al, since it is highly abundant, technologically well-accessible, and offers a high theoretical energy density.

From a crystallographic point of view, diffusion or ionic conduction in crystalline oxygen-containing materials can be described as jumps of a mobile ion between interconnected crystallographic sites in oxygen-coordinated voids. Based on this theory, Voronoi-Dirichlet partitioning can quickly determine structure-immanent void-systems of large crystal structure databases (e.g. the ICSD). By standardising the approach using already well-known Na ion conductors, the method becomes a high-throughput identification tool. The results can be filtered using a range of crystallographic and economic parameters to point towards the highest-promise materials.

The algorithm can be extended by applying bond-valence energy calculations to the most-promising materials and further DFT simulations. The channel that offers the smallest energy changes to an ion will be the conduction path and determines the global electrochemical parameters of the compound.

This work presents the results of the screening for Na ion conductors, incl. high-potential candidates and the first results of the identification of promising oxidic Al-ion conductors involving all 3 approaches and their comparison.

This work is funded by the German BMBF (CryPhysConcept: 03EK3029A; R2R Battery: 03SF0542A), BMUB (BaSta: 0325563B), the Russian Megagrant (14.B25.31.0005), and the RSF (16-13-10158).

MS03-05
In-situ electric field dependent synchrotron x-ray diffraction study characterizing extrinsic and intrinsic contributions to the electromechanical response of high performance piezoceramic (1-*x*)$PbTiO_3$-(*x*)$BiScO_3$

L. Kodumudi Venkataraman[1], D. K. Khatua[2], C. M. Fancher[3], J. L.Jones[3], R. Ranjan[2]
[1]Technische Universität Darmstadt, Institute of Materials Science, Darmstadt, Germany
[2]Indian Institute of Science, Department of Materials Engineering, Bangalore, India
[3]North Carolina State University, Department of Materials Science and Engineering, Raleigh, United States

Ferroelectric systems exhibiting a morphotropic phase boundary (MPB), wherein two ferroelectric phases coexist in equilibrium, have been studied extensively because of their enhanced electromechanical properties. The two main theories proposed to explain the origin of electromechanical response at the MPB are (i) Polarization rotation within the unit cell and (ii) formation of a high density of low energy domain walls. These theories are essentially based on the intrinsic (lattice) and extrinsic (mainly from displacement of domain walls) contributions to the electromechanical response. Advances in structural characterization in the past two decades have led to several reports on the structure and phase transformations at the MPB using *in-situ* and *ex-situ* diffraction techniques. However, a comprehensive study of the intrinsic and extrinsic contributions using structural studies has eluded researchers due to the structural complexity arising from overlapping Bragg profiles of the constituent phases at the MPB.

The MPB based solid solution (1-*x*)$PbTiO_3$-(*x*)$BiScO_3$ is a high temperature, high performance (T_c=450°C and d_{33}=460pC/N) piezoelectric material. In the present study, we have investigated this system using *in-situ* electric field dependent high energy synchrotron x-ray diffraction. The lattice strains, domain switching fractions and field induced phase transformation were characterized for a non-MPB and a MPB composition. The MPB composition exhibits reduced domain switching and lattice strain in comparison with the non-MPB composition and is shown to have additional contributions from field induced phase transformation. The results signify that the origin of large piezoelectric response at the MPB compositions is most likely related to the field induced phase transformations.

MS03-06

Simulation of beryllium crystal under hydrogen pressure

M. Selzer[1], D. Schneider[1], B. Nestler[1]
[1]Karlsruhe Institut of Technology, Institut of Applied Materials, Karlsruhe, Germany

In the effective storage of hydrogen beryllium is a material with good properties. One of the failure mechanisms in materials is the nucleation and growth of micropores. The shape of micropores in thermodynamic equilibrium depends on the lattice anisotropy of the particular material. The surface energy contribution of the particular symmetry planes correspondingly leads to the unique equilibrium shapes of these micropores. In this work we investigate the resulting equilibrium shapes of micropores for a hexagonal lattice using the phase-field method. By varying the interfacial energies of the particular symmetry plane our results show a change in the shape of micropores.

Figure 1

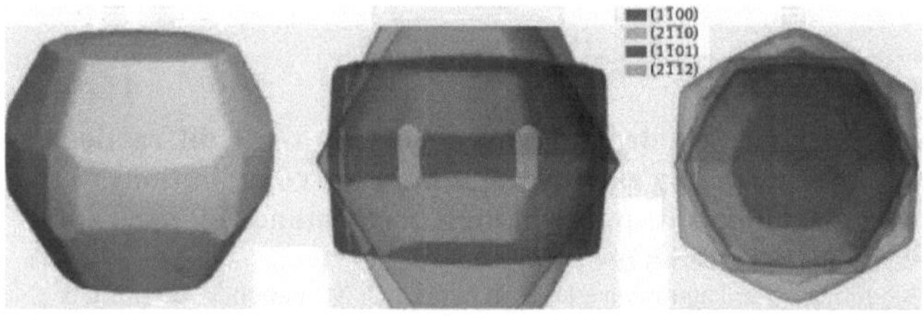

MS04 - Crystal physics

MS04-01
Strontium titanate in the electric field: pyroelectric, piezoelectric and tetragonal

H. Stöcker[1], J. Hanzig[1], E. Mehner[1], S. Jachalke[1], C. Richter[1], M. Zschornak[1], B. Khanbabaee[2], S. Gorfman[2], D. C. Meyer[1]
[1]TU Bergakademie Freiberg, Institute of Experimental Physics, Freiberg, Germany
[2]Universität Siegen, Department of Physics, Siegen, Germany

Pure strontium titanate exhibits a cubic perovskite-type structure at room temperature, but this changes in an external electric field of 1 MV/m. In situ X-ray diffraction shows an increase of the out-of-plane lattice parameter at the positive electrode. Since oxygen vacancies are the most prominent and most mobile defect species in strontium titanate, they play a crucial role also for these structural changes. Their migration away from the positive electrode leads to the formation of a migration-induced field-stabilized polar (MFP) phase. As the name suggests, it can only be observed in the electric field and disappears instantly otherwise. This severely complicates studying the supposed polar properties. Nevertheless, it was possible to prove pyroelectricity and piezoelectricity by a modified Sharp-Garn technique and stroboscopic in situ X-ray diffraction, respectively. Furthermore, the tetragonal structure of this phase, present only in a layer near the anode, was solved by a new resonant diffraction technique.

Measurements of thermally stimulated and pyroelectric currents reveal a pyroelectric coefficient of approx. 30 μC/m²K. From symmetry considerations, space group *P*4*mm* can be derived for the MFP phase of $SrTiO_3$. The piezoelectricity of this phase is shown by stroboscopic time-resolved X-ray diffraction under alternating electric field yielding effective piezoelectric coefficients between 60 and 100 pC/N. Direct structural evidence of the polar character of the new phase was obtained using a new technique of X-ray crystal structure analysis. It is based on targeted suppression of selected Bragg reflections by tuning the resonant scattering power of atoms through X-ray energy variation. Under this condition, the reflection is very sensitive to slight displacements of atoms involved in destructive interference. The obtained tetragonal structure of the MFP phase corresponds to that of $BaTiO_3$ with displacements of Ti and O by -1.3 pm and +3.7 pm, respectively.

The observed field-induced structure change in $SrTiO_3$ may lead to the development of novel pyroelectric and piezoelectric materials derived from initially centrosymmetric compounds.

MS04-03
Anelastic relaxation effects in langasite-type crystal species

C. Hirschle[1], J. Schreuer[1]
[1]Ruhr-Universität Bochum, Institut für Geologie, Mineralogie und Geophysik, Bochum, Germany

Single crystals with langasite-type structure are widely used for high-temperature piezoelectric applications. However, all langasite-type crystal species investigated so far suffer from pronounced acoustic wave attenuation at elevated temperatures which limits their use for example in sensing applications.[1,2] Our work focuses on the investigation of the structural background of the ultrasound attenuation, as there is no consensus on its origin in literature.[1,2]

Resonant ultrasound spectroscopy was used on LGS ($La_3Ga_5SiO_{14}$) and LGT ($La_3Ta_{0.5}Ga_{5.5}O_{14}$) samples to quantify the attenuation via the inverse quality factor Q^{-1} of selected eigenmodes of corresponding samples from room temperature to about 1600 K. The eigenfrequencies were also used to determine the elastic and piezoelectric properties of the samples as they are closely related to many attenuation mechanisms.

Superpositions of two to three attenuation peaks in $Q^{-1}(T)$ are visible for both crystal species. The maxima of these peaks seem to shift to higher temperatures with higher corresponding resonance frequency. The color of the LGT sample shifted from colorless to orange during the heating in air and the attenuation in the sample changed in magnitude following the color change.

Models for Q^{-1} based on point defect processes were fit to the observed Q^{-1}. The refined activation energies are compatible with point defect relaxations and agree well with the activation energies for conduction in these materials.[3] Our observations suggest that both phenomena probably involve the same processes of oxygen vacancy movement.

References:
[1] J. Schreuer and C. Thybaut, IEEE Ultras. Symp., 695–698 (2005).
[2] H. Fritze, *J. Electroceram.* **26**, 122–161 (2011).
[3] Y. Suhak, M. Schulz, D. Richter, and H. Fritze, *Solid State Phenom.* **230**, 267–272 (2015).

MS04-04
Mechanical properties of zircon and the effect of radiation damage

T. Beirau[1], W. D. Nix[2], U. Bismayer[3], L. A. Boatner[4], H. Pöllmann[1], R. C. Ewing[5]
[1]University of Halle, Mineralogy/Geochemistry, Halle, Germany
[2]Stanford University, Department of Materials Science and Engineering, Stanford, United States
[3]University of Hamburg, Department of Earth Sciences, Hamburg, Germany
[4]Oak Ridge National Laboratory, Materials Science and Technology Division, Oak Ridge, United States
[5]Stanford University, Department of Geological Sciences, Stanford, United States

New insights will be presented into the relationship between radiation-dose-dependent structural damage due to natural U and Th impurities and the anisotropic mechanical properties (Poissons ratio, elastic modulus and hardness) of zircon (Beirau et al. 2016). Nanoindentation measurements and calculations, based on elastic stiffness constants (Özkan 1976), revealed a general radiation-induced decrease in hardness and stiffness and an increase in the Poissons ratio of ~ 50 %, respectively, with increasing dose. It was possible to follow the radiation-induced amorphization process to the point that the mechanical properties are isotropic. This work shows that the radiation-dose-dependent changes of the mechanical properties of the mineral zircon can be directly correlated with the amorphous fraction as determined by previous investigations with local and global probes (Ríos et al. 2000, Farnan and Salje 2001, Zhang and Salje 2001). The excellent agreement, revealed by the different methods, indicates a large influence of structural and even local phenomena on the macroscopic mechanical properties. Therefore, this study indicates the importance of acquiring better knowledge about the mechanical long-term stability of radiation damaged materials.

References:
[1] Beirau, T., Nix, W.D., Bismayer, U., Boatner, L.A., Isaacson, S.G. and Ewing, R.C. (2016) Anisotropic mechanical properties of

zircon and the effect of radiation damage. Phys Chem Minerals 43:627–638
[2]Özkan, H. (1976) Effect of nuclear radiation on the elastic moduli of zircon. J Appl Phys 47:4772–4779
[3]Farnan, I. and Salje, E.K.H. (2001) The degree and nature of radiation damage in zircon observed by 29Si nuclear magnetic resonance. J Appl Phys 89:2084–2090
[4]Zhang, M. and Salje, E.K.H. (2001) Infrared spectroscopic analysis of zircon: radiation damage and the metamict state. J Phys Condens Matter 13:3057–3071
[5]Ríos, S., Salje, E.K.H., Zhang, M. and Ewing, R.C. (2000) Amorphization in zircon: evidence for direct impact damage. J Phys Condens Matter 12:2401–2412

MS04-05

High pressure elastic properties of $FeCO_3$ and $MgCO_3$

M. Stekiel[1], T. Nguyen-Thanh[2], S. Chariton[3], C. McCammon[3], A. Bosak[2], R. Luchitskaia[1], K. Refson[4], V. Milman[5], B. Winkler[1]
[1]Goethe Universität, Geowissenschaften, Frankfurt am Main, Germany
[2]ESRF, Grenoble, France
[3]University of Bayreuth, Bayerisches Geoinstitut, Bayreuth, Germany
[4]University of London Egham, Department of Physics, Surrey, United Kingdom
[5]Dassault Systèmes BIOVIA, Cambridge, United Kingdom

Carbonates play a significant role in Earth's deep carbon cycle, as they are being transported to Earth's interior in subducting slabs [1]. Iron bearing magnesite, $Mg_{(1-x)}Fe_xCO_3$, is the most likely candidate for such a deep mantle carbonate. The $FeCO_3$ end member undergoes a spin transition at 45 GPa, associated with a 10% volume collapse [2], and it was shown that the $Mg_{(1-x)}Fe_xCO_3$ solid solutions undergo such a transition as well [3]. In order to better understand the seismic signature of carbonates in the deep Earth we have studied the elastic properties of the end-member carbonates, $FeCO_3$ and $MgCO_3$, at high pressures. We synthesized $FeCO_3$ and performed inelastic X-ray scattering experiments of $FeCO_3$ in a diamond anvil cell at ID28 at the ESRF, focusing on the spin transition. We measured phonon dispersion curves at pressures up to 50 GPa, and derived the C_{33} and C_{44} coefficients of the elastic stiffness tensor. We have complemented the experiments by DFT calculations on $FeCO_3$ and $MgCO_3$ to determine the full elastic stiffness tensor. The excellent agreement between the computed pressure dependence of elastic stiffness coefficients and reported experimental data for $MgCO_3$ [4] and between our high spin IXS data for $FeCO_3$ confirms the accuracy of our calculations and allows us to constrain the elasticity of the low spin compounds. We find that the elastic stiffness coefficients increase significantly over the spin transition in $FeCO_3$, as shown in figure 1. We have also observed a deviation from the linear behavior of elastic stiffness coefficients in $MgCO_3$ at pressures above 30 GPa, see figure 2. The study is supported by funding from the German Science Foundation (DFG-Research Unit FOR2125) and through the DFG-ANR project WI1232/41.

References:
[1] R.M. Hazen, C.M. Schiffries, *Carbon in Earth*, edited by R. Hazen et al., chapt. 1, pp. 1-6, 2013.
[2] B. Lavina et al., *Geophys. Res. Lett.*, vol. 36, pp. L23306, 2009.
[3] A. Spivak et al., *Phys. Chem. Miner.*, vol. 41, pp. 633, 2014.
[4] J. Yang et al., *Earth Planet. Sc. Lett.*, vol. 392, pp. 292, 2014.

Figure 1. C_{33} (a) and C_{44} (b) elastic stiffness coefficients of $FeCO_3$ measured by IXS (full symbols), compared with calculated values (empty symbols: high-spin, partially empty: low-spin siderite) and ambient pressure experimental data [5]. Solid line is a fit to the experimental data. Shaded area indicates the pressure range of the spin transition [2,3].

Figure 2. Calculated elastic stiffness coefficients of $MgCO_3$ (empty symbols) compared to experimental data ("+") [4]. Solid lines are fit to the experimental data.

Figure 1

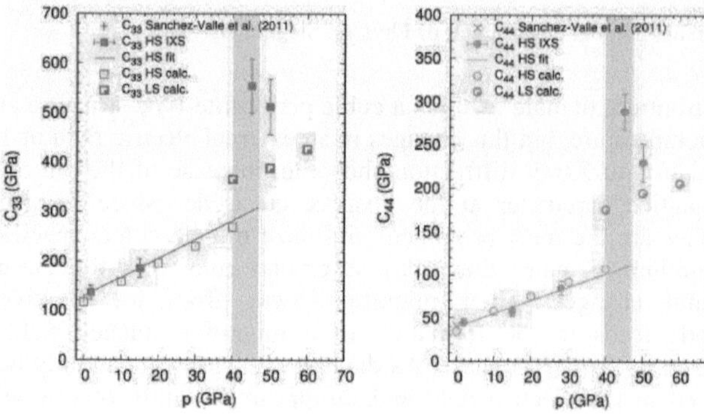

Figure 2

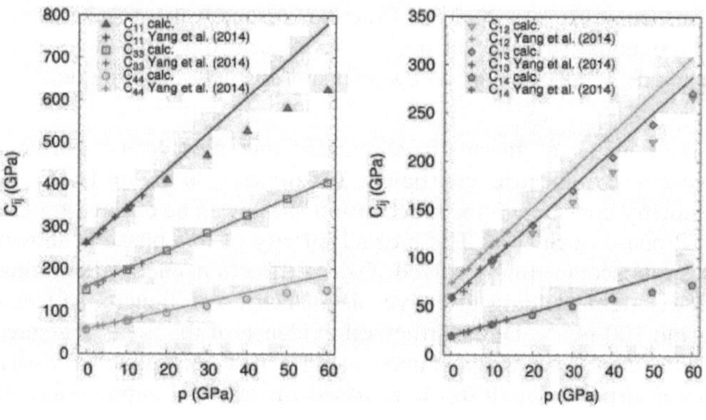

MS04-06

NEW EoS AND NEW METHODS IN THE EOSFIT7 PROGRAM SUITE

R. Angel[1], M. Alvaro[1], J. Gonzalez-Platas[1], F. Nestola[1]
[1]University of Padova, Geosciences, Padova, Italy

Equations of state (EoS) describe the P-T variation of the elastic properties of materials. The first release of the EosFit7 suite of programs and subroutine library (Angel et al., 2014) provided facilities for performing calculations with P-V and P-V-T EoS and for fitting EoS parameters to P-V, T-V and P-V-T data, and to unit-cell parameters. This allows the variation of the elastic compressibility tensor of a crystal to be determined as a function of P and T. The EosFit7-GUI program (Gonzalez-Platas et al., 2016) is a tool for rapid data analysis and for teaching EoS theory. EosFit7c is a console program with more advanced features, and remains the research and development platform. Both programs are available for Mac-OS, Linux and Windows from www.rossangel.net. We now report significant new developments in the EosFit7 program suite and underlying subroutine library.

We have coded the Adapted Polynomial EoS of Holzapfel (2001) into EosFit7 to address the problem that most finite strain EoS do not predict physically-correct properties at infinite compression. To fit PVT data we have added the Mie-Grüneisen-Debye thermal-pressure EoS. We have also introduced a new parameterisation of isothermal models that provides thermodynamically-correct behaviour of the bulk modulus K and dK/dP variation with temperature.

For many crystals there are direct measurements of their elastic moduli as a function of temperature and pressure. EosFit7c has now been modified so that it can perform weighted fits of bulk modulus and volume data separately or simultaneously, so as to provide better constraints on EoS parameters. The extension of simultaneous fitting to linear data for non-cubic materials requires some assumptions to be made about the thermo-elastic behaviour of crystals, and these will be described.

References:

[1] Angel, Gonzalez-Platas, Alvaro (2014): *Zeit. Krist.*, 229:405-419.
[2]Gonzalez-Platas, Alvaro, Nestola, Angel (2016) *J. Appl. Cryst.*, 49:1377-1382
[3]Holzapfel (2001) *Zeit. Krist.*, 216:473-488

EosFit developments are supported by: ERC, INDIMEDEA project 307322 to Fabrizio Nestola, SIR-MIUR grant MILE DEEp (n° RBSI140351) to Matteo Alvaro, and MINECO and EU-FEDER Projects MAT2013-46649-C4-4-P, MAT2015-71070-REDC and MAT2016-75586-C4-4-P to Javier Gonzalez-Platas

MS05 - Macromolecular machines and switches

MS05-01

Molecular structure of hydrazine synthase – an enzyme complex producing rocket fuel

A. Dietl[1], C. Ferousi[2], W. J. Maalcke[2], A. Menzel[3], S. de Vries[4], J. T. Keltjens[2], M. S. M. Jetten[2,4], B. Kartal[2,5], T. R. M. Barends[1]
[1]Max Planck Institute for Medical Research, Biomolecular Mechanisms, Heidelberg, Germany
[2]Radboud University Nijmegen, Institute for Water and Wetland Research, Nijmegen, Netherlands
[3]Swiss Light Source, Villigen, Switzerland
[4]Delft University of Technology, Department of Biotechnology, Delft, Netherlands
[5]Max Planck Institute for Marine Microbiology, Department of Biogeochemistry, Bremen, Germany

Anaerobic ammonium-oxidizing (anammox) bacteria are major players of the biogeochemical nitrogen cycle and are nowadays frequently applied in environmentally friendly waste water treatment procedures. The anammox process combines nitrite (NO_2^-) and ammonium (NH_4^+) to form dinitrogen gas and water, and has been estimated to be responsible for up to 50% of the loss of fixed nitrogen from the sea. Strikingly, it has been discovered that nitric oxide (NO) and hydrazine (N_2H_4) are central intermediates in the anammox metabolism[1]. Hydrazine is a highly reactive compound with a very low redox potential (E_0'= -750 mV) and is therefore also used as a spacecraft propellant. In order to gain insights into the enzymatic mechanism by which hydrazine is biologically synthesized we determined the 2.7 Å resolution crystal structure of a hydrazine synthase (HZS) multienzyme complex (kuste2859-61) isolated from the anammox organism *Kuenenia stuttgartiensis* and performed its biophysical and spectroscopic characterization[2]. The X-ray structure reveals an elongated dimer of heterotrimers (327 kDa) which was verified as the oligomeric state of the complex in solution using analytical ultracentrifugation (AUC) and small-angle X-ray scattering (SAXS). Each heterotrimer harbors two distinct c-type heme-containing active sites as well as two bis-histidine coordinated hemes c that are likely involved in electron transfer. Additionally, the proposed active sites are connected by a system of tunnels. Based on the crystal structure a two-step reaction mechanism for hydrazine synthesis can be proposed: At first nitric oxide is reduced to hydroxylamine (NH_2OH) in a three-electron reduction at the active site of the γ-subunit. The resulting hydroxylamine diffuses through the tunnel to the second active center in the α-subunit, followed by its condensation with ammonia, which finally yields hydrazine.

References:
[1] B. Kartal et al.; FEMS Microbiol Rev **(2013)**, 37, 428
[2] A. Dietl et al.; Nature **(2015)**, 527, 394

MS05-02

Structural Analysis of the Tuberous Sclerosis (TSC) Complex involved in Regulation of Cellular Growth

R. Zech[1], P. Hansmann[1], D. Kümmel[1]
[1]Universität Osnabrück, FB5: Biology/Chemistry, Osnabrück, Germany

The TOR (target of rapamycin) kinase complex is a master regulator of cellular growth and requires, among other inputs, growth factor signaling for its full activation. This signal is conveyed by the small GTPase Rheb, which acts as a molecular switch that cycle between an inactive (GDP) and an active (GTP) state. This function requires activator (guanine nucleotide exchange factors, GEFs) and inactivator (GTPase activating proteins, GAPs) proteins that are responsible for the conversion between the "on" and "off" states. Under resting conditions, Rheb is kept inactive by its GAP, the TSC (tuberous sclerosis) complex. When growth factors are present, the TSC complex is phosphorylated and switch off, thus Rheb and TORC1 become active. Importantly, mutations in the TSC complex cause the genetic disease tuberous sclerosis, which is characterized by the formation of benign tumors in skin, heart, lung and the brain that lead to organ failure. The mechanism of Rheb regulation by the TSC complex and the molecular basis of tuberous sclerosis pathogenesis are poorly understood. We are working on the elucidation of these aspects by the structural and biochemical characterization of the TSC complex. To this end, we have crystallized domains of TSC (Figure 1) and analyzed the influence of pathogenic mutations on structure and function of these proteins.

Figure 1

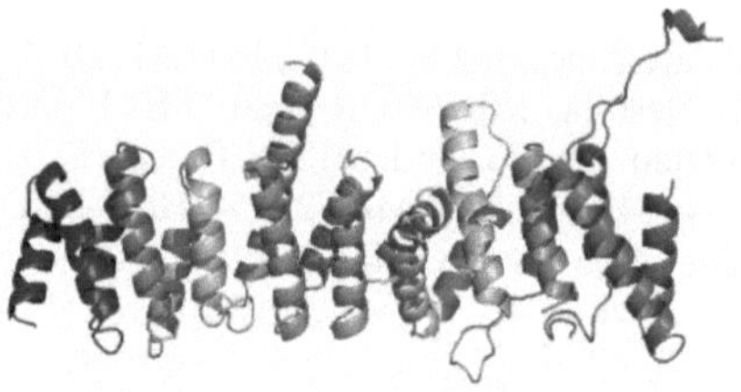

MS05-03

Reaction Control in Fatty Acid Megasynthases

J. Gajewski[1,2], F. Buelens[3], R. Pavlovic[2], M. Fischer[1,2], E. Boles[2], H. Grubmüller[3], M. Grininger[1,2]
[1]Buchmann Institute of Molecular Life Sciences, Frankfurt, Germany
[2]Goethe University Frankfurt, Frankfurt, Germany
[3]Max-Planck-Institute of Biophysical Chemistry, Göttingen, Germany

Fatty acid synthases (FAS) synthesize fatty acids (FA) by repeatedly elongating and modifying a growing acyl chain until a specific length is attained. Baker"s yeast FAS is comprised by two polypeptides, the 207 kDa α-chain and the 229 kDa β-chain, that assemble to a barrel-shaped $\alpha_6\beta_6$-heterododecameric complex of 2.6. MDa that encloses two reaction chambers, each formed by a β-dome and the α-wheel[1]. As multienzymes performing compartmentalized synthesis at high efficiency, FAS are interesting molecules for biotechnological purposes. We engineered yeast FAS for the biosynthesis of technologically relevant small molecules, guided by a combined *in vitro* and *in silico* approach. For example, we performed a highly rational and minimal invasive protein engineering approach for the synthesis of short FA, which are important platform compounds e.g. in biofuel production.

Besides the exploration of the synthetic capability of FAS, we aim to build a foundation for efficient protein engineering, with the specific goal of harnessing evolutionarily related megadalton-scale polyketide synthases (PKS) for the tailored production of bioactive natural compounds[2].

References:
[1] Maier, T., Leibundgut, M., Boehringer, D. & Ban, N. Structure and function of eukaryotic fatty acid synthases. *Q. Rev. Biophys.* **43**, 373-422 (2010).
[2] Gajewski, J., Buelens, F., Serdjukow, S., Janßen, M., Cortina, N., Grubmüller, H. & Grininger, M. Engineering Fatty Acid Synthases for Directed Polyketide Production, *Nat. Chem. Biol.* (accepted).

MS05-04

The inhibition mechanism of human 20S proteasomes enables next-generation inhibitor design

F. Henneberg[1], J. Schrader[1], R. A. Mata[2], K. Tittmann[3], T. R. Schneider[4], H. Stark[1], G. Bourenkov[4], A. Chari[1]
[1]MPI for Biophysical Chemistry, Structural Dynamics, Göttingen, Germany
[2]Institut für Physikalische Chemie, Georg-August-Universität, Göttingen, Germany
[3]Göttinger Zentrum für Molekulare Biowissenschaften und Schwann-Schleiden- Forschungszentrum, Georg-August-Universität Göttingen, Göttingen, Germany
[4]European Molecular Biology Laboratory (EMBL), Hamburg Outstation c/o Deutsches Elektronen Synchrotron (DESY), Hamburg, Germany

Introduction

Protein degradation is mediated via two major pathways, the lysosomal proteolysis and the Ubiquitin-Proteasome Pathway. Gene expression and cell proliferation are controlled by the Ubiquitin-Proteasome Pathway. The 19S lid of the proteasome recognizes and removes the poly-ubiquitin signal and transfers the protein into the 20S proteasome, where it is chopped into peptide fragments. Several crystal structures from different organisms have elucidated the general architecture of the 20S proteasome. However, the attained resolutions of these structures did not allow direct visualization of the mechanisms of peptide hydrolysis and inhibition chemistry.

Objectives

To gain a better understanding of the molecular mechanisms underlying peptide hydrolysis and inhibition chemistry of the 20S proteasome, it is essential to improve the resolution of 20S proteasome crystal structures.

Materials & methods

New protein purification, an optimized crystallization condition, a post-crystallization treatment, and technical improvements at P14 (DESY) were introduced to obtain crystals diffracting to high resolution.

Results

The introduction of the chromatographic free purification method resulted in high protein yields. The structure of the native human 20S proteasome was determined at an unprecedented resolution of 1.8 angstroms. Additionally, six inhibitor-proteasome complex structures were elucidated at resolutions between 1.9 and 2.1 angstroms. Collectively, the high-resolution structures provide new insights into the catalytic mechanisms of inhibition and necessitate a revised description of the proteasome active site.

Conclusion

The optimization of the purification, crystallization and data collection lead to high resolution structure of the 20S proteasome. Knowledge about inhibition mechanisms provides insights into peptide hydrolysis and can guide strategies for the development of next-generation proteasome-based cancer therapeutics. As a perspective, the established protocol are could be used to set up a high throughput pipeline to screen for new inhibitors of the 20S proteasome.

MS06 - Materials for energy storage and conversion

MS06-01

Comparative Analysis of Diffraction Experiments for Lithium Insertion Materials

W. R. Brant[1,2], S. Schmid[1]
[1]The University of Sydney, School of Chemistry, Sydney, Australia
[2]Uppsala University, Department of Materials Chemistry, Ångström Laboratory, Uppsala, Sweden

A comparative study of *ex-situ* and *in-operando* X-ray diffraction experiments using the fast lithium ion conductor $Li_{0.18}Sr_{0.66}Ti_{0.5}Nb_{0.5}O_3$ will be presented. *Ex-situ* analysis of synchrotron X-ray diffraction data suggests that a single phase material exists for all discharges to as low as 0.422 V. For samples with higher lithium content, it is possible to determine the lithium position from the X-ray data. However, *in-operando* X-ray diffraction reveals a kinetically driven two phase region on cycling below 1 V. Monitoring the change in unit cell dimension during electrochemical cycling showed a reduction in the rate of unit cell expansion part way through the first discharge and during the second discharge, caused by a drop in lithium diffusion into the bulk material for higher lithium contents. A more significant change is a jump in the unit cell expansion once the lithium content exceeds one lithium ion per vacant site, caused by damping of octahedral rotations. This provides a link between lithium content and octahedral rotations. Using *in-operando* diffraction may therefore enable to determine the strength of octahedral rotations in defect perovskites and allow correlations with the large variance of ionic conductivities in these materials.

References:
[1] W. R. Brant, D. Li, Q. Gu & S. Schmid, *J. Power Sources* **302**, 126 – 134 (2016). Comparison of *ex-situ* and *operando* X-ray diffraction techniques for investigating lithium insertion defect perovskite $Li_{0.18}Sr_{0.66}Ti_{0.5}Nb_{0.5}O_3$.
[2] W. R. Brant, S. Schmid, A. Kuhn, J. Hester, M. Avdeev, M. Sale & Q. Gu, *ChemPhysChem.* **13**, 2293-2296 (2012). Rapid lithium insertion and location of mobile lithium in the defect perovskite $Li_{2y}Sr_{1-x-y}Ti_{1-2x}Nb_{2x}O_3$.

MS06-02

Structural characterisation of $(Ag_{1-x}Cu_x)_2ZnSnSe_4$ by neutron diffraction.

G. Gurieva[1], A. Franz[1], S. Schorr[1,2]
[1]Helmholtz-Zentrum Berlin, Berlin, Germany
[2]Free University Berlin, Berlin, Germany

$Cu_2ZnSnSe_4$ solar cells, containing only abundant elements, with Ag alloying recently reached efficiency of 10.2% [1]. The open-circuit voltage in CZTSSe devices is believed to be limited by absorber band tailing caused by the exceptionally high density of Cu/Zn antisites. By replacing Cu in CZTSe with Ag, whose covalent radius is 15% larger than that of Cu and Zn, the density of I–II antisite defects (e.g Cu-Zn disorder) is predicted to drop.

$Ag_2ZnSnSe_4$ and $Cu_2ZnSnSe_4$ are quaternary semiconductors belonging to the adamantine compound family. It was shown by neutron diffraction that CZTSe crystallizes in the kesterite type structure (space group I-4) [2]. So far, only X-ray diffraction used for structural characterization of $Ag_2ZnSnSe_4$ was reported in the literature, and it suggests that $Ag_2ZnSnSe_4$ also shows the kesterite type structure [3]. The differentiation between the isoelectronic cations Cu^+ and Zn^{2+} which is the key to the information on Cu-Zn disorder in kesterite type semiconductors is not possible using X-ray diffraction due to their similar scattering factors. Neutrons diffraction can solve this problem; the coherent scattering lengths are sufficiently different for these cations [4].

A detailed structural analysis of $(Ag_{1-x}Cu_x)_2ZnSnSe_4$, grown by solid state reaction, was performed by neutron diffraction at the fine resolution neutron powder diffractometer E9 at BER II (λ=1.7982(1) Å, RT). Rietveld refinement of diffraction data using the FullProf suite software [5] lead to accurate values of lattice constants, and site occupancy factors. The latter have given insights into the cation distribution within the crystal structure of $(Ag_{1-x}Cu_x)_2ZnSnSe_4$ solid solutions with different x values. The correlated information about changes in lattice parameters and cation site occupancies, details on the existing intrinsic point defects and their amounts will be discussed.

References:
[1] T. Gershon, et al Adv. Energy Mater. 2016, 1502468
[2] S.Schorr, Solar Energy Materials and Solar Cells, 95 (2011)1482.
[3] W. Gong et al, phys. stat. sol C., 1-4(2015)
[4] V.F. Sears, Neutron News 3(3) (1992) 26.
[5] Juan Rodriguez-Carvajal and Thierry Roisnel, www.ill.eu/sites/fullprof

MS06-03

Investigations on Oxygen Exchange Rate of Ion Implanted Single Crystalline Strontium Titanate for Air Electrode Application

M. Stöber[1], C. Cherkouk[1], J. Walter[1], M. Schelter[2], J. Zosel[2], S. Prucnal[3], T. Leisegang[1], D. C. Meyer[1], M. Mertig[2]
[1]TU Bergakademie Freiberg, Institut für Experimentelle Physik, Freiberg, Germany
[2]Kurt Schwabe Institute, Meinsberg, Germany
[3]Helmholtz-Zentrum, Institute of Ion Beam Physics and Materials Research, Dresden-Rossendorf, Germany

Energy storage systems which incorporate oxygen from the sorrounding atmosphere through a metal-air electrode into a dielectric material are of recent interest for sustainable power distribution and storage. The efficiency and durability of the modified crystaline surfaces, at which the oxygen reduction reaction (ORR) takes place, is most often a limiting factor.

Metal ion implantation and subsequent annealing can be used to overcome these limits. Single crystaline strontium titanate ($SrTiO_3$) was used due to its high conductivity for oxygen ions. The oxygen exchange rate has been measured by means of Coloumetric Titration. The characteristics of the modified surfaces were investigated using X-ray Photoelectron Spectroscopy (XPS) and Transmission Electron Microscopy (TEM). Furthermore, a theoretical model is suggested in order to obtain the effective rate constants from fitting an iterative solution of the diffusion equation to Coloumetric titration curves depending on oxygen partial pressure and temperature.

The results have shown an increases in exchange rate and quantity as well due to metallic defects at and near the surface of the $SrTiO_3$ single crystals. TEM investigations have revealed the thickness of the modified layer to be about 180 nm. No signs of metal nano clusters have been observed, even though the implanted ions are in a metallic state which has been confirmed using XPS.

We obtained preliminary results, indicating a high potential for an increase of effective rate parameters. The material system will be further investigated with a focus on correlations between structure and catalytic activity. This work is financially supported by the German BMBF (CryPhysConcept: 03EK3029A; R2RBattery: 03SF0542A)

Figure 1

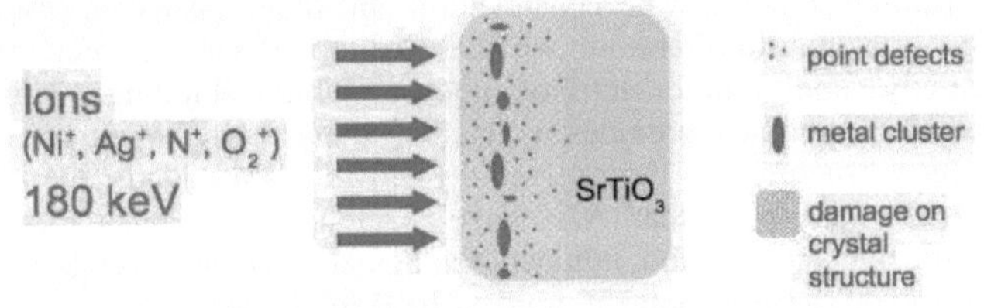

MS06-04

Crystal structure analysis and *in-situ* diffraction studies of new complex aluminum hydrides as hydrogen carriers

C. Weidenthaler[1], T. Bernert[1], M. Felderhoff[1]
[1]Max-Planck-Institut für Kohlenforschung, Mülheim an der Ruhr, Germany

Although complex aluminum hydrides were intensively investigated as potential hydrogen storage materials over the last 20 years, not all of their properties are completely understood [1]. The detailed knowledge of their intrinsic properties will contribute to a deeper understanding of these interesting materials and may explain for example why some materials are reversible while others are not.

In this study complex aluminum hydrides containing elements of group I with higher atomic numbers have been synthesized. The crystal structures of $RbAlH_4$, $CsAlH_4$ and of mixed compounds with sodium will be discussed [2, 3]. Both $RbAlH_4$ and $CsAlH_4$ can be prepared by direct synthesis from the metals under high hydrogen pressure in a ball-mill. An alternative synthesis method is milling of alkali chlorides with commercial $NaAlH_4$ in a planetary ball-mill. $CsAlH_4$ can crystallize in two polymorphic structures which both are stable at room temperature. Both crystal structures were analyzed by X-ray and TOF neutron diffraction and verified by DFT calculations.

The structures can be transformed from one into the other either by temperature increase or by mechanochemical treatment. Unexpectedly, the transformation is rather influenced by the addition of small amounts of catalysts than by the significant reduction of crystallite size. As third compound, mixed aluminum hydrides containing both cations, Rb and Na or Cs and Na in the crystal structure can be synthesized by the mechanochemical route. The crystal structures were analyzed by X-ray diffraction methods and the decomposition pathways were monitored by in-house laboratory *in-situ* diffraction studies. The experimental results were combined with thermal analysis and rehydrogenation experiments. The results obtained on phase transformations, the influence of catalysts and the decomposition pathways provided more information about the structure-property relationship of complex aluminum hydrides.

References:

[1] B. Bogdanović, M. Schwickardi, J. Alloys Compd. 1 (1997) 253.
[2] D. Krech, B. Zibrowius, C. Weidenthaler, M. Felderhoff, Eur. J. Inorg. Chem. (2014) 5683.
[3] T. Bernert, D. Krech, W. Kockelmann, M. Felderhoff, T.J. Frankcombe, C. Weidenthaler, Eur. J. Inorg. Chem. (2015) 5545.

MS07 - Fragment based ligand screening

MS07-01
Addressing Protein Targets with Small Molecule Fragments and Self-optimized Inhibitors by Crystallography

G. Klebe[1]
[1]Univ. of Marburg, Inst. Pharm. Chemistry, Marburg, Germany

In recent years, fragment-based lead discovery has become increasingly popular to identify first lead candidates for relevant drug targets. One advantage of the usage of fragments is their low molecular weight (< 250 g/mol), which leaves sufficient space for subsequent chemical optimization before the pharmaceutically reasonable limit of 500-600 g/mol is reached. However, the inherent low-binding affinity of fragment-like molecules (typically mM to µM range) poses a challenge to current biophysical and biochemical screening methods, particularly whether typically applied pre-screening methods are able to reliably detect all putative fragments that can subsequently be crystallized with the target protein, or whether a direct fragment screening on protein crystals is the superior strategy. We therefore started to design a 361-entry fragment library, and tested this collection against the aspartic protease endothiapepsin. Initially, we screened all compounds by six biophysical methods such as STD-NMR, ESI-MS, thermal shift analysis, micro-thermophoresis, high-concentration enzyme and reporter displacement assay to identify putative binders. Disappointingly, each individual screening method revealed varying hit rates with only a minimal overlap of commonly shared hits. We then performed a crystallographic screening of the entire fragment library directly on crystals. Remarkably, 71 crystal structure could be determined out of the 361 entries and nearly half of the hits were missed by any of the biophysical screening methods. We therefore strongly advocate to screen fragments directly on crystals instead of applying elaborated pre-filters, as the determination of fragment-bound crystal structures is pivotal for any subsequent design project to evolve a fragment into a drug candidate. Remarkably, we detected two structures, which showed a much larger and likely more potent ligand bound to the protein. In these cases, the fragments had reacted under the crystallization conditions to novel products, which were picked by the protein as potent binders. We therefore used the information from our discovered fragment hit sample and started to use dynamic combinatorial chemistry to discover novel leads for endothiapepsin. As hydrazides showed up as promising fragments, we selected the reaction to form acylhydrazones as these molecules assemble under equilibrium conditions in acidic medium. Via two cycles of subsequent dynamic library design, a 54 nM binding hit of could be identified.

MS07-02
The landscape of EPHA2 inhibition

D. Kudlinzki[1], S. Heinzlmeir[2]
[1]DKFZ, DKTK Frankfurt/Mainz, Frankfurt, Germany
[2]TU München, Chair of Proteomics and Bioanalytics, München, Germany

EPH receptors belong to the largest family of receptor tyrosine kinases. They are involved in various developmental and cell-cell communication events. EPHA2, a member of this family, has emerged as an interesting therapeutic target, because malfunctioning is correlated with various diseases (disorders of cardiovascular and nervous system, cancer or pathogen infections). So far, investigation of EPHA2 function and therapeutic targeting of EPHA2 driven pathologies are hampered by the lack of appropriate and selective inhibitors. We established selectivity profiles of 235 clinical kinase inhibitors using a chemical proteomics screen and identified various EPHA2 inhibitors. The binding properties of these off-target inhibitors were investigated by different methods. Protein crystallography was used to delineate the structural determinants of EPHA2 inhibition by numerous clinically approved kinase inhibitors and tool compounds. The identified interaction sites were categorized according to their localization, kinome-wide conservation and their impact on drug target selection. Furthermore, we analyzed the effects on structural plasticity and kinase activity upon inhibitor binding. This combined approach provides a detailed analysis at the interface between structure and drug selectivity profiling. The landscape of EPHA2 inhibition that we have established enables drug repurposing studies and initiated medicinal chemistry programs focusing on the development of novel EPH inhibitors.

References:
[1] Heinzlmeir S, Kudlinzki D, Sreeramulu S, Klaeger S, Gande SL, Linhard V, Wilhelm M, Qiao H, Helm D, Ruprecht B, Saxena K, Médard G, Schwalbe H, Kuster B. ACS Chem Biol. 2016, 11:3400-3411.
[2] Gande SL, Saxena K, Sreeramulu S, Linhard V, Kudlinzki D, Heinzlmeir S, Reichert AJ, Skerra A, Kuster B, Schwalbe H. Chembiochem. 2016, 17:2257-2263.

Fig 1:
Spatial positioning of key, potency and selectivity residues within the EPHA2 drug binding pocket. Key residues: found in inhibitor target proteins, not conserved in kinome; potency residues: highly conserved, add selectivity; selectivity residues: low conserved in binding pocket.

Fig 2:
Increased dynamics of kinase key motifs induced by the inhibitor binding. DFG-in inhibitors support the active K646/E663 salt bridge formation, whereas DFG-out ligands do not (inactive).

Figure 1

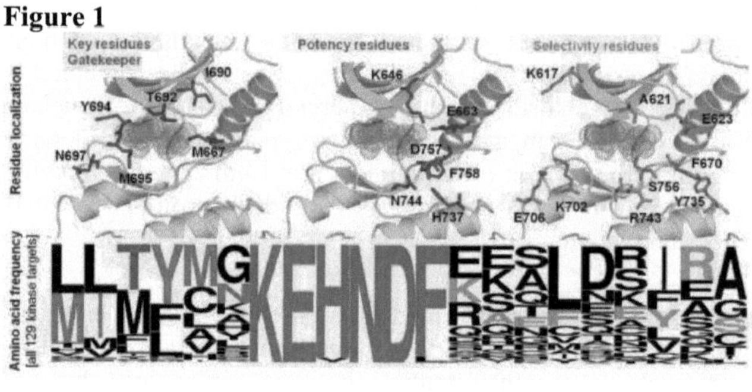

Figure 2

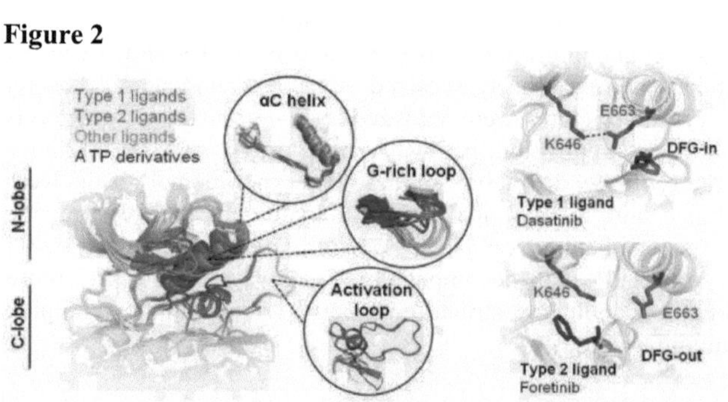

MS07-03

Fragment Based Screening to target the Ubiquitin Activating Enzyme for Cancer Therapy

M. Misra[1], F. Huschmann[2,3], M. Kuhn[4], C. Sotriffer[4], M. Weiss[2], H. Schindelin[1]
[1]Rudolf Virchow Zentrum, University of Wuerzburg, Structural Biology, Wuerzburg, Germany
[2]Helmholtz Zentrum Berlin (HZB) für Materialien und Energie, Macromolecular Crystallography (HZB-MX), Berlin, Germany
[3]Philipps University, Department of Pharmaceutical Chemistry, Marburg, Germany
[4]Institue of Pharmacy and Food Chemistry, Würzburg, Germany

Ubiquitylation, a very common post-translational modification, regulates an array of cellular processes in eukaryotes including proteasomal degradation of target proteins, DNA repair, autophagy, vesicular trafficking and the immune response. Given its important role for normal cellular functioning, it is especially crucial for malignant cells to maintain their higher metabolic rate. The approval of Bortezomib, an inhibitor of the 20 S proteasome, by the Food and Drug Administration, U.S.A., for the treatment of multiple myeloma and mantle cell lymphoma underlined the significance of the Ubiquitin Proteasome System (UPS) as an avenue for cancer therapy. Enzymes present upstream of the proteasome have been targeted to unravel their efficacy in malignant cells. The class of enzymes standing at the apex of the ubiquitylation pathway are the ubiquitin activating enzymes (E1). In humans, there are two such enzymes that activate ubiquitin in an ATP-dependent manner and later transfer the ubiquitin to the E2 class of enzymes in a transthioesterification reaction. E1 enzymes have been shown to be promising targets for the treatment of advanced solid tumors where proteasome inhibitors so far have been ineffective.

In this study, we have utilized fragment based screening to develop novel lead compounds that can inhibit the ubiquitin activating enzyme 1 (Uba1) by screening the 100 compound fragment library developed at the HZB. We combined X-ray crystallography with thermal shift assays as well as activity assays to identify compounds that bind to Uba1 and modulate its activity. We would like to further evolve these fragments using structure based drug design approaches to generate high affinity drug-like compounds that can selectively inhibit Uba1.

MS07-04

Unravelling the nature of the druggable pockets by chemical fragments.

A. Chaikuad[1]
[1]Goethe University Frankfurt, Institute for Pharmaceutical Chemistry and Buchmann Institute for Life Sciences, Frankfurt am Main, Germany

Development of an inhibitor for a target protein typically follows a chemical framework using known chemotypes that have been proven to bind other related proteins in the family. Despite limiting the initial search efforts, the lack of selectivity of the hits is often an obstacle, leading to a requirement of tedious chemistry optimization. Recently, chemical fragment approach have gained a high interest in drug discovery. Not only providing starting chemical scaffolds, an ability of these privileged scaffolds to bind multiple pockets allows also an examination of various medicinal chemical space of the target proteins. Here, we demonstrate the use of fragments towards identification and understanding the specific structural features within the druggable pockets guiding selective inhibitor design. Examples presented here cover two distinct protein families: acetyl-lysine protein interaction modules of the bromodomain family and protein kinases. The presented fragment campaigns revealed not only benefits developing bromodomain inhibitors, but probed the steric nature of the protein kinase backpocket which is a key for selectivity when targeting the highly similar ATP binding sites of kinases.

MS07-05

Fragment-screening studies on the ectonucleotidase CD73

E. R. Scaletti[1], F. Huschmann[2], U. Mueller[2], M. S. Weiss[2], N. Sträter[1]
[1]University of Leipzig, Leipzig, Germany
[2]Helmholtz-Zentrum Berlin für Materialien und Energie, BESSY II, Berlin, Germany

The homodimeric enzyme CD73 is attached via a GPI-anchor at the C-terminus to the cell membrane. It catalyzes the hydrolysis of AMP and is the primary source of extracellular adenosine. The effects of adenosine are often opposite to those of ATP and CD73 activity has an important regulatory role in processes such as inflammation, hypoxia and certain types of cancer.[1]

Structures of CD73 revealed open and closed conformations for the enzyme.[2,3] Each subunit of CD73 is composed of a C- and N-terminal domain and the conformational change is achieved by a large (~100°) rotation of the N-terminal domain. We are studying the inhibitory action of nucleotides on CD73. ATP and ADP are natural inhibitors, which in the immune system reinforce the immunostimulatory action of released ATP by reducing the concentration of adenosine in the early stages of inflammation via CD73 inhibition. Nucleotide derivatives display a high potency and are stable towards degradation, but they also have disadvantages as potential drugs for cancer immunotherapy. They are not orally available and may display unspecific inhibition of other nucleotide-binding proteins.

To aid in the development of novel lead structures based on non-nucleotide inhibitors, we initiated a limited fragment-screening campaign. A construct of human CD73 which produces crystals of the enzyme in the open conformation was purified from refolded inclusion bodies. A small fragment library containing 107 small molecules was screened *via* crystallography. Individual fragments were soaked into open form CD73 crystals and high-resolution datasets were collected at the BESSY light source. Following structure solution and refinement seven non-nucleotide type ligands were identified. These low affinity fragments have binding modes distinct to that observed for known nucleotide substrates/inhibitors of CD73. We also observe fragment binding at the interface between the C- and N-terminal domains of the enzyme. These low affinity fragments may be starting points for the development of allosteric inhibitors blocking the domain motion of the enzyme.

References:

[1] Zimmermann H; Zebisch M; Sträter N (2012), *Purinergic Signal.* 8:437-502.
[2] Knapp K; Zebisch M; Pippel,J; El-Tayeb A; Müller CE; Sträter N (2012), *Structure 20*, 2161-2173.
[3] Heuts DPHM; Weissenborn MJ; Olkov RV; Shaw AM; Levy C; and Scrutton NS (2012). *ChemBioChem 13*, 2384-2391.

MS08 - Phase transitions and dynamic phenomena

MS08-01

Phase transitions and new structures at low temperature in $Ba_{1-x}Sr_xCuSi_2O_6$ spin dimer compounds

N. van Well[1], P. Puphal[2], B. Wehinger[1,3], M. Kubus[1,4], J. Scheffer[1], S. van Smaalen[5], C. Rüegg[1,3]
[1]Paul Scherrer Institute, LNS, Villigen, Switzerland
[2]Goethe-University, Physikalisches Institut, Frankfurt am Main, Germany
[3]University of Geneva, Department of Quantum Matter Physics, Geneva, Switzerland
[4]University of Bern, Department of Chemistry and Biochemistry, Bern, Switzerland
[5]University of Bayreuth, Laboratory of Crystallography, Bayreuth, Germany

$BaCuSi_2O_6$ is a quasi-two dimensional spin dimer system and a model material for studying Bose-Einstein condensation of magnons in high magnetic fields [1]. It can i.e. be grown with the flux method or with oxygen partial pressure. The new $Ba_{1-x}Sr_xCuSi_2O_6$ mixed system (Fig. a) can only be grown with $x \leq 0.3$ using enhanced oxygen partial pressure [2]. The compounds of this system, and $BaCuSi_2O_6$, both grown with the two methods, have the same tetragonal structure (*I*41/*acd*) at RT. If they are grown with oxygen partial pressure, they show no structural phase transition. This tetragonal average structure is stable down to LT of at least 1.5K. $BaCuSi_2O_6$ crystals, grown by using flux (Fig. b), also show a tetragonal structure (*I*41/*acd*) at RT, but reveal a phase transition from tetragonal to orthorhombic structure (*I*bam) between 80K and 100K with two types of dimer layers [3]. Oxygen partial pressure acts as control parameter for the growth process, and results in the differences of the structure and the physical properties of these compounds. The last results from the synchrotron single crystal diffraction study at LT for these compounds at the BM01A of ESRF will be presented, which show a twinning of $Ba_{1-x}Sr_xCuSi_2O_6$ (Fig. c and d). The first results of the measurements under pressure (0.85 kbar) at LT at the ID27 of ESRF show that the phase transition, as described above, is shifted to a higher temperature at around 200K. To solve the structure, two alternatives have been applied: structure determination with four twins, or, with two twins and two incommensurable vectors. A detailed understanding of the crystal structure, depending on the oxygen content, will allow the study of the spin dynamics of field-induced order states in this model magnetic compound of high current interest with only one type of dimer layers, which shows the same distance between the Cu atoms in the structure.

Figure a) Crystal of $Ba_{0.7}Sr_{0.3}CuSi_2O_6$ grown under oxygen partial pressure, b) $BaCuSi_2O_6$ crystal grown with $LiBO_2$ flux, c) Detector picture of the orthorhombic spitting at 90K with incommensurable reflexes, d) the same reflexes without splitting at 290K

References:
[1] Jaime, M. et al., Phys. Rev. Lett. 2004, 93, 087203
[2] van Well N et al., Crystal Growth & Design 16, 3416 (2016). DOI: 10.1021/acs.cgd.6b00399
[3] Sheptyakov, D. V. et al, Phys. Rev. 2012, B86, 014433

Figure 1

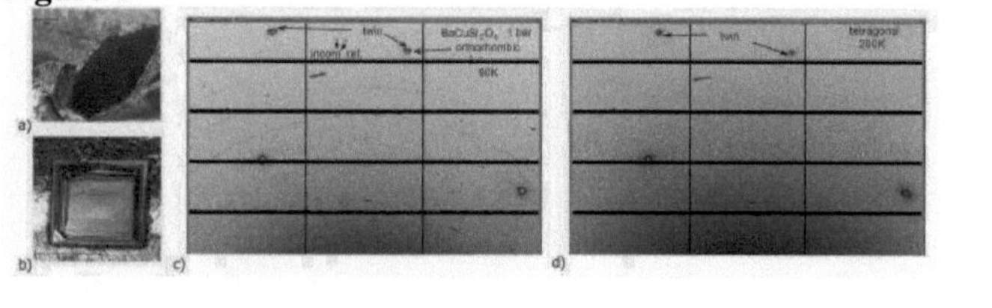

MS08-02

Polymorphic phase-transitions of $CsGaQ_2$ (Q = S, Se)

D. Friedrich[1], M. Schlosser[1], A. Pfitzner[1]
[1]Universität Regensburg, Institut für Anorganische Chemie , Regensburg, Germany

Ternary group 13 chalcogenometallates $M_xT_yQ_z$ containing alkali metal cations (M = alkali metal, T = triel, Q = chalcogen) crystallize in a large variety of different crystal structure types.[1] The anionic structures of all these compounds are formed by linked TQ_4 tetrahedra. Condensation of the small building blocks leads to more complex one-, two- or three-dimensional oligomeric or polymeric anions. The crystal structures of the chalcogenotrielates contain these anions embedded in a cationic surrounding of the alkali metal cations.

The *MTQ*2 family members $CsGaS_2$ and $CsGaSe_2$ both crystallize in two different polymorphic modifications.[2,3] At elevated temperatures, a phase transition from $CsGaQ_2$-*mC*64 to $CsGaQ_2$-*mC*16 (Q = S, Se) takes place. The phase-transition is accompanied by a complete structural reconstruction from two-dimensional layers $^2_\infty[Ga_4Q_8^{4-}]$ in the low temperature phase $CsGaQ_2$-*mC*64 to one-dimensional chains $^1_\infty[GaQ_2^-]$ in the high-temperature phase $CsGaQ_2$-*mC*16 (Figure 1). Whereas the temperature-induced phase-transition of the selenide is reversible, the formation of $CsGaS_2$-*mC*16 is irreversible. By application of high pressure (> 5 GPa), however, the low-temperature polymorph $CsGaS_2$-*mC*64 can be recovered. For $CsGaTe_2$ only data about the *mC*64 polymorph are known to date.[4]

Figure 1: Schematic representation of the reversible phase-transitions in $CsGaQ_2$ (Q = S, Se) featuring the anionic substructures in both polymorphs and the crystal shape, as well as the transition pathways.

The high-temperature phase-transitions of both compounds were further studied *in situ* by high-temperature X-ray powder diffraction. Diffraction experiments performed under isothermal conditions were used to study the phase transition kinetics, crystallization behavior and activation energies. The obtained data were analyzed by using the JMAK-theory of crystal growth[5] and different growth mechanisms found in literature.[6]

References:
[1] B. Krebs, *Angew. Chem.*, **1983**, *95*, 113-134.
[2] D. Friedrich, M. Schlosser, A. Pfitzner, *Crystal Growth & Design* **2016**, *16*, 3983-3992.
[3] D. Friedrich, M. Schlosser, R. Weihrich, A. Pfitzner, *Inorganic Chemistry Frontiers* **2016**, DOI:10.1039/C6QI00462H.
[4] M. Avrami, *J. Chem. Phys.* **1939**, *7*, 1103-1112; M. Avrami, *J. Chem. Phys.* **1940**, *8*, 212-224; M. Avrami, *J. Chem. Phys.* **1941**, *9*, 177-184.
[5] J. H. Sharp, G. W. Brindley, N. N. Achar, *J. Am. Ceram. Soc.* **1966**, *49*, 379-382.
[6] E. J. Wu, M. A. Pell, T. M. Fuelberth, J. A. Ibers, *Z. Kristallogr. - NCS* **1997**, *212*, 91.

Figure 1

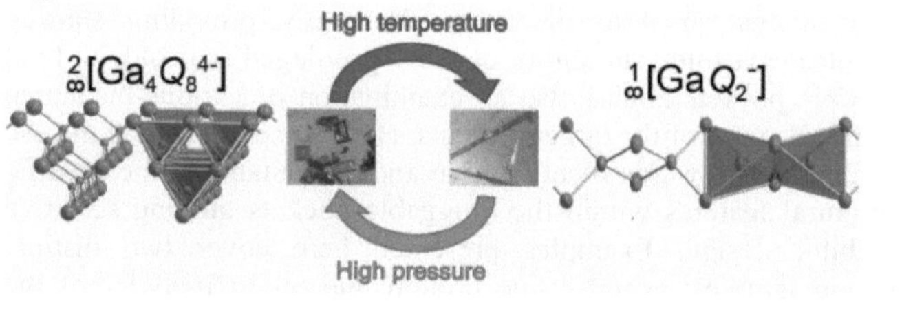

MS08-03

Kinetic effects in piezoelectric actuator materials

M. Hinterstein[1], M. Knapp[1], A. Studer[2], M. Etter[3], A. Schoekel[1], H. Choe[4], S. Gorfman[4], M. J. Hoffmann[1]
[1]Karlsruhe Institute of Technology, Institute for Applied Materials, Karlsruhe, Germany
[2]Australian Nuclear Science and Technology Organisation, Bragg Institute, Sydney, Australia
[3]Deutsches Elektronensynchrotron, Hamburg, Germany
[4]University of Siegen, Department of Physics, Siegen, Germany

Piezoelectric ceramics exhibit the remarkable property to couple elastic strain and polarization. Especially actuators rely on high electric fields to generate high strains and forces. The two most important characteristics of this class of materials are macroscopic strain and piezoelectric coefficient. Despite extensive studies and elaborated measurement techniques, the correlation between macroscopic strain and structural response is still not fully understood.

Most of the relevant systems found up to now are compositions close to phase boundaries linking highly correlated phases. Apart from the well-known field induced structural responses such as domain switching and the converse piezoelectric effect we recently identified field induced phase transitions in different systems as an additional poling mechanism. With *in situ* X-ray and neutron diffraction we were able to identify a range of kinetic effects over almost 9 orders of magnitude in time. These effects range from creep in the low μHz range to complex phase transitions in the range of kHz. These effects were only accessible with a combination of highly sophisticated quantitative structure and texture analysis and stroboscopic powder diffraction.

MS08-04

Zintl phases hydride LnTtHx (Ln = La, Nd) (Tt = Si, Ge, Sn)

A. Werwein[1], H. Auer[1], L. Kuske[1], H. Kohlmann[1]
[1]Universität Leipzig, Institut für Anorganik, Leipzig, Germany

Zintl phases are polar intermetallics composing of group one or two metal and a group 13 to 16 element. Zintl phases react in various ways with hydrogen. The fundamental concept is the oxidation of the polyanion. The formed hydride ion can be incorporated in voids of the crystal structure (interstitial hydride) or bonded covalently to the polyanion (polyanionic hydride) [1]. Due to the oxidation the connectivity can be changed. Especially the Zintl phases which crystallize in the CrB structure type AeTt (Ae = Ca, Sr, Ba Tt = Si, Ge, Sn) show interesting hydrogenation behavior, since both motifs were observed [2,3].

Other compounds which crystallize in the CrB structure type are LnTt [4], which also exhibit the FeB structure type depending on lanthanide. Considering structural and physical properties they may be described by an ionic formula Ln3+Tt2-e-, where the polyanion [Tt2-]n form chains.

The title compounds were prepared by arc melting of the elements and studied during the hydrogenation by *in situ* thermal analysis under hydrogen pressures of 5 MPa and temperatures up to 450°C. The title compounds react with hydrogen at temperatures between 200 and 300 oC. X-ray powder diffraction showed that the compounds form filled FeB and CrB structure types. The compounds with the FeB structure type further show a phase transition. While at low temperatures the cell volume increases, at higher temperature there is a change from the filled FeB to the filled CrB structure type. The compound LaGe (FeB) first reacts to the filled FeB type (a = 8.2710(4) Å, b = 4.1487(2) Å, c = 6.3282(3) Å) and then to a filled CrB structure type (a = 4.35225(10) Å, b = 12.2368(3) Å, c = 4.12452(9) Å). Neutron diffraction studies on the deuteride revealed that the hydrogen incorporated the tetrahedral voids in both structure types. The hydrogen site is completely filled for the FeB type and occupied with 85 % for the CrB type. The compound NdGe (CrB structure type) also incorporated hydrogen in tetrahedral voids and forms the phase NdGeD (a = 4.21969(14) Å , b = 11.8697(4) Å, c = 4.03851(14) Å). The other title compounds LaSi, NdSi (FeB) and LaSn (CrB) react in the same way.

References:
[1] U. Häußermann, V. F. Kranak, K. Puhakainen, *Struct. Bond.* **2011**, *139*, 143-162
[2] H. Wu, W. Zhou, T.J. Udovic, J.J. Rush, T. Yildirim, *Phys. Rev. B*, **2006**, *74*, 224101-1-224101-5
[3] H. Auer, R. Guehne, M. Bertmer, S. Weber, P. Wenderoth, T. C. Hansen, J. Haase, H. Kohlmann, *Inorg. Chem., accepted*
[4] A. E. Dwight, J. W. Downey, R. A. jr.Conner, *Acta Cryst.* **1967**, *23*, 860-862

Figure 1

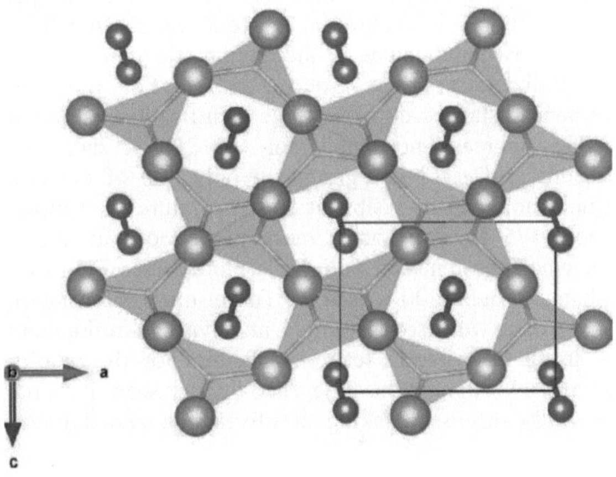

Figure 2

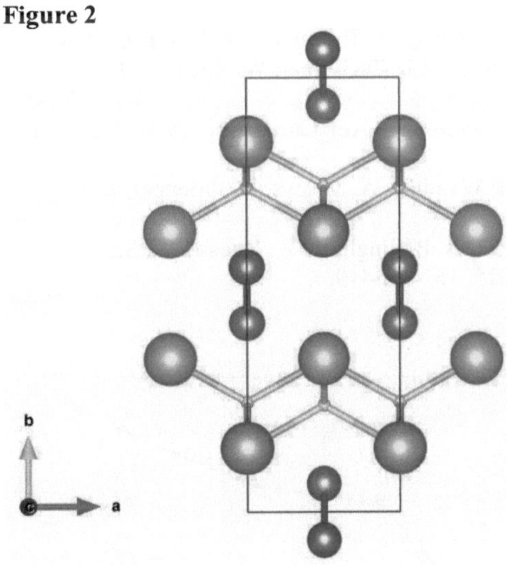

MS08-05

Temperature-dependent structural changes in a Jacobsen catalyst.X-ray induced spin-crossover, phase transition or just rotational disorder?

J. Ballay[1], M. Lutz[2], O. Janka[3], R. Pöttgen[3], C. Czekelius[4], B. Dittrich[1]
[1]Heinrich-Heine Universität Düsseldorf, Anorganische Chemie und Strukturchemie II, Düsseldorf, Germany
[2]Utrecht University, Bijvoet Center for Biomolecular Research; Crystal and Structural Chemistry, Utrecht, Netherlands
[3]Westfälischen Wilhelms-Universität Münster, Institut für Anorganische und Analytische Chemie, Münster, Germany
[4]Heinrich-Heine Universität Düsseldorf, Institut für Organische Chemie und Makromolekulare Chemie, Düsseldorf, Germany

Chiral Jacobsen catalysts are widely used for enantioselective epoxidation of prochiral olefines [1]. In a Jacobsen catalyst the central Manganese atom is in a +3 oxidation state (*d*4-electron configuration). We studied N,N'-bis(3,5-di-tert-butylsalicylidene)-1,2-cyclohexanediaminomanganese(III) chloride in more detail by quantum chemical computations and by single-crystal X-ray diffraction. We also find that, based on magnetic measurements in the temperature range from 3 to 300 K, our powdered sample was in a low-spin state within the investigated temperature region. X-ray data were measured at temperatures of 100, 104, 112, 125, 150, 175, 200, 250 and 300 K to comparably high resolutions around 0.66 Angstoem-1 throughout, most of them (104 - 250 K) on the same specimen at the SLS synchrotron in sequence, starting from 250 K. There are two molecules in the asymmeric unit and the molecule crystallizes in space group $P2_12_12_1$. For most of temperatures bond distances do not change significantly, although the Uiso values increase noticeably for the 200 K data set. Rotational disorder of tert-butyl groups in only one of the two asymmetric-unit molecules is visible at all temperatures, but might be static below 175 K; it becomes more pronounced at higher temperatures; when the unit-cell volume is plotted (see Figure) a significant change around 200 K is seen. To get more information from the X-ray data we also performed an invariom refinement followed by using tailor-made scattering factors for the whole-molecule [2] and investigated the difference density seen. Possible explanations will be discussed, taking all information available into account.

References:
[1] E. N. Jacobsen, W. Zhang, A. R. Muci, J. R. Ecker, L. Deng, J. Am. Chem. Soc. 1991, 113, 7063-7064; W. Zhang, J. L. Loebach,
S. R. Wilson, E. N . Jacobsen, J. Am. Chem. Soc. 1990, 112, 2801-2803.
[2] B. Dittrich, C. M. Wandtke, A. Meents, K. Proepper, K. C. Mondal,
P. P. Samuel, N. A. Sk, A. P. Singh, H. W. Roesky, N. Sidhu, ChemPhysChem, 2015, 16, 412-419.

Figure 1

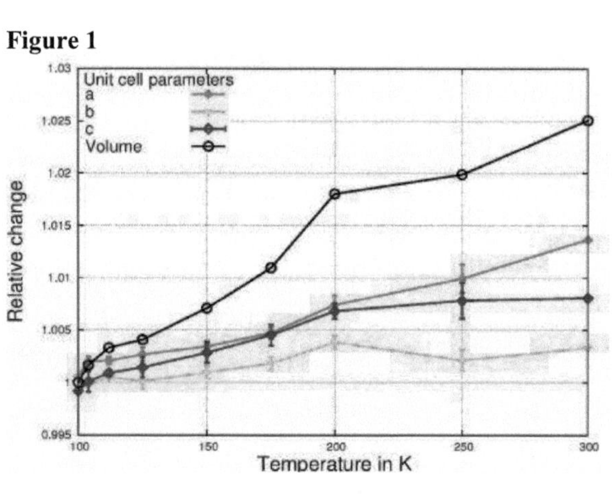

MS08-06

Structure and Thermal Stability of Boehmite derived Transition Aluminas

M. Rudolph[1], M. Motylenko[1], T. Zienert[1], H. Stöcker[2], D. Rafaja[1]
[1]TU Bergakademie Freiberg, Institute of Materials Science, Freiberg, Germany
[2]TU Bergakademie Freiberg, Institute of Experimental Physics, Freiberg, Germany

One of the most interesting intermediary aluminum oxides accompanying the transition from boehmite to corundum is γ-Al_2O_3 that is assumed to facilitate the segregation of non-metallic inclusions from metallic melts during the casting process. Therefore, this metastable alumina is considered as an important constituent of active and reactive surface coatings on ceramic filters.

However, γ-Al_2O_3 is only stable in a certain temperature range. At higher temperatures, it first transforms to other transition phases with less favorable properties and later to the thermodynamically stable corundum. For this reason, the understanding of the structural changes as well as their kinetics during the transition process is essential.

In this study, the structural changes occurring during the conversion of boehmite over the metastable phases γ-, δ- and θ-Al_2O_3 to the stable corundum were investigated using *in situ* and *ex situ* X-ray diffraction (XRD) as well as transmission electron microscopy (TEM) and thermal analysis (DTA/TG). For this purpose, highly crystalline boehmite powder was calcined at temperatures between 300 and 1400 °C with exposure times of one and 20 hours.

TEM and XRD investigations confirmed the crystallographic relationships between individual phases reported in [1–4], but also revealed some discrepancies in the accepted structural descriptions of the transition phases. Furthermore, it could be shown that the transition process from highly defective γ-Al_2O_3 towards the final transition phase θ-Al_2O_3 starts at temperatures exceeding 600 °C and proceeds continuously. Therefore, the transition from γ- to θ-Al_2O_3 proceeds rather over a δ-state than over a distinct δ-phase through a continuous reordering of aluminum ions over a longer range within the preserved oxygen sublattice. This process is kinetically inhibited by a limited mobility of cations at lower temperatures. Consequently, the typical structural changes were postponed towards higher temperatures for shorter calcination times.

References:
[1] Zhou R.S. and Snyder R.L. *Acta Cryst.,* 1991, B47, 617. [2] Repelin Y. and Husson E. *Mat. Res. Bull.,* 1990, 25, 611. [3] Levin I. and Brandon D. *J. Am. Ceram. Soc.,* 1998, 81, 1995. [4] Paglia G., Buckley C.E., Rohl A.L., Hart R.D., Winter K., Studer A.J., Hunter B.A. and Hanna J.V. *Chem. Mater.,* 2004, 16, 220.

MS09 - Hot Biostructures

MS09-02
A RecQ helicase in disguise:
How the unconventional Structure of the human RecQ4 Helicase defines its special position in the RecQ helicase family.

S. Kaiser[1], F. Sauer[1], C. Kisker[1]
[1]Rudolf-Virchow-Center, Structural Biology, Wuerzburg, Germany

The human RecQ4 helicase is a member of the evolutionary conserved RecQ family, an important class of enzymes, dedicated to preserve genomic integrity by operating in DNA repair, telomere maintenance and DNA replication. While reduced RecQ4 activity is associated with cancer predisposition and premature aging, RecQ4-upregulation correlates with carcinogenesis and metastasis. Within the RecQ family, RecQ4 assumes an exceptional position, lacking characteristic domains, which are common for all other RecQ enzymes. We have recently solved the first crystal structure of human RecQ4, encompassing the conserved ATPase core and a novel C-terminal domain, featuring a zinc-binding site and two distinct types of winged-helix motifs. Our mutational analysis suggests a DNA-strand separation mechanism for human RecQ4, which is more closely related to bacterial RecQ helicases than to its human family members. In summary, our structural and functional analysis offers a first glimpse into the molecular details of RecQ4 helicase function and provides the groundwork to examine the complex phenotypes of RecQ4-associated diseases.

MS09-03
Structural Studies on Ectonucleotidases Involved in Purinergic Signaling

C. Döhler[1], M. Zebisch[1], N. Sträter[1]
[1]University of Leipzig, Institute of Bioanalytical Chemistry, Leipzig, Germany

Nucleotide pyrophosphatses/phosphodiesterases (NPPs) are a family of ectophosphodiesterases comprising 7 members in vertebrates [1]. NPPs are glycoproteins and able to hydrolyze a wide range of molecules involved in different signaling pathways (e.g. in purinergic signaling). Whereas NPP1 and 3 are specific for nucleotides and dinucleotides, the natural substrates of NPP2, NPP5 and NPP7 are phospholipids. NPP1-3 include besides the catalytic domain a nuclease-like domain, which has no catalytic activity. Furthermore at the N-terminus of NPP1-3 two consecutive cysteinerich somatomedin B (SMB)like domains are located, which are involved in substrate binding (NPP2) and membrane anchoring (NPP1 and 3). NPP4-7 are only contain the catalytic domain. Apart from NPP2 all NPP family members are membrane associated [2]. Based on their involvement in many physiological functions and diseases NPPs are regarded as attractive drug targets. We aim to determine crystal structures of these proteins to characterize the structural basis of substrate specificity and the catalytic mechanisms. Structures of NPP1 and NPP2 from vertebrates revealed first insights in domain arrangement and ligand binding [3, 4]. Nevertheless for further investigations of the catalytic mechanism of NPP enzymes high resolution structures in complex with substrates or substrate analogs are needed. We report on structural studies of rat NPP, in unliganded form and in complex with various substrates.

References:

[1] Zimmermann H., Zebisch M., Sträter N. (2012). Cellular function and molecular structure of ecto-nucleotidases. Purinergic Signal. 8:437-502.
[2] Stefan C., Jansen S., Bollen M., (2006). Modulation of purinergic signalling by NPP-type ectophosphodiesterases. Purinergic Signal *2*, 361-70
Nishimasu H., Okudaira S., Hama K., Mihara E., Dohmae N., Inoue A., Ishitani R., Takagi J., Aoki J., Nureki O., (2011) Crystal structure of autotaxin and insight into GPCR activation by lipid mediators. Nature Structural and Molecular Biology *18*, 205-12
Katoa K., Nishimasua H., Okudairab S., Miharac E., Ishitania R., Takagic J., Aokib J., Nureki O., (2012). Crystal structure of Enpp1, an extracellular glycoprotein involved in bone mineralization and insulin signalling. PNAS *109*, 16876-81

MS09-04
***Escherichia coli* HrpB – a helicase containing a unique C-terminal region**

A. J. Pietrzyk-Brzezinska[1], M. C. Wahl[1]
[1]Fu Berlin, Department of Biology, Chemistry, Pharmacy, Berlin, Germany

Introduction
Bacteria harbor one or several RNA helicases, which are enzymes that harvest the chemical energy of NTP hydrolysis to effect changes in the conformation of RNAs or protein-RNA complexes. Among them, poorly investigated members of the DExH/RHA family of superfamily 2 helicases may play important roles as co- or post-transcriptional gene regulators, allowing adaptation of bacteria to changing environments and stress conditions.

Objectives
The aim of this study was the structural investigation of the *Escherichia coli* DEFH-box RNA helicase HrpB and its structure-guided functional analysis.

Materials & Methods
HrpB was produced recombinantly and crystallized using the vapor diffusion method. The crystal structure of HrpB was solved using SeMet-derivatized HrpB crystals and single-wavelength anomalous dispersion. The activities of HrpB were investigated using various biochemical, biophysical and molecular biological approaches.

Results
We determined the crystal structure of HrpB at 2.23 Å resolution. The protein can be divided into six domains: two RecA-like domains, a winged helix domain (WH), a helical bundle domain (HB), an oligonucleotide/oligosaccharide-binding fold (OB) and a large C-terminal region (CTR). While a helicase core comprising RecA1, RecA2, WH, HB and OB domains has previously been seen in some eukaryotic DExH/RHA helicases, the CTR represents a novel expansion that is unique to HrpB proteins. To elucidate the functional role of the CTR, we are presently conducting activity assays using structure-guided truncation variants of the enzyme, monitoring nucleotide binding, nucleic acid binding, nucleotide hydrolysis and nucleic acid unwinding. We also constructed an *E. coli hrpB* deletion strain and are monitoring effects of the manipulation on growth and stress tolerance.

Conclusion
The crystal structure of HrpB represents the first structure of a bacterial DExH/RHA family RNA helicase, allowing structural comparisons to known eukaryotic relatives. It revealed a novel helicase-associated CTR, which may modulate the enzymes helicase mechanism, its interaction with substrates or its association with other proteins.

MS10 - Electron Crystallography

MS10-01
Structural characterisation of porous functional materials using a combination of electron diffraction tomography, HRTEM and powder X-ray diffraction

H. Zhao[1], B. Barton[1], I. Großkreuz[2], H. Gies[2], T. Rhauderwiek[3], N. Stock[3], U. Kolb[1]
[1]Institut für Anorganische Chemie und Analytische Chemie , Johannes Gutenberg-Universität Mainz , Mainz, Germany
[2]Department of Geology, Mineralogy and Geophysics, Ruhr University Bochum, Bochum, Germany
[3]Institute for Inorganic Chemistry, Christian-Albrechts-University Kiel, Kiel, Germany

Micro- and mesoporous materials with high surface area and large pore volume, such as zeolites and metal-organic framework (MOFs), have attracted much attention due to their potential applications for catalysis, gas storage, drug delivery, etc. The structural investigation of these materials is a key step to understand their physical properties for optimization in applications and to design novel materials. As compared to the established crystal structure solution approach by X-ray powder diffraction, electron diffraction (ED) has advantages for structures with large unit-cell, disorder, superstructure, pseudo symmetry etc. Automated electron diffraction tomography (ADT) combines nano diffraction, tilt series acquisition and electron beam precession to derive a three-dimensional atomic structure from a single nano crystal [1]. In recent years, a number of *ab initio* structure solutions were successfully performed based on ADT data. For complex materials, the crystal structure is best studied by a combination of electron diffraction, HRTEM and X-ray diffraction.

The interlayer-expanded microporous titanosilicate (Ti-IEZ-RUB-36) shows catalytic activity for the oxidation of small molecules and bulky molecules [2]. An orthorhombic lattice with a = 24.0726 Å, b = 14.0251 Å, c = 7.4118 Å in space group $Pnm2_1$ was obtained from the indexing of the powder X-ray diffraction data and also from the reconstruction of three-dimensional ED data. Strong diffuse streaks (Fig.1) were observed in the ED data, which indicates the layer stacking of *bc* plane along the a direction. The *ab initio* structure solution was performed using direct methods from ADT data.

The crystal structure of a Zirconium metal-organic framework ZrMOF was investigated by ADT. A large tetragonal unit cell with $a = b$ = 45.32 Å, c = 8.16 Å was obtained from reconstructed three-dimensional ED data. Symmetry analysis results in the space group $I4_1/acd$. Metal atoms were clearly observed in the structure solution against ADT data. Structure refinement and HRTEM analysis are ongoing.

References:
[1] U. Kolb, E. Mugnaioli and T. E. Gorelik, Cryst. Res. Technol., 2011, 46, 542–554.
[2] F. S. Xiao, B. Xie, H. Zhang, L. Wang, X. Meng, W. Zhang, X. Bao, B. Yilmaz, U. Müller, H. Gies, H. Imai, T. Tatsumi and D. De Vos, ChemCatChem, 2011, 3, 1442–1446.

Figure 1

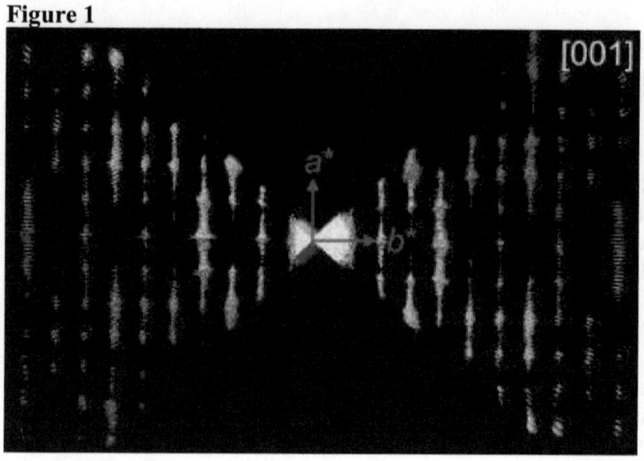

MS10-02
The Rotation Method in Electron Crystallography

T. Gruene[1], J. P. Abrahams[1,2], E. van Genderen[1], M. Clabbers[2], T. Blum[2]
[1]Paul Scherrer Institut, Laboratory of Biomolecular Research, Villigen PSI, Switzerland
[2]University of Basel, C-CINA, Basel, Switzerland

The rotation method is the *de facto* standard for structure determination of organic and macromolecular compounds with single crystal X-ray diffraction[1]. Encouraged by recent development in hardware and data processing methods, the rotation method is now also being applied in electron diffraction[2-4]. The group of authors for this contribution focus on the improvement of hardware and methods in order to make electron diffraction a complementary method for structure determination. One initial milestones is the Timepix/ Medipix family of detectors. This hybrid pixel detectors were used for the first diffraction data recorded from a 3D protein crystal[5]. We now have two Timepix cameras installed at a FEI Polara and a FEI Talos microscope. We have determined the structure from various radiation sensitive materials like crystals pharmaceuticals, zeolites, and proteins. Our data provides insight into which aspects of the data collection and data processing process deserve deserve most attention in order to improve the reliability of the models. The latest results from our work include the assessment of correcting lens distortions in order to produce more precise unit cell dimensions, and the structure of pharmaceutical compounds where the data are strong enough to refine hydrogen atom positions - in X-ray crystallography, their positions are usually calculated as riding atoms. These results, as shown in Fig. 1 where direct methods assign all but one atom type correctly, imply that current data quality is much better than traditional quality indicators from X-ray crystallography and provide guide lines for the construction of a dedicated electron diffraction instrument.

References:
[1] Arndt & AJ Wonacott, "The Rotation Method in Crystallography" (1977)
[2] TE Gorelik, J van de Streek, AFM Kilbinger, G Bryunklaus, U Kolb; Acta Cryst
(2012), B68, 171-181
[3] W Wan, J Sun, J Su, S Hovmoeller, X Zou; J Appl. Cryst. (2013), 46, 1863-1873
[4] M Gemmi, P Oleynikov, Z. Kristallogr. -- Cryst. Mater. (2013), 51-58
[5] D Georgieva, J Jansen, I Sikharulidze, L Jiang, HW Zandbergen, JP Abrahams;
J. Instrum (2011), Vol. 6, C01033

Figure 1

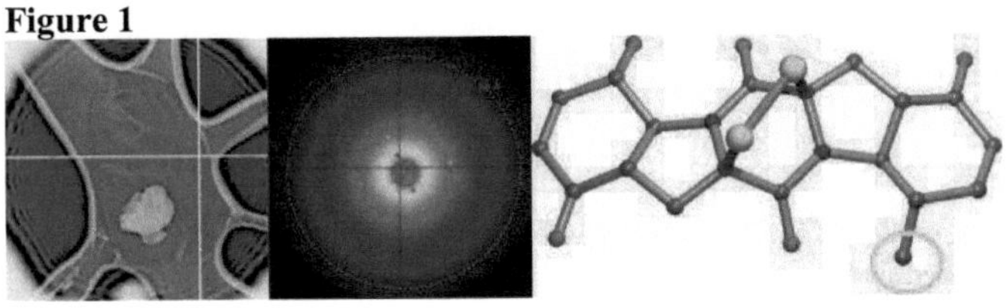

MS10-03
Crystal structure of a disordered organic pigment determined by electron diffraction

T. E. Gorelik[1], C. Czech[1], S. M. Hammer[1], M. U. Schmidt[1]
[1]Johann Wolfgang Goethe-University Frankfurt am Main, Institute of Inorganic and Analytcal Chemistry, Frankfurt am Main, Germany

Quinacridone is an industrial organic pigment used for coloration of paints, plastics and printing inks. In its application the pigment is not dissolved but used as a fine crystalline powder. The molecule can crystallize in four polymorphic forms with different colour and stability. The crystal structure of three polymorphs were known, while the structure of the aII form remained unresolved so far. The fine crystals size of the aII polymorph hinders the structure analysis using single crystal or powder X-ray diffraction techniques.

Due to the instrumental flexibility of Transmission Electron Microscopes (TEMs) electron diffraction data of quality suitable for a structure analysis can be collected from crystals of a few tens of nanometres in size. This makes electron diffraction a unique method accessing the crystal structure of nanocrystals. The robustness and reproducibility of structure analysis from electron diffraction data was numerously demonstrated during the last decade.

As TEMs are mostly designed to work with electron-beam stable materials, the main challenge of collecting electron diffraction from an organic crystal is the optimization of the experimental conditions in order to reduce the beam damage. Apart from the practical issues in data collection, diffraction data of aII quinacridone crystals showed intensive diffuse scattering.

Despite the listed issues, the structure of aII quinacridone was solved from electron diffraction data. The structure solution path together with the explanation of the diffuse scattering will be presented. The relation of the newly obtained structure to the previously solved polymorphs will be discussed.

MS10-04
TEM investigation of $CuMn_2O_4$ spinel catalyst upon reduction and subsequent reoxidation

M. Trapp[1], M. M. Müller[1], H. J. Kleebe[1]
[1]TU Darmstadt, Institute of Applied Geosciences, Darmstadt, Germany

1. Introduction
A novel catalyst based on copper-manganese spinel was developed for the application in vehicle engine three-way-catalysts (TWC) in the recent past[1]. The effects of reducing and oxidizing exhaust species on this material, which are relevant for the long-term usage, were investigated indirectly and macroscopically so far but not directly on a microscopic scale. In order to gain more insight in this material and the proceeding reactions, it was subjected to an investigation by transmission electron microscopy (TEM).

2. Objectives
TEM was performed on a sample series of a pure $CuMn_2O_4$ spinel catalyst model system in order to study the material itself as well as the effects of its reduction by carbon monoxide and its subsequent reoxidation by oxygen.

3. Materials & methods
Fresh, reduced and reoxidized $CuMn_2O_4$ powder samples were prepared for TEM. The reduction was performed by CO at 600°C for 5, 10, 20 and 120 min, whereas the reoxidation was done by O_2 for 120 min. Conventional TEM imaging was combined with probe-corrected STEM, electron diffraction, EDS and image simulations.

4. Results
A phase transition from $CuMn_2O_4$ spinel into separated MnO and Cu-rich phases, which are CuO or Cu depending on the reduction period, was observed in the reduced samples. Furthermore, a previously unknown surface coverage of all MnO grains by epitaxially grown, nanocrystalline Mn_3O_4 was found. The phase separation was completely reversed and pure $CuMn_2O_4$ spinel was reconstituted in the reoxidized samples. SAED imaging showed superlattice reflections to be a characteristic feature of the fresh spinel whereas absent in the reoxidized spinel.

5. Conclusion
A complete reduction into separated phases with a subsequent, likewise complete reoxidation of $CuMn_2O_4$ spinel is possible under the conditions applied. With respect to the close relationship between $CuMn_2O_4$ and Mn_3O_4, the nanocrystals on the MnO surface are assumed to act as seed crystals for the spinel recrystallization. The superlattice reflections were assigned to a $P4_332$ supercell formed by cation ordering, which is destroyed by the reduction and not reconstituted in the reoxidized spinel.

References:
[1] S. Golden, Z. Nazarpoor, and M. Launois, "Novel Mixed Metal Oxide Structure for Next Generation Three-Way Catalysts." in. SAE International, 2015.

MS10-05
Radial distribution function imaging: a new TEM method to analyze multiphase amorphous materials

X. Mu[1,2], D. Wang[1,3], T. Feng[4], C. Kübel[1,2,3]
[1]Karlsruhe Institute of Technology (KIT), Institut für Nanotechnology (INT), Eggenstein-Leopoldshafen, Germany
[2]Karlsruhe Institute of Technology (KIT), Helmholtz-Institute Ulm for Electrochemical Energy Storage (HIU), Ulm, Germany
[3]Karlsruhe Institute of Technology (KIT), Karlsruhe Nano Micro Facility (KNMF), Eggenstein-Leopoldshafen, Germany
[4]Nanjing University of Science and Technology (NJUST), Herbert Gleiter Institute of Nanoscience, Nanjing, China

Interpreting the atomic structure of amorphous materials is a challenge, especially in case of heterogeneous nanoglasses, which have fueled interests because of their unusual structure and properties [1]. Only few experimental means offer a way to characterize disordered structures. Atomic radial distribution function (RDF) is one of the important tools, which was first applied to X-ray diffraction of organic solids and extended to electron diffraction for inorganic glasses [2]. The RDF describes the probability to find atomic pairs as a function of pair separation and thus provides short/medium-range structural information [2]. However, traditional diffraction experiments only provide an average over large sample areas. Critical local variations can be hidden in the averaged signal.

Here, we demonstrate a newly developed scanning transmission electron microscopy (STEM) method, RDF-imaging [3], combining electron diffraction in STEM with RDF and hyperspectral analysis for structural mapping of multiphase amorphous materials with nanometer resolution. A 4-dimensional diffraction map is acquired by recording diffraction patterns in STEM with quasi parallel nano-beam and ~1 nm spot size. RDFs are calculated from all diffraction patterns to construct a 3D data cube of RDFs. These can be analyzed by hyperspectral techniques to obtain phase maps of multiphase amorphous materials. The structure of each phase can be analyzed based on the RDFs in terms of bond distance, angle and coordination number.

Application to amorphous ZrO2/ZrFe multilayers will be shown as example. In addition to the amorphous ZrO2 and ZrFe phases, a 3-4 nm interface layer between ZrO2 and ZrFe was identified by RDF mapping, which could not be identified by traditional TEM techniques such as EELS and EDX maps. By analyzing the RDF, the atomic structure of the interfacial layers could be understood. It possesses the same atomic packing as amorphous ZrO2 phase but with a 0.04 Å shrinkage of the average bond distance. The change in bond distance could be due to O depletion and Fe incorporation.

References:
[1] R Witte et al., Appl Phys Lett **103** (2013), p. 073106.
[2] D J H Cockayne, D R Mckenzie, Acta Cryst. A 44 (1988), p. 870.
[3] X Mu et al., Ultramicroscopy 168 (2016), p. 1.

MS11 - Structural biology meets chemistry

MS11-01
The fine-tuned machinery of O2-tolerant [NiFe] hydrogenase

A. Schmidt[1], P. Scheerer[1], J. Kalms[1]
[1]Charite - Universitaetsklinikum Berlin, Institut fuer medizinische Physik und Biophysik, Berlin, Germany

Hydrogenases are metalloenzymes catalyzing the interconversion of H2 2H+ + 2e-. Hydrogenases exist in all domains of life, but only a small subgroup of [NiFe] hydrogenases evolved the feature of H2 conversion under aerobic conditions. For enabling the aerobic H2 oxidation in [NiFe] hydrogenases, multiple adaptable pathways have been evolved. Structural investigations on this biological machine may lead to new developments in renewable energy technologies [1].

The membrane-bound [NiFe] hydrogenase (MBH) of *Ralstonia eutropha (R.e.)* is one of the best investigated O2-tolerant hydrogenases. Several crystal structures of *R.e.*MBH as wildtype or with multiple substitutions in different redox states reveal a fine-tuned interplay between pathways and channels that lead to a perfect transport of reagents to/from the active site [2, 3]. For H2 oxidation the active site requires the delivery of H2 via a hydrophobic gas channel. H2 oxidation liberates electrons which are guided via an electron pathway to an electron acceptor. Subsequently, the electrons enter the quinone pool of the respiratory chain as reduction power for the cell [2]. Under aerobic conditions additionally the active site has to reduce the attacking O2 + 4e- + 4H+ → 2H2O. Consequently, the electron pathway, consisting of three [FeS] clusters, has to operate bidirectional. Therefore, a unique [4Fe3S] cluster proximal to the active site undergoes a redox-dependent reversible transformation. For H+ delivery several pathways have been investigated and introduce new questions that might be answered by investigative methods e.g. neutron diffraction. The gas channel that is supplying also the inhibitory O2 has been adapted especially in quantity and size to remain the hydrogenase activity for the system [4]. H2O produced under O2 reduction is released through a water channel. This complex system is still not completely understood and moreover sensitive to X-rays. Consequently a near radiation-damage free technique, the free-electron laser, has been used to gain further insights into the functionality of this enzyme.

References:
[1] Fritsch et al (2013) *NatRevMicrobiol.* 2, 106-114.
[2] Fritsch & Scheerer et al (2011) *Nature* 479 (7372), 249-52.
[3] Frielingsdorf et al (2014) *Nat Chem Biol.* 10, 378-385.
[4] Kalms et al (2016) *Angew Chem Int Ed Engl.* 55(18), 5586-90

Figure 1

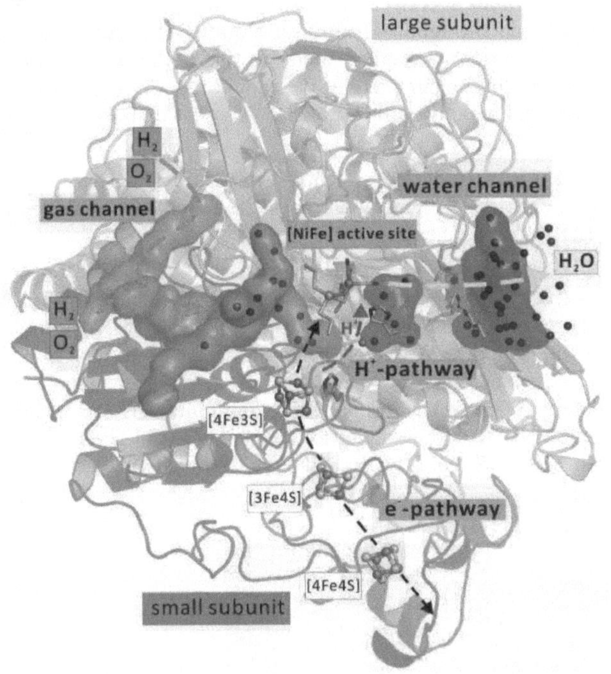

MS11-02
BIODIFF - a neutron diffractometer optimized for crystals with large unit cells: New developments and recent application examples

T. E. Schrader[1], A. Ostermann[2], M. Monkenbusch[3], B. Laatsch[4], P. Jüttner[2], W. Petry[2], D. Richter[1,3]
[1]Forschungszentrum Jülich GmbH, Jülich Centre for Neutron Science, Garching, Germany
[2]Technische Universität München, Heinz Maier-Leibnitz Zentrum (MLZ), Garching, Germany
[3]Forschungszentrum Jülich GmbH, Institute for Complex Systems ICS, Jülich, Germany
[4]Forschungszentrum Jülich GmbH, Engineering and Technology (ZEA-1), Jülich, Germany

The neutron single crystal diffractometer BIODIFF at the research reactor Heinz Maier-Leibnitz (FRM II) is especially designed to collect data from crystals with large unit cells. The main field of application is the structural analysis of proteins, especially the determination of hydrogen atom positions. BIODIFF is a joint project of the Jülich Centre for Neutron Science (JCNS) and the FRM II. BIODIFF is designed as a monochromatic instrument with a narrow wavelength spread of less than 3 %. To cover a large solid angle the main detector of BIODIFF consists of a neutron imaging plate in a cylindrical geometry with online read-out capability. An optical CCD-camera pointing at the sample position is used to quickly align the sample with respect to the neutron beam. The main advantage of BIODIFF is the possibility to adapt the wavelength to the size of the unit cell of the sample crystal while operating with a clean monochromatic beam that keeps the background level low. BIODFF is equipped with a standard Oxford Cryosystem "Cryostream 700+" which allows measurements in the temperature range from 90 K up to 500 K. A new kappa goniometer head was added recently (see Figure 1). This allows an automated tilting of the crystal in order to increase the completeness of the data set when recording another set of frames in the tilted geometry. Typical scientific questions addressed are the determination of protonation states of amino acid side chains in proteins and the characterization of the hydrogen bonding networks between the protein active centre and an inhibitor or substrate. Figure 2 shows one example of a typical application of BIODIFF as a neutron protein crystallography instrument: a deuterium ion is found in the middle between the carboxy group of an amino acid side chain and the imidazol nitrogen atom of the product of the catalyzed reaction rendering them to have the same acidity constant. Furthur examples to illustrate the power of neutron protein crystallography will be given in this contribution.

Figure 1 An Oxford Cryosystems "Cryostream 700 Plus" being aligned to the crystal pin mounted on the new kappa goniometer head (ARINAX, France).

Figure 2 A deuterium atom in between an amino acid side chain and the product as seen by a data set from the BIODIFF instrument. The difference Fo-Fc nuclear density omit map in blue of the shared D^+ ion and of the N9 nitrogen and deuteron atom contoured to 3.5σ. The Fo-Fc X-ray omit map of the adenine molecule and D198, contoured to 3.0σ is shown in green. Figure taken from PNAS, vol. 113, pp. 13756–13761 (2016).

Figure 1

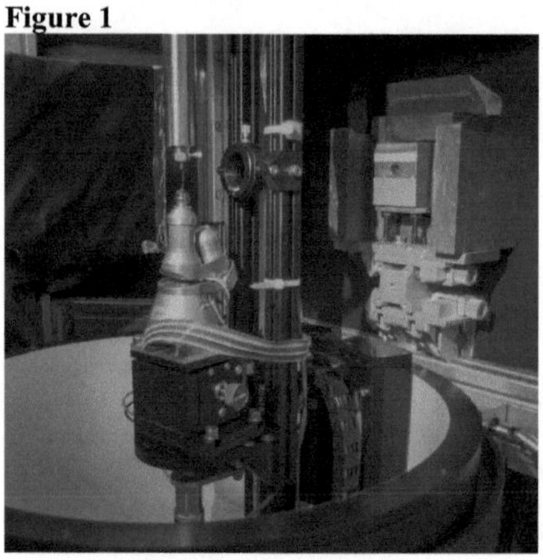

Figure 2

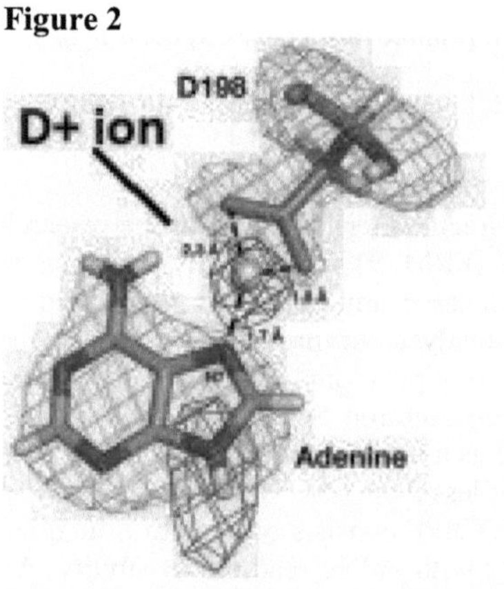

MS11-03

Structural basis for the formation of (*2S,4R*)-methyl-proline within the biosynthesis of griselimycins, natural products with anti-tuberculosis activity

P. Lukat[1], Y. Katsuyama[2,3], S. Wenzel[3], M. Brönstrup[1], R. Müller[3], W. Blankenfeldt[1]
[1]Helmholtz-Centre for Infection Research, SFPR - Structure & Function of Proteins, Braunschweig, Germany
[2]University of Tokyo, Department of Biotechnology, Graduate School of Agricultural and Life Sciences, Tokyo, Japan
[3]Helmholtz-Institute for Pharmaceutical Research, Saarbrücken, Germany

Recently, griselimycins (GMs) were described as natural products exhibiting good anti-tuberculosis activity[1]. Their depsidecapeptide scaffold contains the non-proteinogenic amino acid (*2S,4R*)-4-methyl-proline. While only two of these residues are incorporated into the major component produced by fermentation cultures of the natural producer *Streptomyces* strain, a minor species is secreted that contains a third (*2S,4R*)-4-methyl-proline residue. This third (*2S,4R*)-4-methyl-proline is highly important for the metabolic stability of GMs, as the methyl group at this position blocks oxidative degradation of the drug candidate[1]. Due to this finding, we set out to elucidate the molecular basis of (*2S,4R*)-4-methyl-proline biosynthesis and its incorporation into GMs. Within the GM biosynthetic gene cluster, we identified a sub-operon responsible for 4-methyl-proline formation. We were able to show that 4-methyl-proline is generated from l-leucine in four steps analogous to a previously reported 4-methyl-proline biosynthesis pathway in cyanobacteria[2-3] with the key difference that in the latter pathway (*2S,4S*)-4-methyl-proline is formed instead of the (*2S,4R*)-diastereomer. For GM biosynthesis, we identified the the leucine hydroxylase GriE to catalyze the stereospecifity-determining step within (*2S,4R*)-4-methyl-proline formation. We were able to determine the crystal structures of GriE in complex with its substrates and reaction products at 1.8 and 1.5 Å resolution, providing insight into the stereospecificity of (*2S,4R*)-4-methyl-proline generation.

References:

[1] A. Kling, P. Lukat, D. V. Almeida, A. Bauer, E. Fontaine, S. Sordello, N. Zaburannyi, J. Herrmann, S. C. Wenzel, C. Konig, N. C. Ammerman, M. B. Barrio, K. Borchers, F. Bordon-Pallier, M. Bronstrup, G. Courtemanche, M. Gerlitz, M. Geslin, P. Hammann, D. W. Heinz, H. Hoffmann, S. Klieber, M. Kohlmann, M. Kurz, C. Lair, H. Matter, E. Nuermberger, S. Tyagi, L. Fraisse, J. H. Grosset, S. Lagrange, R. Muller, *Science* **2015**, *348*, 1106-1112.
[2] H. Luesch, D. Hoffmann, J. M. Hevel, J. E. Becker, T. Golakoti, R. E. Moore, *J Org Chem* **2003**, *68*, 83-91.
[3] M. Hibi, T. Kawashima, P. M. Sokolov, S. V. Smirnov, T. Kodera, M. Sugiyama, S. Shimizu, K. Yokozeki, J. Ogawa, *Appl Microbiol Biotechnol* **2012**.

MS11-04
Elucidation of photochemical mechanisms by combination of protein crystallography, spectroscopy and computational chemistry

T. Domratcheva[1]
[1]Max Planck Institute for Medical Research, Department of Biomolecular Mechanisms, Heidelberg, Germany

Crystal structures of proteins provide critically important information for elucidation of biomolecular mechanisms. In particular, the structures stimulate spectroscopy experiments characterizing proteins during their functioning. The obtained spectral signatures can be further linked to the protein function with the help of computational-chemistry models which are based on the crystal structures. In the past decade, the approach combining structural, spectroscopy and computational studies led to elucidation of the role of hydrogen bonds in photosensory flavoproteins that function as photoreceptors. Flavoproteins responsive to wavelengths at the maximum of the solar spectrum undergo photoactivation by redox reactions of the flavin chromophore. Crystal structures of these photoreceptors revealed hydrogen-bonding interactions that are involved in spectral and redox tuning of the flavin chromophore. Infrared spectroscopy studies supplemented with selective and global isotope labeling identified spectral signatures of these hydrogen bonds upon photoactivation [1, 2]. We developed quantum-chemical models that proved to be indispensable in linking the identified spectral signatures to the active-site structures. Furthermore, our computational models elucidated that hydrogen bonds specifically stabilize the 1-e reduced flavin. We demonstrated that such stabilization is central in DNA repair enzyme photolyase [1], bacterial photoreceptor BLUF [2], and putative magnetoreceptor cryptochrome [3].

References:
[1] Mahaputra Wijaya, M. *et al.* Single hydrogen bond donation from flavin N5 to proximal asparagine ensures FAD reduction in DNA photolyase. *J. Am. Chem. Soc.* 138, 4368–4376 (2016).
[2] Domratcheva, T. *et al.* Evidence for tautomerisation of glutamine in BLUF blue light receptors by vibrational spectroscopy and computational chemistry. *Sci. Rep.* **6**, 22669; doi: 10.1038/srep22669 (2016).
[3] Solov"yov, I.A., Domratcheva, T. & Schulten, K. Separation of photo-induced radical pair in cryptochrome to a functionally critical distance. *Sci. Rep.* **4**, 3845; doi: 10.1038/srep03845 (2014).

MS12 - Spectroscopy as supporting tool in crystallography

MS12-01
The First Ni(II) Complex of 4-Methoxypicolinic Acid: Synthesis, Crystal Structure, Infrared and UV-Vis spectra: A combined Experimental and DFT Study

Î TAMER[1]
[1]Sakarya University, Physics, Sakarya, Turkey

A novel Ni(II) complex of 4-methoxypicolinic acid (4-mepicH), [Ni(4-mepic)$_2$(H_2O)$_2$], was synthesized and its crystal structure was fully determined by X-Ray diffraction technique. Figure 1 demonstrates that donor O and N atoms coordinate to central Ni(II) ion, forming five-membered chelate ring. Two aqua ligands coordinated to the central ion contribute to octahedral coordination sphere of central Ni(II) ion. FT-IR spectrum for Ni(II) complex was recorded in the range of 4000-600 cm^{-1}. The crystal structure was also verified by FT-IR spectum. The wavenumber interwal between symmetric and asymmetric COO- stretching vibrations demonstrated that the carboxylate group coordinates to Ni(II) ion as a monodentate ligand. The UV-Vis spectrum for Ni(II) complex was recorded in ethanol and DMSO solvent. In Uv-Vis spectrum of Ni(II) complex, the ligand-ligand and ligand-metal charge trasnfer peaks have been observed. The calculated natural charge found to be smaller than that of free Ni(II), indicating that the electronic charge transfer from 4-mepic to Ni(II) ion is greater than that of opposite direction. In order to provide a deep understanding concerning the relation between crystal structure and spectroscopic properties, density functional theory calculations were performed for the synthesized complex.

Figure 1. The crystal structure of [Ni(4-mepic)$_2$(H_2O)$_2$] complex

Figure 1

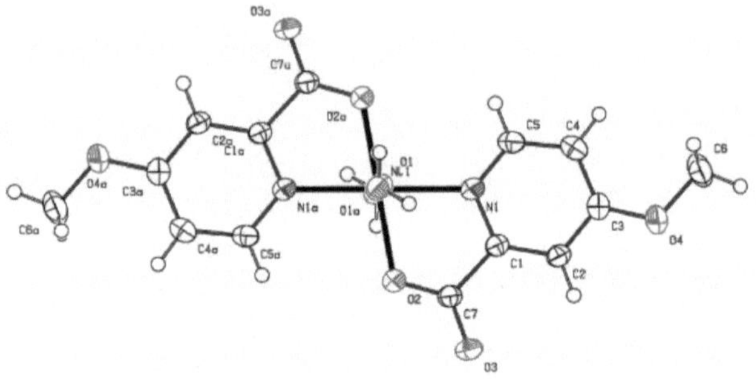

MS12-02
Crystal structure solution and phase characterization of α/β-Ni(OH)$_2$-like basic nickel chlorides

S. Bette[1], R. E. Dinnebier[1], D. Freyer[2]
[1]Max-Planck-Institute for Solid State Research, Stuttgart, Germany
[2]TU Bergakademie Freiberg, Institute for Inorganic Chemistry, Freiberg, Germany

Question

Brucite like β-Ni(OH)$_2$ and its LDH phases, "α-Ni(OH)$_2$", with incorporated water and/or anions are important electrode materials and the basis of modern nickel metal hydride (NiMH) technologies.[1] Basic nickel chlorides, i.e. nickel(II) chloride hydroxides, are closely related to these materials, but they have only been poorly characterized yet and all crystal structures are unknown. [2,3]

The present work focuses on the characterization and crystal structure elucidation of basic nickel chlorides.

Methods

Various nickel(II) chloride hydroxides, NiCl(OH), $NiCl_x(OH)_{2-x}$, $NiCl(OH)_3$, $Ni_3Cl(OH)_5 \cdot 4\,H_2O$, $Ni_3Cl_{2+x}(OH)_{4-x} \cdot 4\,H_2O$ (x = 0.1) and $Ni_3Cl_{2+x}(OH)_{4-x} \cdot 2\,H_2O$ (x = 0.26, 0.48, 0.82) were synthesized.[4-8] The phase compositions were derived from chemical analyses and the crystal structures solved from XRPD data.

Results

The crystal structures of all basic nickel chlorides consist of layers or chains of edge sharing, distorted $NiCl_{6x}(OH/H_2O)_{6-6x}$ octahedra. In the structures of $Ni_3Cl_{2+x}(OH)_{4-x} \cdot 4\,H_2O$[4] and $Ni_3Cl_{2+x}(OH)_{4-x} \cdot 2\,H_2O$[7] these octahedra form Ni(OH)$_2$ like sheets interrupted by cation vacancies (Fig. 1). The relationship to β-Ni(OH)$_2$ is reflected by their spectroscopic and magnetic properties.

The variable Cl^- to OH^- ratio of most phases originates from occupationally disordered Cl/O sites. From the XRPD data the distribution of OH^- and H_2O within the crystal structures of nickel(II) chloride hydroxide hydrates cannot be estimated directly. Distinct water molecules were localized by comprehensive IR-spectroscopy.[4]

Conclusions

Eight novel basic nickel chlorides were synthesized and characterized for the first time. Structure solution from XRPD data and complementary analyses provided a deep crystallographic understanding of these phases.

References:
[1] Hall et al. *Proc. R. Soc. A* **2014**, *471*, 20140792-1-20140792-65.
[2] Feitknecht, Collet, *Helv. Chim. Acta* **1939**, *22*, 1428-1444.
[3] Lv et al. *Electrochim. Acta.***2016**, *216*, 35-43.
[4] Bette et al. *Inorg. Chem.* **2014**, *53*, 4316-4325.
[5] Bette et al. *J. Solid State Chem.* **2015**, *228*, 131-140.
[6] Bette et al. *J. Appl. Crystallogr.* **2015**, *48*, 1706-1718.
[7] Bette et al. *Eur. J. Inorg. Chem.* **2016**, *12*, 1875-1885.
[8] Bette et al. *Eur. J. Inorg. Chem.* **2017**, *accepted for publication.*

Fig 1. Comparison of β-Ni(OH)$_2$ like layers.

Figure 1

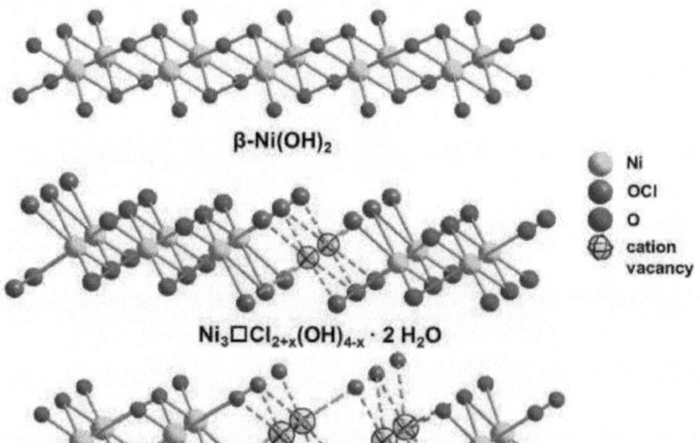

MS12-03

Improving QM/MM and MO/MO computations of displacement parameters incrystal structures and comparison to measured IR spectra forpredicting both external and internal modes of vibration

B. Dittrich[1], D. R. Dombrowski[1], C. A. Sever[1], J. Lübben[1]
[1]Heinrich-Heine Universität Düsseldorf, Anorganische Chemie und Strukturchemie II, Düsseldorf, Germany

The overall scale of anisotropic displacement parameters (ADPs) for crystal structures determined by single crystal x-ray or neutron diffraction is usually not well known. Disagreements concerning their absolute scale between the same structure measured at the same temperature are more common than good agreement for the few cases were both neutron as well as x-ray diffraction data are available. The need for empirical scaling procedures [1] illustrates this unsatisfactory situation. Availability of a procedure that is independent of experimental sources of error like extinction or absorption, and that is un-demanding in terms of the computational effort, would therefore be highly desirable. Experimental IR spectra can help to calibrate ab initio computations, whose computed IR frequencies can be converted to ADPs. Here periodic DFT computations start to provide the desired results [2], albeit at high computational cost. In this contribution we will report on our efforts to further develop QM/MM or MO/MO methodology with the aim to provide an computationally less demanding but still accurate alternative to compute atomic displacement parameters. QM/MM or MO/MO computations can already provide internal modes of vibration that, combined with a TLS fit to non-hydrogen ADPs, reliably provide estimates of ADPs for hydrogen atoms [3]. More recently prediction of the correct temperature dependence from pre-calculated fragment contributions was also achieved [4]. Since our earlier attempts to predict all-atom ADPs by the same means required empirical scaling, and since we were not able to take temperature dependence into account properly we extended and modified source code and procedures in the program BAERLAUCH [5]. Experimental IR/THz spectra will be compared to our theoretical predictions and we will report ADP results of the amino acid L-homoserine and the carbapenem antibiotic imipenem monohydrate. For the latter structure multi-temperature x-ray data were collected.

References:

[1] Blessing, R. H. (1995). Acta Cryst. B, 51, 816–823.
[2] Madsen, A. Ø., Civalleri, B., Ferrabone, M., Pascale, F. & Erba, A. (2013). Acta Cryst. A, 69, 309–321; Deringer, V. L., Stoffel, R. P., Togo, A., Eck, B., Meven, M. & Dronskowski, R. (2014). Cryst. Eng. Comm. 16, 10907–10915.
[3] Whitten, A. E. & Spackman, M. A. (2006). Acta Cryst. B, 62, 875–888.
[4] Lübben, J., Bourhis, L. J. & Dittrich, B. (2015). J. Appl. Cryst. 48, 1785–1793.
[5] Dittrich, B., Pfitzenreuter, S. & Hübschle, C. B. (2012). Acta Cryst. A, 68, 110–116.

MS12-04

Complex substitution in Intermetallics

V. Peters[1], A. Korthaus[1], O. Pecher[2,1,3], F. Haarmann[2,1]
[1]RWTH Aachen, Institut für Anorganische Chemie, Aachen, Germany
[2]MPI-CPfS, Chemische Metallkunde, Dresden, Germany
[3]University of Cambridge, Department of Chemistry, Cambridge, United Kingdom

A combination of X-ray diffraction, quantum mechanical calculations, and NMR spectroscopy was recently presented to investigate the local atomic environments in Intermetallic Phases (IPs)[1,2]. Even the minor influence of cationic substitution on the anionic sublattice was resolved for $Sr_{1-x}Ba_xGa_2$[3]. A substitution of Ga atoms in the solid solutions of $M(Al_{1-x}Ga_x)_4$ with M = Sr, Ba results in a preferred occupation of one of the two positions of the anionic sublattice within the $BaAl_4$ type structure. This causes a minimum of the lattice parameter *a* for x = 0.5[4].

The IPs $M_{1-x}Ga_{2+3x}$ possess homogeneity ranges with $0.069 \leq x \leq 0.135$ at 750 °C for M = Ca[5] and $0 \leq x \leq 0.056$ at 400 °C for M = Sr[6]. According to X-ray diffraction a complex substitution of the cations by Ga_3 units is realized in the AlB_2 type structures. Both materials possess anisotropic metallic conductivity. Thus, their crystallites can be aligned in the magnetic field resulting in improved experimental resolution for M = Ca and a complex orientation dependent NMR line shape for M = Sr. These were used in $^{69;71}Ga$ NMR experiments to study the Ga environments. QM calculations confirm a significant influence of the Ga_3 groups on the AlB_2 type host lattice.

References:

[1] F. Haarmann, K. Koch, D. Grüner, W. Schnelle, O. Pecher, R. Cardoso-Gil, H. Borrmann, H. Rosner, Yu. Grin. Chem. Eur. J. 15, 1673 (2009).
[2] F. Haarmann, K. Koch, P. Jeglič, O. Pecher, H. Rosner, Yu. Grin. Chem. Eur. J. 17, 7560 (2011).
[3] O. Pecher, B. Mausolf, K. Lamberts, D. Oligschläger, C. Niewieszol (née Merkens), U. Englert, F. Haarmann. Chem. Eur. J. 21, 13971 (2015).
[4] O. Pecher, B. Mausolf, V. Peters, K. Lamberts, A. Korthaus, F. Haarmann. Chem. Eur. J. 22, 17833 (2016).
[5] O. Pecher. NMR-Spektroskopie an intermetallischen Phasen in den Systemen EA Al Ga mit EA = Ca, Sr und Ba. Dissertation, RWTH Aachen University (2013).
[6] F. Haarmann, Yu. Prots. Z. anorg. allg. Chem. 632, 2135 (2006).

Figure 1

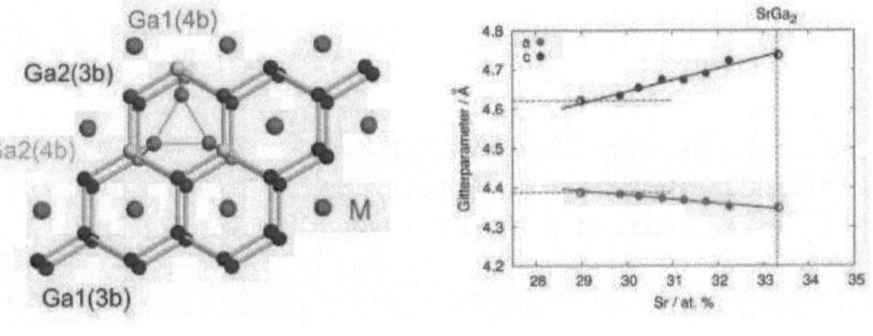

MS12-05

Pressure induced phase transition in Eu doped $CaCO_3$ detected by fluorescence spectroscopy

C. J. Fruhner[1], L. Bayarjargal[1], N. Schrodt[1], R. Luchitskaia[1], B. Winkler[1]
[1]Goethe University, Crystallography, Frankfurt am Main, Germany

The phase transition from calcite to aragonite has been the topic of numerous studies and, at ambient *T*, proceeds via intermediate phases (calcite-II --> III --> IIIb) [1]. A pressure induced formation of "aragonitic" amorphous $CaCO_3$ was described by [2].

The goal of this study is to employ fluorescence spectroscopy to further elucidate the calcite to aragonite phase transition and to demonstrate the viability of this approach for the study of phase relations of carbonates at high *p* and *T*. Fluorescence of Eu^{3+} can be employed for site symmetry detection [3].

High-*p* experiments were carried out in diamond anvil cells loaded with Eu-doped samples, ruby as pressure marker and argon as pressure transmitting medium. Fluorescence spectra and lifetimes were recorded between 0 – 25 GPa at ambient *T*.

The pressure induced shift of one emission peak of aragonite:Eu^{3+} and of calcite:Eu^{3+} was determined during pressure increase and decrease (Fig. 1). In aragonite there is no evidence for a phase transition, as there is only a linear, reversible pressure-induced fluorescence shift. In contrast, the fluorescence in calcite:Eu^{3+} shows initially a non-linear pressure-dependence on pressure increase. At $p > 12$ GPa, the pressure dependence changes irreversibly. Upon pressure release, the fluorescence energies are those of aragonite. Additionally, a Pair-Distribution-Function analysis using diffraction data collected on beamline P02.2 @ PETRA III (Fig. 2) of the recovered sample shows a local structure similar to that of aragonite.

We have shown that Eu^{3+} fluorescence allows us to detect a pressure-induced phase transition to aragonite at 12 – 15 GPa. The formation of calcite II and III/IIIb has not been detected, which is consistent with the similarity of the local environments in calcite I, II and III.

Financial support from the DFG (Projects BA 4020 and WI 1232) is gratefully acknowledged.

Figure 1 Pressure-dependence of the fluorescence. Filled symbols: P increase, open symbols: P decrease; left: line: linear fit right: upper line: Guide to the eye, bottom line: linear fit

Figure 2 PDF analysis of simulated calcite and aragonite compared to experimental results

References
[1] Merlini et al., *Min. Mag.* **78** 225 (2014).
[2] Fernandez-Martinez et al, Angew. Chem. **52** 8354 (2013).
[3] M. M. Fernandes, et al. *J. Colloid Interface Sci.* **321** 323 (2008).

Figure 1

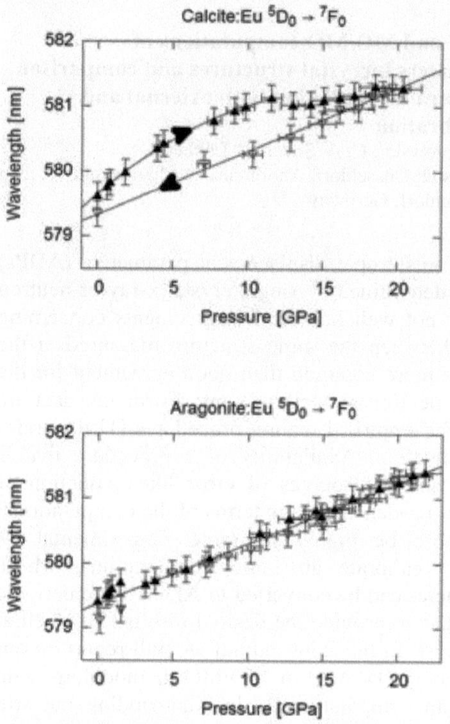

Figure 2

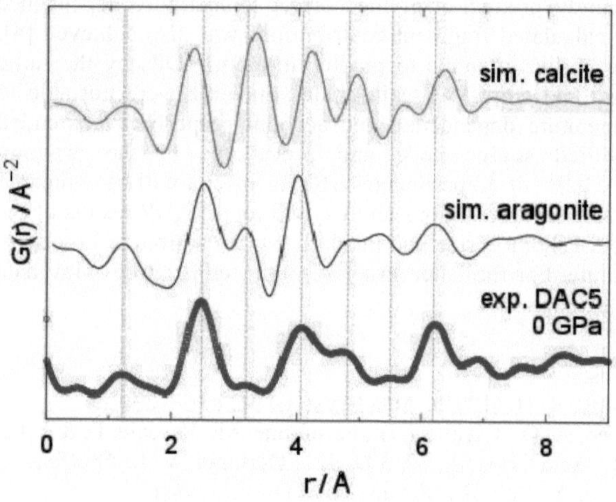

MS12-06

Microfocusing Source Based Beam Solution for Analytical Instruments

P. U. Pennartz[1]
[1]Rigaku Innovative Technologies, Eschweiler, Germany

Rigaku Innovative Technologies (RIT) has developed a full line of microfocusing sealed tube X-ray generators for x-ray analytical instruments. These instruments include XRD and XRF based analytical tools and metrology tools. Microfocusing sources offer high flux and well defined beam characteristics for XRD applications. RITs beam solution also offers ultra-low background by monochromatic excitation, especially with the elimination of x-ray diffraction noise. The performance of a beam solution depends on the system design, choice of sources, multilayer coating and optic precision. In this presentation, we will show you the performance of these elements and examples of their applications. We will especially emphasize the importance of the optical system design and the engineering capability to achieve the most closely coupled optic, resulting in the highest brilliance and flexibility of any source designed for XRD or small-spot XRF.

Many different optic designs are possible, which includes single corner graded multilayer optics often referred to as CMF or Montel optics, 4 corner graded multilayer optics to achieve even higher brilliance, precision Johansson monochromators, and specialty multilayer-crystal hybrid optics to achieve either a larger area or more narrow bandpass. With a modular design concept, MicroMax-003F and MicroMax-002+ can be readily configured for a solution for SAXS, Crystallography, XRR, High-Pressure, small spot XRF or other specialized applications such as flood-field use such as 2d-detector testing.

MS13 - Molecular Crystallography at the Limits

MS13-01
ab initio ORTEP: fundamentals, function, tentative future

R. Dronskowski[1]
[1]RWTH Aachen University, Chemistry, Aachen, Germany

Thermal properties of solid-state materials are a fundamental topic of study with important implications for crystallography. Thus, anisotropic displacement parameters (ADPs) are routinely used to quantify the thermal motion of atoms in crystals, as exemplified by Johnson"s ingenious ORTEP drawing routine invented back in 1965 [1] but still valid in 2017. While ORTEPs are most effectively derived from high-resolution neutron diffraction, modern electronic-structure methods have also enabled their first-principles prediction using periodic density-functional theory. This is performed by also including thermal vibrations (phonons) of the crystal lattice, typically described within the harmonic or the quasi-harmonic approximation [2], on top of the well-defined electronic ground state [3] reached by state-of-the-art exchange-correlation functionals also including dispersion interactions. This very method is easily illustrated by various examples dealing with – but not restricted to – molecular crystals such as small nitrogen-containing molecules [4], crystalline pentachloropyridine [5], classical transition-metal carbonyls [6], and prototypical elemental solids such as α-sulfur [7]. As a general rule, theory leads to excellent agreement with experiment, with important implications for realistic materials modeling at finite temperatures and semi-automated computer routines [8] but also with respect to the paramount importance of experimental validation using state-of-the-art measurements.

References:

[1] C. K. Johnson: OR-TEP: A FORTRAN Thermal-Ellipsoid Plot Program for Crystal Structure Illustrations, ORNL-3794 (Rev.), Union Carbide Corporation, Oak Ridge National Laboratory, June 1965.
[2] R. Stoffel, C. Wessel, M.-W. Lumey, R. Dronskowski, *Angew. Chem. Int. Ed.* **2010**, *49*, 5242.
[3] R. Dronskowski: *Computational Chemistry of Solid State Materials*, Wiley-VCH, Weinheim, New York 2005.
[4] V. L. Deringer, R. P. Stoffel, A. Togo, B. Eck, M. Meven, R. Dronskowski, *CrystEngComm* **2014**, *16*, 10907.
[5] J. George, A. Wang, V. L. Deringer, R. Wang, R. Dronskowski, U. Englert, *CrystEngComm* **2015**, *17*, 7414.
[6] V. L. Deringer, A. Wang, J. George, R. Dronskowski, U. Englert, *Dalton Trans.* **2016**, *45*, 13680.
[7] J. George, V. L. Deringer, A. Wang, P. Müller, U. Englert, R. Dronskowski, *J. Chem. Phys.* **2016**, *145*, 234512-1.
[8] www.ellipsoids.de

MS13-02
Diffuse Single Crystal Scattering corrected for Molecular Formfactor Effects

E. Schmidt[1], R. Neder[1]
[1]Friedrich-Alexander-Universität Erlangen-Nürnberg, Lehrstuhl für Kristallografie und Strukturphysik, Erlangen, Germany

Introduction

The determination of the average structure of a crystalline material via analysis of Bragg scattering data has long since become a routine method, at least for materials with moderate complexity. No technique equivalent to direct methods has been established to date for the analysis of disordered crystalline materials. For full data analysis the weaker diffuse scattering in between the Bragg peaks has to be considered.

Objectives

We show that chemical short range order (SRO) in two component molecular crystals can be solved directly by separating the influence of the molecular form factor from the diffraction pattern. This is demonstrated by analyzing the diffuse scattering of tris-tert-butyl-1,3,5-benzene tricarboxamide.

Materials & Methods

A theoretical model for SRO in two component systems has been established by Warren et al. [1]:

$$I_{SRO}(h_1h_2h_3) = Nm_Am_B|f_A-f_B|^2(1+\sum_{lmn} \alpha_{lmn} \cos(2\pi(h_1l+h_2m+h_3n)))$$

Here I_{SRO} is the diffuse intensity due to SRO, which is a function of the reciprocal space coordinates h_1, h_2 and h_3. The crystal consists of N atoms/molecules of species A and B. m_A, m_B, and f_A, f_B are the respective concentrations and atomic/molecular form factors of the species. The summation over lmn is a summation over all interatomic vectors, and the α_{lmn} are the Warren-Cowley SRO parameters. For simple materials with a primitive unit-cell where only one site is occupied, the diffuse scattering can simply be divided by the absolute molecular form factor difference squared $|f_A-f_B|^2$. A consecutive projection into one reciprocal unit cell allows a linear regression algorithm to obtain the α_{lmn}.

Results

The suggested algorithm was tested on the diffuse scattering of tris-tert-butyl-1,3,5-benzene tricarboxamide. The data was taken from [2] with kind permission. The results obtained by the division and projection algorithm are in very well agreement with the results from the 3DΔ-PDF analysis performed with YELL [3].

Conclusion

We show that single crystal diffuse scattering can be corrected for molecular form factor effects by dividing the data by the absolute form factor difference squared. This is a fast and robust way of analyzing SRO.

References:

[1] Warren, B.E., Averbach, B.L. & Roberts, B.W. (1951). J. Appl. Phys. 1493–1496, 2–78.
[2] Simonov, A., Weber, T. & Steurer, W. (2014). J. Appl. Cryst. 47, 2011-2018.
[3] Simonov, A., Weber, T. & Steurer, W. (2014). J. Appl. Cryst. 47, 1146-1152.

MS13-03
How Far Can We See? Atomic Resolution Structure of Giant Supramolecules

A. Virovets[1,2], E. Peresypkina[1,2], M. Scheer[1,2]
[1]University of Regensburg, Institute of Inorganic Chemistry, Regensburg, Germany
[2]Institute of Inorganic Chemistry SB RAS, Novosibirsk, Russian Federation

One of the ultimate goals of the single crystal X-ray structural analysis is to obtain as fine and detailed structure of the molecule as possible. For small molecules, not only precise geometrical characteristics but also charge density distribution can be obtained from high-angle diffraction experiment. On the other hand, giant molecular structures containing thousands non-H atoms in the asymmetric unit allow the researcher to work only with the diffraction data of limited quality.

Since 2003, we have been investigating the chemistry of pentaphosphaferrocene, $[Cp^RFe(\eta^5\text{-}P_5)]$ ($Cp^R = \eta^5\text{-}C_5R_5$, R = Me, CH_2Ph, PhC_4H_9). Phosporous atoms of planar P_5-rings coordinate Cu^+ and Ag^+ cations resulting in self-assembled spherical supramolecules of 2.1 – 4.6 nm in diameter [1-5]. Recently we have found out that $[Cp^RTa(\eta^4\text{-}P_4)(CO)_2]$ can also be used as bulding block for giant supramolecules [6]. The central metal-phosporous core with fullerene or fullerene-like topologies sometimes possesses a central cavity, which is able to encapsulate and store various guest molecules [1].

Here we discuss our recent results with particular focus on the rather less-debated problematics of the atomic resolution for the giant non-biological molecules using our own experience in comparison with literature examples on giant clusters. We will address the questions: How can high-resolution data be obtained? How do they help us to understand the chemical aspects of the self-assembly process? How information from other physical methods and knowledge of structural chemistry can help us to verify and to improve the structural model? What information can be obtained from low-angle data?

We thank ERC grant AdG339072-SELFPHOS for support. The research was partly done at the light source PETRA III at DESY.

References:
[1] E. Peresypkina, C. Heindl, A. Virovets, M. Scheer (2016) *Structure and Bonding*, online, DOI 10.1007/430_2016_2.
[2] J. Bai, A. Virovets, M. Scheer (2003), *Science*, **300**, 781.
[3] S. Heinl, E. Peresypkina, J. Sutter, M. Scheer (2015) *Angew. Chem. Int. Ed.*, **54**, 13431.
[4] C. Heindl, E. Peresypkina, A. Virovets et al (2015) *J. Am. Chem. Soc.*, *137*, 10938.
[5] E. Peresypkina, A. Virovets, M. Scheer (2016) *Crystal Growth & Des.*, **16**, 2335.
[6] F. Dielmann, E. Peresypkina, B. Krämer et al (2016) *Angew. Chem. Int. Ed.*, **55**, 14833.

Figure 1

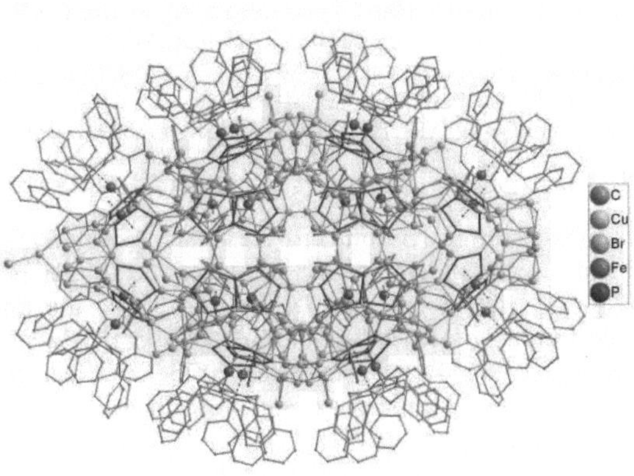

MS13-04
Caviplexes: Complexes of Macrocyclic NHC-Ligands with defined Cavities

A. Pöthig[1]
[1]Technical University of Munich, Catalysis Research Center, Garching, Germany

Two different classes of complexes of the same macrocyclic ligand platform, both exhibiting a structurally well-defined second coordination sphere, are presented. Square planar coordinating metals like nickel (II) form dinuclear saddle-shaped complexes.[1] Due to attractive Coulomb interactions these saddle-shaped compounds can assemble to so-called capsoplexes via encapsulation of guest ions.[2] In contrast, linearly coordinating transition metals as silver(I) and gold(I) form octanuclear pillarplexes.[3] These metallocavitands with very defined tubular cavities are able to exclusively host linear molecules, the gold(I) complexes exhibit luminescent behavior, and their solubility can easily be adjusted by simple anion exchange reactions. The synthesis and characterization of these novel complex classes as well as catalytic and supramolecular applications are presented.

References:
[1] P. J. Altmann, C. Jandl, A. Pöthig*, *Dalton Trans.*, **2015**, *44*, 11278-11281.
[2] P. J. Altmann, A. Pöthig*, *Chem. Commun.*, **2016**, *52*, 9089-9092.
[3] P. J. Altmann, A. Pöthig*, *J. Am. Chem. Soc.*, **2016**, *138* (40), 13171–13174.

Figure 1

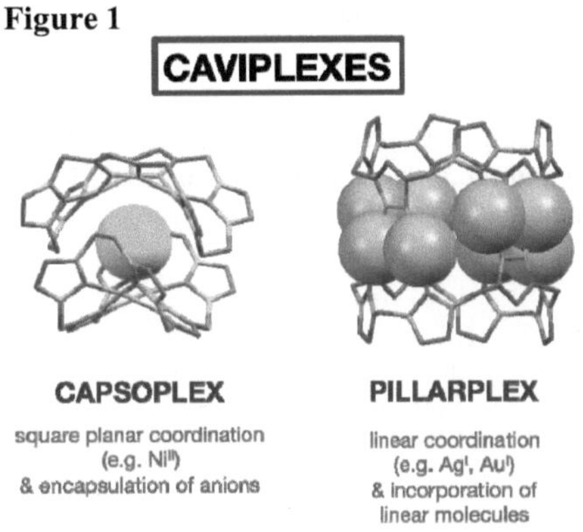

MS13-05

Facilities for Macromolecular Crystallography at the HZB

M. Gerlach[1], C. Feiler[1], R. Förster[1], C. Gless[1], N. Grabicki[1], T. Hauß[1], M. Hellmig[1], F. Huschmann[1], A. Kastner[1], P. Malecki[1], K. Röwer[1], L. Schmuckermaier[1], M. Steffien[1], P. Wilk[1], M. S. Weiss[1]
[1]Helmholtz-Zentrum Berlin, Macromolecular Crystallography, Berlin, Germany

The Helmholtz-Zentrum Berlin operates three beamlines for macromolecular crystallography at the electron storage ring BESSY II [1,2]. BL14.1 and BL14.2 are tunable in the photon energy range from 5 to 15.5 keV, while BL14.3 is a fixed-energy side station (13.8 keV). They feature state-of-the-art experimental stations and ancillary facilities, serving more than 100 research groups across Europe. More than 2200 protein structures measured at BESSY II have resulted in PDB depositions so far, and with more than 450 PDB depositions in 2016, they are currently among the most productive MX-stations in Europe.

The experimental endstation of BL14.1 provides high degree of automation and is equipped with a very large Pilatus 6M detector, a CATS sample changer robot and an MD2 multi-axis goniometer. In 2016, BL14.2 underwent a comprehensive upgrade to increase the performance in terms of sample throughput, allowing for large automated fragment-screening campaigns in the near future. It now features a Pilatus3S 2M detector, a G-Rob sample changer and a piezo-controlled nanodiffractometer (Figure 1). Its large sample dewar can accommodate up to 294 samples, supporting both, SPINE- and UNIPUCKS. A high-resolution UV/Vis microspectrophotometer can be employed in order to perform spectro-photometric measurements during diffraction data collection, e.g. to account for radiation damage (Figure 2).

All standard data collection procedures, such as SAD and MAD, are possible, as well as long-wavelength measurements and element identification using an X-ray fluorescence detector. Further experimental possibilities are radiation induced phasing experiments using a pulsed UV-laser, *in situ* crystal screening and controlled crystal dehydration using an HC1 dehydration device [3]. Furthermore, the HZB-MX group operates an S1 BioLab which supports the complete workflow from protein purification to crystallization.

References:
[1] U. Mueller et al., *J. Synch. Rad.* **19**, 442
[2] U. Mueller et al., *Eur. Phys. J. Plus* **130**, 141
[3] M. W. Bowler et al (2015). *Cryst. Growth Des.* **15**, 1043

Figure 1: The new endstation of BL14.2 with the G-ROB sample changer (right), the nanodiffractometer (center) and the PILATUS3S 2M detector (left).

Figure 2: Absorbance spectra of the redox indicator Resazurin in the oxidized state (blue) and the reduced state (red) measured with the microspectrophotometer.

Figure 1

Figure 2

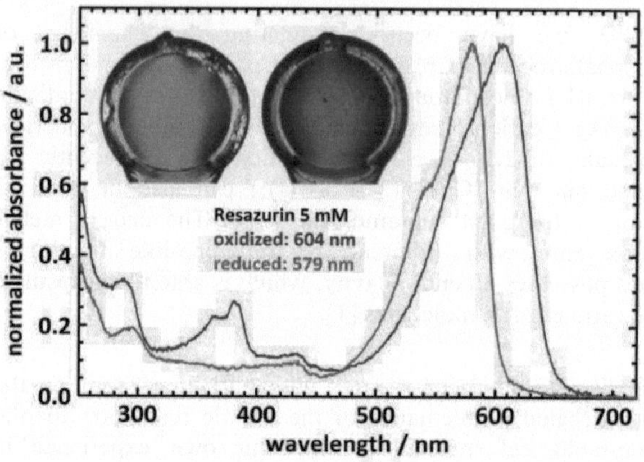

MS14 - in situ/in operando studies of energy materials

MS14-01
In situ diffraction 100

H. Kohlmann[1]
[1]University Leipzig, Leipzig, Germany

100 years ago, in 1917, Hull performed X-ray powder diffraction experiments on iron at variable temperatures up to 1000°C [1], which we may regard as the first *in situ* diffraction experiment. Ever since X-ray and later neutron diffraction evolved and are nowadays widely used and convenient tools for materials characterization and structure determination. Considerable progress was made in the development of radiation sources and diffraction instrumentation, enabling time dependent *in situ* studies of a wide range of processes today [2-4].

In this contribution we will define "*in situ* diffraction" and give a brief historical sketch of this field. The potential of the method will be demonstrated by recent examples of *in situ* X-ray and neutron powder diffraction experiments on hydrogen storage, magnetic and battery materials, metastable oxides as photocatalysts and the HDDR process (hydrogenation-decomposition-desorption-recombination). Special emphasis will be given to advanced sample environment and time-resolved studies of chemical reactions and phase transitions in the solid state.

References:
[1] A. W. Hull, *Phys. Rev.* **1917**, *10*, 661-697.
[2] T. C. Hansen, H. Kohlmann, *Z. Anorg. Allg. Chem.* **2014**, *640*, 3044-3063.
[3] B. R. S. Hansen, DK. T. Moller, M. Paskevicius, A.-C. Dippel, P. Walter, C. J. Webb, C. Pistidda, N. Bergemann, M. Dornheim, T. Klassen, J.-E. Jorgensen, T. R. Jensen, *J. Appl. Crystallogr.* **2015**, *48*, 1234-1241.
[4] D. R. Neuville, L. Hennet, P. Florian, D. de Ligny, *Rev. Miner. Geochem.* **2014**, *78*, 779-800.

MS14-02
Uncommon phase transformation in $LiRhO_2$ upon Li-extraction

D. Mikhailova[1], O. Karakulina[2], D. Batuk[2], A. Abakumov[3], J. Hadermann[2], A. Tsirlin[4], H. Ehrenberg[1]
[1]Karlsruhe Institute of Technology, Karlsruhe, Germany
[2]EMAT, Antwerp, Belgium
[3]Skoltech, Moscow, Russian Federation
[4]University of Augsburg, Augsburg, Germany

Layered $Li(M,Li)O_2$ (M – transition metal) ordered rock salt-type structures are used in advanced metal-ion batteries as one of the best hosts for the reversible intercalation of Li ions [1]. Besides the conventional redox reaction involving oxidation/reduction of the M cation upon Li extraction/insertion, partial oxygen oxidation increases the capacity while maintaining the oxidized oxygen species in the lattice through high covalency of the M-O bonding [2]. The typical degradation mechanism of the $Li(M,Li)O_2$ electrodes involves partially irreversible M cation migration towards the Li positions resulting in gradual capacity/voltage fade. In the present work we report about a quite different, unique structural behaviour of a layered LiRhO2 compound upon chemical or electrochemical Li extraction.

When charged above 4.0 V, $LiRhO_2$ exhibits massive, concerted and partially reversible migration of Rh cations towards empty octahedral positions in the Li layers resulting in the $Li_yRh_3O_6$ phase with a rutile-ramsdellite intergrowth structure with rutile-like [1x1] and bigger ramsdellite-like [2x1] tunnels (Fig. 1), as it was shown by TEM and synchrotron powder diffraction methods. The Rh cation migration is accompanied by an anionic redox process, and oxygen dimers with short O-O distances of 2.26 Å are formed. Upon lithiation of the fully delithiated material, a fraction of the $Li_yRh_3O_6$ phase transforms back into the layered $Li_{1-x}RhO_2$ structure. Surprisingly, the transformed structure can accept significantly more Li compared to the layered $LiRhO_2$ due to the population of both ramsdellite and rutile-type tunnels with Li atoms. The novel $Li_yRh_3O_6$ structure is remarkably stable upon electrochemical cycling illustrating that proper structural implementation of the oxidized oxygen species can open a pathway towards deliberate employment of the anion redox chemistry in high capacity/high voltage positive electrodes for metal-ion batteries.

Figure 1. Upon chemical or electrochemical oxidation, layered $LiRhO_2$ shows a unique structural transformation, which involves both cation migration and oxidation of oxygen resulting in a stable tunnel-like rutile-ramsdellite intergrowth $Li_yRh_3O_6$ structure with short O-O distances that are compatible with the connectivity of RhO6 octahedra.

References:
[1] Lithium Batteries: Science and Technology, Edited by G.-A. Nazri, G. Pistoia, Springer Science + Business Media, 2009, 708 p.
[2] M. Oishi et al. *J. Power Sources* 2015, 276, 89-94.

Figure 1

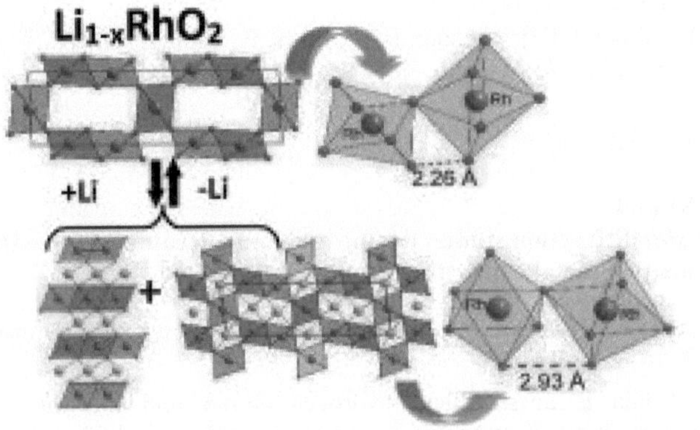

S14-03

INVESTIGATING THE FORMATION OF NANOSIZED Pt-M (M = Ni, Co) ALLOYS BY IN-SITU PAIR DISTRIBUTION FUNCTION ANALYSIS

S. ORTATATLI[1], C. WEIDENTHALER[1], J. KNOSSALLA[1]
[1]Max-Planck-Institut für Kohlenforschung, Heterogeneous Catalysis, Mülheim an der Ruhr, Germany

Alloying Pt with other metals is an eligible way to decrease the amount of noble metal used for the catalyst preparation employed in proton exchange membrane fuel cells (PEMFCs). Carbon supported Pt-alloys are often pre-annealed at elevated temperatures under reducing or inert atmosphere to enhance the catalytic activity before being employed in a PEMFC[1]. However, the ordering phenomena during annealing treatments are not fully understood. In order to gain comprehensive understanding on the alloy formation, ordering phenomena, and, dealloying, pair distribution function (PDF) analysis is the most promising approach to elucidate the structural changes at nanoscale. Although Pt-based alloys have been studied intensively in the literature for application as electrocatalysts[2], there is only limited knowledge on the local structure of these alloys and on the alloying procedure based on PDF analysis[3]. In this work, the formation of Pt-M (M = Ni, Co) nanoalloys starting from Pt-M precursors impregnated into hollow graphitic spheres was monitored by means of *in-situ* PDF analysis using synchrotron radiation. Upon heating a disordered PtCo alloy was formed, followed by increased ordering during cooling. For the PtNi alloy neither during heating nor cooling full ordering was obtained.

References
[1] Stephens, I. E., Bondarenko, A. S., Grønbjerg, U., Rossmeisl, J., Chorkendorff, I., *Energy Environ. Sci.*, 5, **2012**, 6744.
[2] Jia, Q., Liang, W., Bates, M. K., Mani, P., Lee, W., Mukarjee, S., *ACS Nano*. 9, **2015**, 387.
[3] Redmond, E. L., Setzler, B. P., Juhas, P., Billinge, S. J. L., Fuller, T. F., *Electrochem. Solid St.*, 15 (5), **2012**, B72-B74.

MS14-04

In-situ diffraction studies during ammonia decomposition with transition metal catalysts

J. Tseng[1], C. Weidenthaler[1]
[1]Max-Planck-Institut für Kohlenforschung, Mülheim an der Ruhr, Germany

Ammonia is an excellent hydrogen carrier and the catalytic decomposition of ammonia is one way to generate hydrogen for fuel cell applications without formation of CO_x. The most active catalyst so far is Ru on carbon nanotubes, but bimetallic compounds or metal nitrides are also under investigation as potential catalysts since noble metals are rather expensive and limited in availability. In this work we study different transition metal catalysts with the focus on Fe-based catalysts during the decomposition of ammonia. Starting from transition metal oxides, the catalyst formation was studied by *in-situ* X-ray diffraction under reaction conditions using in-house laboratory instruments. Crystallographic studies with respect to phase changes, crystal structure variations, and microstructure properties have been performed [1,2]. The behavior under reaction conditions and the catalytic activity can significantly differ for each catalyst system: while some transition metal oxides reduce during the reaction with ammonia to the metals, others form nitrides. In case of iron oxides precursors, nitrides are formed during the reaction with ammonia while for cobalt oxides precursors metals form after activation [3,4]. Molybdenum-based catalysts are a very good example that various factors govern activity. Structure changes as well as changes of specific surface areas, and defect concentrations have to be considered [1,2]. In this work we also present studies on the use of bimetallic catalysts (Fe-Co, Fe-Ni) for ammonia decomposition. The formation of metal nitrides, metals and/or alloys during the reactions is monitored by *in-situ* XRD. The structural changes and detailed reaction pathways were studied by the evaluation of the diffraction data.

References:
[1] Tagliazucca et al., J. Catal. 2013, 305, 277.
[2] Tagliazucca et al., Phys. Chem. Chem. Phys. 2014, 16(13), 6182.
[3] Feyen et al., Chem.-Eur. J. 2011, 17, 598.
[4] Gu et al. J. Phys. Chem. C. 2015, 30, 17102.

MS15 - New Crystal Structures and Structure Systematics

MS15-01
$[LiAl_2(OH)_6]X·mH_2O$ LDH containing different carboxylic anions of various chain sizes

A. Niksch[1]
[1]Institut für Geowissenschaften und Geographie, Mineralogie/Geochemie, Halle, Germany

The stoichiometry of layered double hydroxides can normally be formulated as $[M^{z+}{}_{1-x}M^{3+}{}_x(OH)_2]^{p+}[(A^{n-})_{p/n}·mH_2O]$ with z = 2 or 1, M = (mono-), bi- and trivalent metallic elements, A = organic or inorganic anion and m = amount of interlayer H2O depending on the temperature, relative humidity and hydration level. A Li^+ / Al^{3+} containing LDH presents a unique case with mono- and trivalent cations and a fixed ratio of Li : Al = 1 : 2 **[1]**. Synthesized $[LiAl_2(OH)_6][Cl·0.5H_2O]$ precursor are able to exchange the interlayer anion with organic anions of different sizes or charges. The positively charged main layer remains stable and is not capable of ion exchange.

The carboxylic acid containing Li-LDH were synthesized by anion exchange of a $[LiAL_2(OH)_6][Cl·0,5H_2O]$ ($P6_3/m$) precursor with the relevant Li-carboxylate hydrate. These organic salts were synthesized by the neutralization and recrystallization of an organic acid with LiOH (Tab. 1). For the anion exchange, 1 g precursor was added to a 4 - 5 M solution of the relevant organic salt. After stirring at 90 °C for approximately 24 h the product was filtered, washed with 60ml distilled water and dried (R 35 %). The $[LiAL_2(OH)_6][C_xH_y(O_z/S)·mH_2O]$ LDHs were characterized by XRD, HT XRD, TGA-DSC, IR-spectroscopy and SEM (EDX). It could be proved, that a complete anion exchange for all organic anions was executed with no trace of precursor Cl^- anions left. By increasing the size of the organic anion, the lattice parameter ***c*** is also increasing while the lattice parameter ***a***, ***b*** remain constant. The symmetry remains always hexagonal. Investigations with HT XRD show almost always (except $[LiAl_2(OH)_6][C_7H_7O_3S]$) an increase of the lattice parameter ***c*** at higher temperatures (100 - 200 °C) which is attributable to a change in the orientation of the organic anion. Independent of the carboxylate anion size the replacement of Cl^- with aromatic and aliphatic anions in $[LiAl_2(OH)_6][X·mH_2O]$ LDHs is possible.

References:
[1] A. V. Besserguenev et. al, Synthesis and structure of the Gibbsite Intercalation Compunds $[LiAl_2(OH)_6]X$ {X=Cl, Br, NO3} and $[LiAl_2(OH)_6]Cl·H_2O$ Using Synchrotron X-ray and Neutron Powder Diffracion, C. Mater, 1997

Figure 1

Tab. 1: organic salts used for anion exchange

aliphatic molecules	aromatic molecules
Li-Formiate·nH2O	Li-Benzoate·nH2O
Li-Acetate·nH2O	Li-Phenyl acetate·nH2O
Li-Glycolate·nH2O	Li-Phenyl propionate·nH2O
Li-Propionate·nH2O	Li-Phenyl butyrate·nH2O
Li-Butyrate·nH2O	Li-Phenyl valerate·nH2O
Li-Valerate·nH2O	Li-Benzenesulfonate·nH2O
Li-Methansulfonate·nH2O	Li-Toluol-4-sulfonate·nH2O
Li-Ethansulfonate·nH2O	Li-Phthalate·nH2O
Li-Glutarate·nH2O	Li-Isophthalate·nH2O
Li-Malonate·nH2O	Li-Terephthalate·nH2O
Li-Oxalate·nH2O	
Li-Succinate·nH2O	

MS15-02
Crystal Structures of Organolithium Compounds from Powder Data

A. Bodach[1], R. Hebestreit[1], L. Fink[1], M. Bolte[1], M. U. Schmidt[1]
[1]Goethe University Frankfurt, Chemistry, Frankfurt am Main, Germany

Since their development organolithium compounds have been applied many times very successfully in organic and organometallic chemistry, even on industrial scale.[1] Until now, only a few structures of pure/lewis-base free organolithium compounds have been reported, whereas the fundamental understanding of chemistry relies on structures, most of them determined from X-ray diffraction.

With modified Schlenk techniques, instead of glove boxes, highly reactive powder samples can be prepared in flame sealed borosilicate glass capillaries under severe inert gas conditions. These samples were measured on high resolution laboratory diffractometers and annealed/protected under nitrogen gas, if necessary.

Preparations and crystal structure determinations of several organolithium compounds are presented here: *p*-tolyllithium forms a dimer (Fig.1), which is connected to an infinite zigzag chain by Li-π interactions, which is isotypic to PhLi[2] and MesLi[3]. The crystal structure of *o*-tolyllithium was determined in Sohncke space group $P2_12_12_1$, where two symmetrically independent molecules form a dimer in *E* configuration. Furthermore, $Li[CH_2P\mathit{t}Bu_2]$, which is a reaction intermediate for the syntheses of *frustrated lewis pairs* "FLP", forms polymeric chains of Li_2C_2-rings and chair-like $Li_2C_2P_2$-rings.[4] Finally, a Li-analogue of the "Turbo-Grignard" (*R*MgBr·LiCl) was isolated as a co-crystal with the composition of $[(LiC_8H_9)_3·MTBE_3·LiBr]$.[5]

References:
[1] M. Schlosser, *Organometallics in Synthesis: A Manual*, Wiley, **1994**; *Third Manual*, Wiley, **2013**. [2] R. E. Dinnebier *et al.*, *Journal of the American Chemical Society* **1998**, *120*, 1430–1433. [3] A. Hübner *et al. Dalton Transactions* **2010**, *39*, 7528–7533. [4] L. Fink *et al., Zeitschrift für anorganische und allgemeine Chemie* **2016**, *642*, 282–287. [5] M. Blasberg *et al.*, Organometallics **2012**, *31*, 1001–1005. A. Krasovskiy *et al., Angewandte Chemie International Edition* **2004**, *43*, 3333–3336.

Fig.1: Packing motif of *p*-tolyllithium, black carbon, grey lithium, white hydrogen.

Figure 1

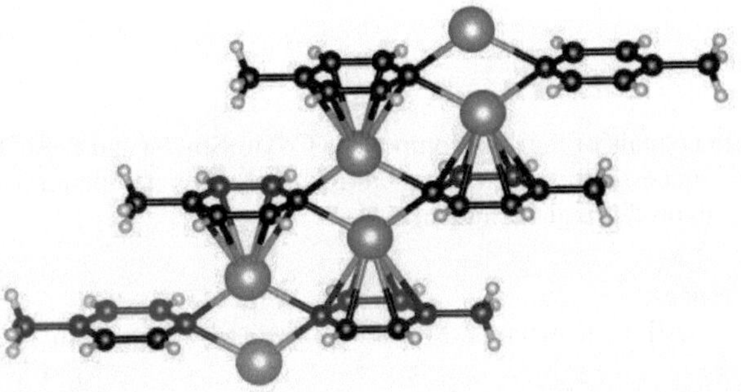

MS15-03

Hydrothermal synthesis of $CoMo_2Sb_2O_{10}$ and $FeSbO_2F_2$.

S. I. Ali[1], R. K. Kremer[2], M. Johnsson[1]
[1]Stockholm University, Materials and Environmental Chemistry (MMK), Stockholm, Sweden
[2]Max Planck Institute for Solid State Research, Stuttgart, Germany

Introduction:

Transition metal oxides and oxohalides containing p-block elements with a streochemically active lone-pair are of interest due to that they often show interesting physical properties such as magnetic frustration and nonlinear optical second harmonic generation.[1-3] The p-block elements prefer to bond to oxygen while transition metal cations bond to both oxygen and halide ions in such an environment. This leads to open structures that very often are layered and where the chance to form non-centrosymmetric compounds is relatively high.

Objectives:

The objective is to find out new transition metal oxide of M-M'-Sb-O -type and oxo-fluorides of M"-Sb-O-F -type (M / M' / M" = Transition metal) by hydrothermal techniques and then investigate their physical properties.

Materials & Methods:

The compounds were synthesized by hydrothermal techniques by mixing MF_2, $M'O_2$ and Sb_2O_3 for M-M'-Sb-O-type and $M'F_2$ and Sb_2O_3 for M"-Sb-O-F-type. Single crystal- and powder X-ray data were collected for structural determination. Thermal gravimetric analyses and magnetic susceptibilities were performed using a TA Discovery and a MPMS SQUID magnetometer.

Results:

The new compounds $CoMo_2Sb_2O_{10}$ crystalizes in the monoclinic space group *C2/c* with unit cell parameters a = 27.392(6) Å, b = 5.774(1) Å, c = 5.002(1) Å, β = 94.214(5)° and the new compound and $FeSbO_2F_2$ in *C2/c* with a = 11.9129(15) Å, b = 4.9605(5) Å, c = 5.5000(6) Å, β = 103.897(7)° respectively. $CoMo_2Sb_2O_{10}$ consists of charge neutral $[CoMo_2Sb_2O_{10}]_n$ layers where each layer is made up of corner sharing chains of $[Co_2O_8]_n$, $[Mo_2O_8]_n$ and $[SbO_2]_n$. $FeSbO_2F_2$ crystal structure consists of $[FeO_2F_2]_n$ sheets and $[SbO_2]_n$ chains. $CoMo_2Sb_2O_{10}$ is stable below 530°C. The magnetic susceptibility $CoMo_2Sb_2O_{10}$ follows the Curie-Weiss law above 50K with a short range antiferromagnetic transition below ca. 5K.

Conclusion:

Single crystals of the new compounds $CoMo_2Sb_2O_{10}$ and $FeSbO_2F_2$ were successfully synthesized at relatively low temperature by using hydrothermal techniques.

References:

[1] S. I. Ali, R. K. Kremer, M. Johnsson, *Inorg. Chem.* **2016,** *55,* 11490.
[2] H. Yu, H. Zhang, P. S. Halasyamani, *Cryst. Growth Des.* **2016,** 16, 1081.
[3] S. Hu, M. Johnsson, *Dalton Trans.* **2012,** 41, 12786.

Keywords: Hydrothermal synthesis, Transition metal Oxides and Oxo-halides, Single crystal X-ray diffraction, Magnetic Properties.

MS15-04

Crystal structure and polymorphism of $NaSrVO_4$: the first $A^IB^{II}X^VO_4$ larnite related structure from X-ray powder data

G. Nenert[1], P. O'Meara[2], T. Degen[1]
[1]PANalytical B. V., Almelo, Netherlands
[2]PANalytical Ltd., Cambridge, United Kingdom

The crystal chemistry of $A^IB^{II}X^VO_4$ (A^I = alkali ion, B^{II} = alkali-earth ion, X^V = P, V, As) is very rich and leads to numerous polymorphic phases which belong to 7 different structures types: olivine, arcanite, glaserite, tridymite, α-K_2SO_4, β-Na_2SO_4 and γ-Na_2SO_4 [1]. Among the various families (X = P, V, As); the phosphates have been the most widely investigated. Besides the purely interest from a crystal chemistry point of view, the research activities related to this family of materials is driven mainly due to their ferroelectric and ferroelastic properties and possible applications as phosphors for LEDs [1,2].

Here, we report on a new structural type within these families. The previously unknown crystal structure of sodium strontium vanadate, $NaSrVO_4$, is determined and refined from laboratory X-ray powder diffraction data. The title compound exhibits the larnite β-Ca_2SiO_4 structure. This is the first report of such chemistry $A^IB^{II}X^VO_4$ which exhibits this structural type. $NaSrVO_4$ exhibits 2 first order phase transitions above room temperature towards respectively an orthorhombic phase and an α-Ca_2SiO_4 related structure.

This sduty demonstrates that the larnite structure can be stabilized with chemical composition which has been so far never considered. This is opening potentially new possibilities to better understand the chemistry of larnite and related phases such as flamite.

References:

[1] V. A. Isupov, *Ferroelectrics*, **2002** *274:1*, 203-283.
[2] S. Choi, Y. J. Yun, S. J. Kim, H.-K. Jung *Optical Letters,* **2013** *38*, 1346-1348.

Figure 1

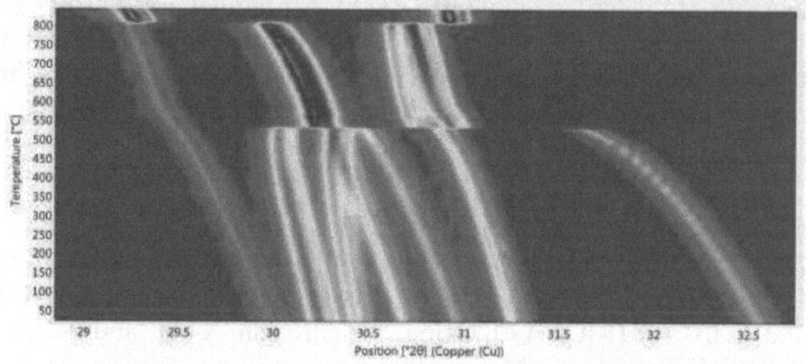

MS15-05

Structural Crystallography of Novel Alkali Neptunyl Selenate and Selenite Compounds: 2D-Sheet Selenates and 1D-Selenite

E. M. Langer[1], O. Walter[2], E. V. Alekseev[1,3]
[1]Forschungszentrum Jülich - IEK-6, Jülich, Germany
[2]European Commission, Directorate for Nuclear Safety and Security - JRC, Karlsruhe, Germany
[3]RWTH Aachen, Institut für Kristallographie, Aachen, Germany

Introduction

Spent nuclear fuel consists of considerable amounts of Np and it is considered to be a very tricky problem for the long-term disposal of nuclear waste, as is also true for the long-lived fission product 79Se. Due to its wide variety of possible oxidation states ranging from +II to +VII for Np and typically +IV and +VI for the occurring high yield fission product Se, this research is of keen importance to assess the long-term storage of nuclear waste in a safe manner.

Objectives

To achieve this, a full structural as well as spectroscopic characterization of novel compounds is essential in further understanding potential forming mechanisms. Using different synthetic methods, novel compounds were to be synthesized and analysed structurally by single crystal X-ray diffraction and spectroscopically by means of Raman and IR-spectra.

Materials & methods

Slow evaporation from solution at ambient conditions as well as hydrothermal synthesis was approached; Np(V)-nitrate solution, Se(IV)O2, H2Se(VI)O4 and alkali metal nitrates and chlorides were used.

Results

We were able to synthesize and characterize two novel neptunyl-selenites as well as three novel neptunyl-selenates. The table in the appendix gives an overview of the found compounds. The isostructural selenites (**1,2**) consist of 1D-chains of square antiprismatic NpO8 polyhedra interconnected by four trigonal pyramidal SeO3 polyhedra. **3** and **4** possesses bipyramidal NpO7 monodentately interconnected with SeO4 tetrahedra with one NpO7(SeO4)5 block and one NpO7(SeO4)3, resulting in almost planar iso-topologic 2D-sheets. **5** consists of pentagonal bipyramidal NpO7 monodentately interconnected with SeO4 tetrahedra on every planar O to form undulated two dimensional layers. In all five structures the alkali metal cations act as counter cations for charge balance and is positioned in the interspaces. Additionally three novel Neptunyl nitrates were found inadvertently in the process (**6-8**).

Conclusion

Considering the variety within the found structures of the Np-Se-O family the major impact of redox chemistry is well visible. The fact that this family was increased by only a few experiments from 15 to 20 known compounds, demonstrates how under-explored this system is, making it even more important for a fundamental understanding of actinide chemistry.

Figure 1

Table 1: List of novel Np containing compounds obtained via slow evaporation.

	Compound	Spacegroup	a [Å]	b [Å]	c [Å]	α [°]	β [°]	γ [°]	V [Å³]	Dimension	Remarks
	Selenites (SeIV)										
(1)	K2[Np(SeO3)2(HSeO3)2](H2O)	P 4/mnc	12,66	12,66	10,21	90	90	90	1635,37	1D	1)
(2)	Rb2[Np(SeO3)2(HSeO3)2](H2O)	P 4/mnc	12,83	12,83	10,25	90	90	90	1687,60	1D	1)
	Selenates (SeVI)										
(3)	K2[(NpO2)2(SeO4)3(H2O)2](H2O)3.5	P 12(1)/m1	11,92	13,38	13,63	90	109,50	90	2049,35	2D	2)
(4)	Rb2[(NpO2)2(SeO4)3(H2O)4](H2O)3	P -1	8,39	11,77	13,21	102,54	106,94	103,01	1158,47	2D	
(5)	Cs2[(NpO2)2SeO4)3]	P -42(1)m	9,87	9,87	8,14	90	90	90	792,66	2D	3)
	Nitrates										
(6)	[(NpO2)(NO3)2(H2O)2] (H2O)4	C mc2(1)	13,21	7,99	11,29	90	90	90	1191,64	1D	
(7)	Rb (NpO2)(NO3)3	R -3m	9,22	9,22	9,42	90	90	120	692,90	1D	4)
(8)	Cs (NpO2)(NO3)3	R -3m	9,50	9,50	9,75	90	90	120	762,25	1D	4)

1) isostructural to each other
2) isostructural to $(H_2O)_2[(UO_2)_2(SeO_4)_3(H_2O)_2](H_2O)_{3.5}$
3) isostructural to $Cs_2[(UO_2)_2SeO_4)_3]$, unpublished but found in previous own work
4) isostructural to each other

MS16 - Nanoscience

MS16-01
Electron tomography as a powerful tool for superstructure crystallography and defect analysis of self-assembled nanoclusters

C. Kübel[1]
[1]KIT, INT, Eggenstein-Leopoldshafen, Germany

Transmission electron microscopy is a powerful tool to characterize the 2D and 3D structure from the atomic to the mesoscale. In particular, electron tomography has been well-established to image the complex 3D structure of materials and in biology with (sub)nanometer resolution. With the high spatial resolution, it is well-suited to characterize the self-assembled superstructures of a wide range of nanoclusters enabling mesoscale crystallography in real-space. This is an interesting alternative to traditional small angle X-ray scattering experiments for superstructure characterization. Moreover, it enables characterization of individual submicron supercrystals even in case of high-defect densities and allows analysis of local structural variations, e.g. at defects or the surface.

In this presentation, I will illustrate the electron tomography based superlattice characterization of different nanoclusters looking into the self-assembly of thiophenyl capped CdS nanoclusters [1] and POM-POSS Janus-type co-cluster connected by different linkers [2].

References:
[1] T.I. Levchenko et al., Chem. Mat., **2015**, *27(10)*, 3666-3682.
[2] C. Ma et al, Angew. Chemie. Int. Ed. Engl, **2015**, *54*, 15699-15704.

Figure Captions

Figure 1: Electron tomographic reconstruction of a self-assembled 2.3 nm CdS nanocluster superlattice (with 5 nm gold particles): a) volume rendering, b) digital slice of individual crystallographic defects (dislocations, voids) in the 3D reconstruction.

Figure 2: a) Low-dose STEM image of self-assembled POM-POSS co-clusters imaged in [100] orientation, b) volume rendering of the electron tomographic reconstruction viewed in the same orientation, c) 3D FFT of the electron tomographic reconstruction showing the different allowed reflections fitting to an ordered bicontinuous double diamond structure – {110} (dark blue), {111} (green), {200} (red), {211} (purple) and {221} (light blue).

Figure 1

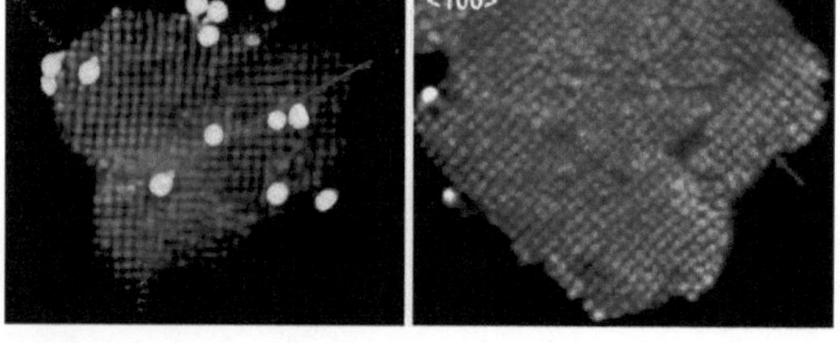

Figure 2

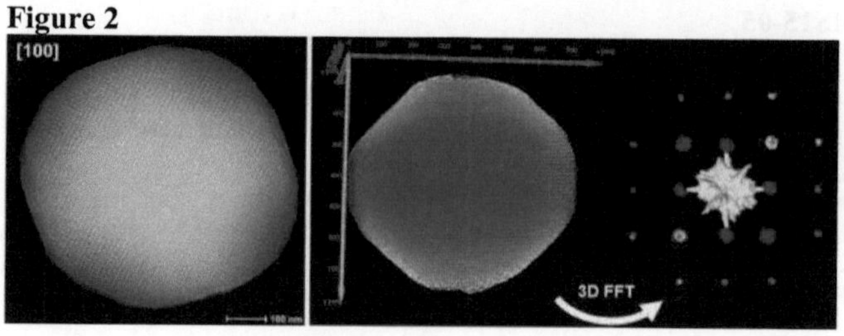

MS16-02
Ferroelectric Domains in Strained $K_{0.75}Na_{0.25}NbO_3$ Epitaxial Films Grown on (110) $TbScO_3$ Probed by Scanning X-Ray Nanodiffraction

M. Schmidbauer[1], M. Hanke[2], A. Kwasniewski[1], D. Braun[1], L. von Helden[1], S. J. Leake[3], J. Schwarzkopf[1]
[1]Leibniz-Institut für Kristallzüchtung, Berlin, Germany
[2]Paul-Drude-Institut für Festkörperelektronik, Berlin, Germany
[3]ESRF The European Synchrotron, Grenoble, France

In ferroelectric materials phase symmetry and structural distortions are strongly coupled to piezoelectric properties. Periodic domain structures are of particular interest from both fundamental and technological point of view. Such periodic polarization modulations on a nanometer scale can be engineered by the use of substrates with suitable (anisotropic) misfit strains. We focus on $K_{0.75}Na_{0.25}NbO_3$ epitaxial layers grown on (110) $TbScO_3$ orthorhombic substrates using metal-organic chemical vapor deposition. A highly regular one dimensional ferroelectric domains pattern is formed (Fig.1a) which extends over several micrometers with a lateral periodicity of about 50 nm. The monoclinic symmetry of the domains is controlled by the anisotropic epitaxial lattice strain, which is highly compressive in one in-plane direction and weakly tensile in the corresponding orthogonal direction.

The ferroelectric domain pattern is analyzed by piezoresponse force microscopy and conventional high-resolution x-ray diffraction (Fig.1c,d). The monoclinic M_A phase is identified [1], which is associated with both a strong vertical and lateral electrical polarization component. A structural variant of a 90° rotated M_A domain pattern is also observed (Fig.1a,b), however, with significantly lower probability in agreement with energy considerations based on linear elasticity theory.

In order to independently investigate the different domain variants scanning x-ray nanodiffraction has been performed by using a focused x-ray beam of about 100 nm probe size. This experiment proves that the 90° variant shows a larger vertical lattice spacing than the 0° domain variant. Calculations based on linear elasticity theory substantiate that this difference is caused by the elastic anisotropy of the $K_{0.75}Na_{0.25}NbO_3$ epitaxial layer.

References:
[1] J. Schwarzkopf, D. Braun, M. Hanke, A. Kwasniewski, J. Sellmann, M. Schmidbauer, *J. Appl. Cryst.* **49**, (2016), 375.

Figure 1

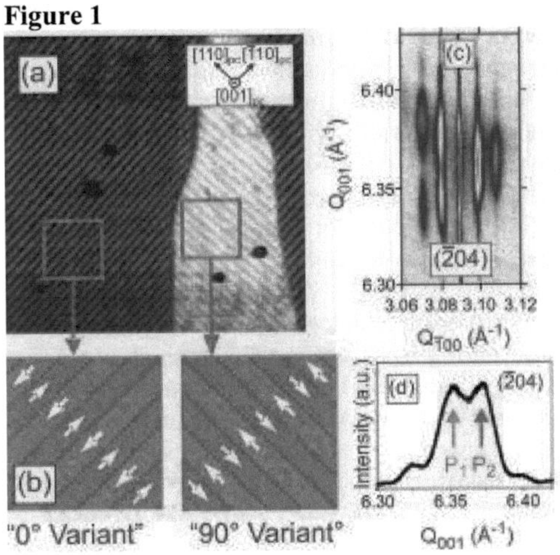

MS16-03

An algorithm to decorate nanoparticle surfaces

R. Neder[1]
[1]Friedrich-Alexander-Universität Erlangen-Nürnberg, Kristallographie und Strukturphysik, Erlangen, Germany

Introduction

Under favorable conditions (small size, neutron scattering) the decoration of nanoparticles by stabilizing molecules can be observed directly. [1] It is not straightforward to simulate and refine a nanoparticle structure with a decorated surface.

Objectives

An algorithm to decorate nanoparticles by any type of stabilizing molecule is developed. Four different binding concepts shall allow flexible application to a wide variety of possibilities.

Materials & Methods

The algorithm has been implemented into the DISCUS suite [2] and uses the general tools present within this software to generate and modify the nanoparticle core. This includes the options to introduce internal and or surface disorder and strain fields as well as flexible particle shapes. The stabilizing molecules are attached in four different bond schemes that include a single bond between core and ligand molecules as well as double and multiple bond pattern between the core and the ligand. The algorithm achieves a fully automatic placement that requires the user only to specify the general concepts like bond type and ligand density.

Results

The algorithm has been tested on neutron PDF data collected for ZnO nanoparticles stabilized by citric acid. The experimental data are well reproduced, specifically the interatomic distance peaks observed at distances that do not correspond to internal distances of the ZnO core nor to those of the citrate molecule. Corresponding results are achieved for gold nanoparticles.

Conclusion

A flexible algorithm to decorate any type of nanoparticle surface with stabilizing molecules under a variety of bonding schemes is presented. Its inclusion into the DISCUS suite allows simulation and refinement of the corresponding nanoparticle structures with respect to powder and PDF data.

References:
[1] Page, K. et al. (2011). J. Appl. Cryst. 44, 327–336.
[2] Neder, R.B., Proffen T. (2008) Diffuse Scattering and Defect Structure Simulations, Oxford University Press

MS16-05

Structural and compositional analysis of isolated core-shell (In,Ga)N/GaN rods based on nanofocus x-ray diffraction and scanning transmission electron microscopy

T. Krause[1], M. Hanke[1], L. Nicolai[1], Z. Cheng[1], M. Niehle[1], A. Trampert[1], M. Kahnt[2,3], G. Falkenberg[2], C. Schroer[2,3], J. Hartmann[4,5], H. Zhou[4], H. Wehmann[4,5], A. Waag[4,5]
[1]Paul-Drude-Institut, Mikrostruktur, Berlin, Germany
[2]Deutsches Elektronen-Synchrotron, Hamburg, Germany
[3]Universität Hamburg, Department Physik, Hamburg, Germany
[4]Technische Universität Braunschweig, Institut für Halbleitertechnik and Laboratory for Emerging Nanometrology, Braunschweig, Germany
[5]Epitaxy Competence Center ec2, Braunschweig, Germany

The incorporation of indium atoms into a GaN matrix offers the possibility to tune the wavelength of the emitted light. This makes the ternary (In,Ga)N alloy embedded into a GaN matrix interesting for a variety of optoelectronic applications. To improve and expand existing group III-nitride based semiconductor technologies researchers have focused their activities on the investigation of low-dimensional structures such as quantum dots or nanowires. For example, 3D core-shell rods are discussed as promising candidates for next generation light emitting diodes. If the structures are grown not in an axial but in a core-shell geometry, the optically active area can be significantly increased compared to planar structures. Moreover, concerning GaN the growth of (In,Ga)N quantum wells on non-polar m-planes is especially interesting because the optical performance is not impaired by polarization potentials and the quantum confined Stark effect. Hence, indium content and quantum well width mainly determine the band gap.

We have used nanofocus x-ray diffraction to investigate structure and local strain field of an isolated GaN micro-rod with a shell, which comprises of a GaN spacer, an (In,Ga)N/GaN double quantum well (QW) and a p-doped GaN cap. Due to the high spatial resolution of the x-ray beam, we are able to investigate several distinct volumes of one individual side-facet. Here, we find a drastic increase in thickness of the outer GaN shell along the rod height. Additionally, we performed high-angle annular dark-field scanning transmission electron microscopy measurements on several rods from the same sample showing that also (In,Ga)N double QW and GaN barrier thicknesses increase strongly along the height. Moreover, plastic relaxation is observed in the top part of the rod. Based on the experimentally obtained structural parameters, we simulated the strain induced deformation using the finite element method which served as input for subsequent kinematic scattering simulations. The simulations reveal a significant increase of elastic in-plane relaxation along the rod height. However, at a certain height the occurrence of plastic relaxation yields a decrease of the elastic strain. Due to the experimentally obtained structural input for the finite element simulations, we are able to translate the elastic relaxation into an indium concentration.

MS17 - Advances and application of neutron and synchrotron radiation

MS17-01
Probing structural distortions with high precision – Resonant X-ray Diffraction with photon energies tuned to destructive interference

M. Zschornak[1], C. Richter[1,2], D. V. Novikov[2], E. Mehner[1], M. Nentwich[1], S. Gorfman[3], J. Hanzig[1], H. Stöcker[1], T. Leisegang[1], D. C. Meyer[1]
[1]TU Bergakademie Freiberg, Institut für Experimentelle Physik, Freiberg, Germany
[2]Deutsches Elektronen-Synchrotron (DESY), Photon Science, Hamburg, Germany
[3]University of Siegen, Department of Physics, Siegen, Germany

The variability of X-ray photon energy at synchrotrons presents unique capabilities to increase the contrast for the characterization of subtle structure deviations. Within Resonant X-ray Diffraction (RXD), tuning independent scattering contributions of selected, resonantly scattering atoms in the unit cell can even promote a total destructive interference of allowed Bragg reflections for a majority of structures. At a certain photon energy the structure factor vanishes, resulting a high contrast several orders of magnitude upon change of structural parameters. This permits to elucidate structural details with very high accuracy. Diminutive deviations of the ideal structure will change this specific energy in the RXD spectra giving access to e.g. static as well as dynamic atomic displacements.

We used this enhanced RXD approach for the first time to determine the crystal structure of a recently found, crystalline phase of strontium titanate $SrTiO_3$ formed under the influence of an electric field [1]. A loss of inversion symmetry had been evident from the experimental detection of pyro- and piezoelectricity for this phase [2, 3], but standard methods had failed to obtain the structural parameters. The measurements were performed in the vicinity of the Sr-*K* absorption edge on reflections with exactly one odd Miller index. With high structural precision of 1 pm the polar character of the tetragonal phase was validated in terms of $BaTiO_3$-like, polar displacements of titanium and oxygen in opposed directions along the applied field.

References:
[1] Hanzig et al., Phys. Rev. B 88, 024104 (2013).
[2] Hanzig et al., New Journal of Physics 17, 023036 (2015).
[3] Khanbabaee et al., Applied Physics Letters 109, 222901 (2016).

MS17-02
High-Pressure Geophysics using Synchrotron Radiations

H. J. Mueller[1], F. R. Schilling[1], C. Scheffzük[1], B. R. I. Mueller[1]
[1]Karlsruher Institut für Technology, AGW, Karlsruhe, Germany

The interpretation of the dynamic processes within our planet relays on highly resolved seismic data from Earths deep interior requires measurements of the physical properties of Earths materials under experimental simulated mantle conditions. Considering this, geophysical high pressure research is faced the technological challenge to increase pressure and sample volume at the same time to be able to perform in situ experiments with geophysical representative complex samples. Recent large volume presses at synchrotron facilities already provide sample volumes 3 to 7 orders of magnitude bigger than in diamond anvil cells far beyond transition zone conditions. The sample size of several cubic millimeters allows elastic wave experiments in the low to medium MHz range. Together with the small and even adjustable temperature gradients over the whole sample, this technique makes anisotropy and grain boundary effects in complex systems accessible for the measurements of elastic and inelastic properties. The real elastic wave propagation have also no limits for opaque and encapsulated samples. Using triple-mode transducers and the data transfer function technique for ultrasonic interferometry makes transient measurements in complex systems under non-equilibrium conditions possible. Ultrasonic interferometry necessarily requires in situ sample deformation measurement by X-radiography. In situ falling sphere viscosimetry require time-resolved X-radiography. Using a D-DIA even the measurement of elastic and inelastic properties in the seismic frequency range is achievable. The paper presents techniques and results of recent experimental high pressure geophysics in the low velocity zones within the Earths mantle.

MS17-03
Feasibility of high-pressure single-crystal neutron diffraction in diamond anvil cells on the diffractometer HEIDI at the Heinz Maier-Leibnitz Zentrum (MLZ) in Garching

A. Grzechnik[1], M. Meven[1], K. Friese[2]
[1]RWTH Aachen University, Institute of Crystallography, Aachen, Germany
[2]Research Centre Jülich, JCNS-2, Jülich, Germany

High-pressure single-crystal x-ray diffraction in diamond anvil cells (DAC) is a routine technique to study structures of crystalline solids. Data in the DAC can easily be measured on every single-crystal x-ray diffractometer and processed with standard software. On the other hand, single-crystal neutron diffraction in opposite anvil cells has been limited to very few studies. The reason is that even at the most advanced neutron sources the flux densities are far below the corresponding photon flux densities at synchrotron and laboratory instruments. Hence, neutron-diffraction experiments are difficult due to the limited sample volume in the DAC. Neutron diffraction is complementary to x-ray diffraction, providing the information for instance on magnetic order and hydrous species. It is then timely to explore the feasibility of neutron measurements in the DAC at hot sources that offer the benefit of various short neutron wavelengths with high fluxes.

We have now started a project to establish single-crystal neutron diffraction measurements in the DAC on the diffractometer HEIDI at the MLZ. Several existing DACs have been tested there to optimize the most efficient scattering geometry. The signal-to-noise ratio improves significantly after minimizing the beam size and when using hot neutrons due to reduced scattering/absorption of the neutron beam by the components of the DAC. Consequently, we have developed optimized DACs for measurements at room and low temperatures.

We have studied a crystal (~0.3 mm^3) of $MnFe_4Si_3$ in the panoramic Ahsbahs-type diamond anvil cell [*Rev. Sci. Instr.* **55**, 99 (1984)] at ambient pressure and at 1 GPa. It has been possible to find reflections of the sample using standard searching routines in the reciprocal space. Several hundred reflections have been collected using a point detector at both pressures. It has also been possible to refine the structural model for $MnFe_4Si_3$ using both data sets with the program Jana2006 to $R_{obs}<10\%$.

Our results so far demonstrate the feasibility of single-crystal neutron diffraction in the DAC with hot neutrons at the MLZ. Further progress will be achieved by upgrading both the monochromator to operate at short wavelengths and the neutron guide to minimize the beam size so that much smaller crystals (<0.1 mm^3) could be investigated.

MS17-04

Determination of coordination site disorder in $LiMn_{2-x}Ti_xO_4$ through combined X-ray diffraction and XAS studies

D. T. Murphy[1], S. Schmid[1]
[1]The University of Sydney, School of Chemistry, Sydney, Australia

Spinel type lithium metal oxides are attractive cathode materials for rechargeable lithium ion batteries. In particular, LiMn2O4 was considered a promising cathode material as it is relatively cheaper, environmentally more friendly and safer to operate than the widely used $LiCoO_2$. Doping of $LiMn_2O_4$ with tetravalent titanium was reported to increase the stability of the structure. The presence of Ti^{4+} is able to suppress the Jahn-Teller effect of Mn^{3+} resulting in a more stable spinel framework and therefore, the cycling ability is significantly improved. In addition, a recent study on nanophase $LiMnTiO_4$ showed that capacities up to 290 mA h g^{-1} are achievable, rendering the material a very attractive electrode material indeed [1].

This presentation focuses on the structural investigation of $LiMn_{2-x}Ti_xO_4$ as prepared through solid state syntheses employing different heating and cooling regimes [2]. The phase behaviour of quenched and slowly cooled $LiMn_{2-x}Ti_xO_4$ was confirmed through variable temperature synchrotron X-ray and neutron powder diffraction measurements. The distribution of Li between tetrahedral and octahedral sites was determined from diffraction data. Due to their very similar X-ray scattering factors, however, analysis of the Mn/Ti distribution in addition required Mn and Ti K-edge X-ray absorption near edge structure spectra. These revealed, e.g. for the x = 1 member, the presence of Mn^{3+} in primarily octahedral and Ti^{4+} in octahedral and tetrahedral environments, with very slight variations depending on the synthesis conditions. Magnetic measurements indicated the dominance of antiferromagnetic interactions in both the slowly cooled and quenched samples below 4.5 K.

References:
[1] Chen, R.; Knapp, M.; Yavuz, M.; Heinzmann, R.; Wang, D.; Ren, S.; Trouillet, V.; Lebedkin, S.; Doyle, S.; Hahn, H.; Ehrenberg, H.; Indris, S. *J. Phys. Chem. C*, **2014**, *118*, 12608.
[2] D. T. Murphy, S. Schmid, J. R. Hester, P. E. R. Blanchard & W. Miiller, *Inorg.Chem.* **2015**, *54*, 4636.

MS17-05

The D19 diffractometer at the ILL: for chemical and protein crystallography

E. Mossou[1]
[1]Institut Laue Langevin, Grenoble, France

D19 is a monochromatic thermal neutron diffractometer for single crystal and fibre diffraction. With a wavelength selection of 0.8-2.4Å, and a large position sensitive detector, it is versatile and provides large solid angle coverage. D19 is optimized for the study of large chemical systems, small proteins and fibrous polymers but is also very efficient for texture analysis and accurate structures of smaller molecules.

In the recent past, D19 has produced remarkable results from a number of protein and chemical systems, yielding high impact publications1,2,3,4,5,6,7. These studies have fully exploited the high flux of the instrument and the large solid angle of the detector. D19 has allowed the highest resolution protein structure ever determined with good completeness. This access to ultra-high resolution helped uncovering a world of structural detail below that visible by X-ray protein crystallography, with highly structured water, hydronium and Zundel ion networks that are likely to be of major importance in protein charge transfer processes. The D19 diffractometer is unmatched for this scale of structural biology, and the availability of perdeuterated protein from the Life Sciences group offers highly optimized sample production.

This talk will focus on presenting the instrument as well as highlight some recent examples of protein and chemical crystallography where data collected on D19 has brought unique insight to the structure. Ongoing instrument and software developments are likely to further broaden the scope of the instrument and allow larger systems to be studied.

References
[1] A.Y. Kovalevsky , L. Hanson , S.Z. Fisher , M. Mustyakimov , S.A. Mason , V.T. Forsyth , M.P. Blakeley , D.A. Keen , T. Wagner , H.L. Carrell , A.K. Katz , J.P. Glusker , P. Langan . Structure. 18(6): 688-99, 2010.
[2] A. Y. Kovalevsky, B. L. Hanson, S. A. Mason, T. Yoshida, S. Z. Fisher, M. Mustyakimov, V. T. Forsyth, M. P. Blakeley, D. A. Keen, P. Langan. Angew. Chem . 50 , 7520-7523 (2011).
[3] M.G. Cuypers, S.A. Mason, M. P. Blakeley, E.P. Mitchell, M. Haertlein, and V. T. Forsyth, Angew. Chem .52 :1022-25, 2013.
[4] P. Langan, A.K. Sangha, T. Wymore, J.M. Parks, Z.K. Yang, B.L. Hanson, Z. Fisher, S.A. Mason,
[5] Guidi T., Gillon B., Mason S.A., Garlatti E., Carretta S., Santini P., Stunault A., Caciuffo R., van Slageren J., Klemke B., Cousson A., Timco G.A., Winpenny R.E.P. Nature Communications **6,** 7061-1-7061-6 (2015)
[6] R. Lee, S.A. Mason, E. Mossou, G. Lamming, M. R. Probert, and J. W. Steed. *ACS Crystal Design and Growth. In press*
[7] M.G. Cuypers, S.A Mason, E. Mossou, M. Haertlein, V.T. Forsyth, E.P. Mitchell. *Scientific Reports*, 6, 31487, 2016

MS17-06

Analysis of VHCF damage in duplex stainless steel using micro-beam X-ray diffraction and a pnCCD detector.

A. Abboud[1], U. Pietsch[1], B. Dönges[2], H. J. Christ[2]
[1]University of Siegen, Department of Physics, Siegen, Germany
[2]University of Siegen, Institut für Werkstofftechnik, Siegen, Germany

During very high cycle fatigue (VHCF) of austenitic-ferritic duplex stainless steel (DSS), dislocations arrange in patterns within the austenite grains leading to crack formation at grain boundaries towards the ferrite phase and consequently inducing mechanical failure. How and when the self-organization processes of dislocations take place have remained elusive. Measurements using an energy dispersive 2D detector (pnCCD) promises to shed light on the subject making it possible to identify individual grains and their respective Bragg reflections. Using the pnCCD, we investigate the arrangement of dislocations in the vicinity of a fatigue crack in VHCL DSS. Using a sub-micrometer sized x-ray beam we scan multiple region of interest in step size of 1um. This allows to determine the crystallographic orientation and local texture simultaneously with high spatial resolution without moving the sample. By studying the energy and spatial evolution of the Laue spots during in situ loading, we capture an insight view into the processes relevant for understanding of the strength and work-hardening of steel alloys.

Poster

Advances and application of neutron and synchrotron radiation

P1
Modern CPAD Detector Technology for Higher Energy X-rays
M. Adam[1], E. Hovestreydt[1], R. Durst[1]
[1] Bruker AXS GmbH, Karlsruhe, Germany

Higher energy X-rays result in a compressed diffraction pattern and enable higher resolution to be achieved, which is particularly advantageous for charge density studies or when the diffraction geometry is restricted, for example by a high-pressure cell. Other advantages include strongly reduced extinction and reduced absorption, which is proportional to about λ3 and strongly reduced extinction.

Recent advances in X-ray source and mirror technology have provided the home-lab market with viable solutions for higher energy X-rays, namely the Ag Kα (0.56089 Å) IμS 3.0 and the In Kα (0.51359 Å) METALJET. These high flux density systems with small, focused beams overcome the decrease in the absolute scattering power of the crystal, which is proportional to λ3.

The higher energy X-rays challenge modern detectors, and especially Si-sensor based HPAD detectors. HPAD detectors suffer from low DQE and parallax effects due the thick sensors employed. The latest generation of CPAD (charge integrating pixel area detector) on the other hand, with their thin and extremely efficient high Z-element scintillators, overcome these challenges, and provide the ideal solution for shorter wavelength experiments.

This presentation will focus on recent advances in X-ray detector hardware and software development, and will highlight experiments where higher energy X-rays are advantageous, as outlined above.

Figure 1

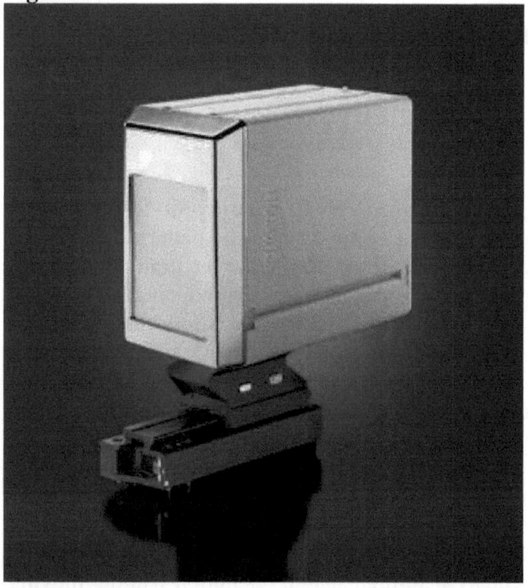

P2
in-situ synchrotron radiation study of Ag-Sn bilayer growth
N. Zotov[1], P. Schützendübe[2], P. Rossi[2], E. J. Mittemeijer[2], S. Ibrahimkutty[3], P. Wochner[3]
[1]University of Stuttgart, Institute of Material Science, Stuttgart, Germany
[2]Max Planck Institute for Intelligent Systems, Stuttgart, Germany
[3]Max Planck Institute for Solid State Reserach, Stuttgart, Germany

Recently it has been shown that the sequence of the intermetallic compounds (IMC) formation in metal bilayers of Ag-Sn and Ag-In, prepared by thermal evaporation, sensitively depends on the stacking sequence [1-2]. The large diffusion coefficients for the Ag-Sn and Ag-In systems lead to rapid interdiffusion even at RT and preclude a detailed investigation of the phase formation and the structure of the interface layer during deposition with ex-situ X-ray diffraction methods.

In-situ time-resolved grazing incidence X-ray diffraction was used in the present contribution in order to study the kinetics of phase formation during deposition of Ag-Sn bilayers. The experiments were performed at the MPI beam line of the ANKA synchrotron radiation facility (photon energy 12 keV) using a specially-build ultra-high vacuum MBE-chamber, equipped with two electron-beam evaporators of type EFM3. The MBE chamber was mounted on a heavy-duty diffractometer in vertical scattering geometry. The 7 cm high Be-window, covering 360°, allows the in-situ study of thin film growth under grazing incidence from practically all azimuthal directions. The deposition rates were about 1 Å/s for both Ag and Sn.

The in-situ experiments show that the first IMC formed is always the Ag_3Sn phase, independently of the stacking sequence of the bilayers (Ag-on-top or Sn-on-top). In Sn-on-top bilayers only the Ag_3Sn phase is observed during the Sn deposition. In the case of Ag-on-top bilayers two growth regimes are observed. For thick Sn bottom layers (tSn > 60 nm) only Ag_3Sn is formed during the whole Ag deposition. For thin Sn bottom layers (t_{Sn} < 60 nm), a second phase Ag_4Sn is formed once the Sn bottom layer is consumed. The kinetics of nucleation and growth of the Ag_3Sn phase indicates most probably diffusion-controlled growth with nucleation rate rapidly decreasing with time

References:
[1] P. Rossi, N. Zotov, E.J. Mittemeijer, Surface and Coatings Technology, 295 (2016) 88-92
[2] P. Rossi, N. Zotov, E.J. Mittemeijer, J Applied Physics 120 (2016) 165308

P3
Temperature-dependent structural and spectroscopic investigations of $(Bi_{1-x}Fe_x)FeO_3$ (x = 0.17 and 0.33) using neutron elastic and light inelastic scattering
A. Kirsch[1,2], M. M. Murshed[1,2], M. Kirkham[3], A. Huq[3], T. M. Gesing[1,2]
[1]Universität Bremen, Institut für Anorganische Chemie und Kristallographie, Bremen, Germany
[2]Universität Bremen, MAPEX Center for Materials and Processes, Bremen, Germany
[3]Oak Ridge National Laboratory, Chemical and Engineering Materials Division, Oak Ridge, Tennessee 3783, United States

BiFeO3 is one of the widely studied multiferroic materials for interesting properties [1] depending on crystal chemistry [2] and crystallite/particle dimensionality [3]. Recently, a 57Fe Mössbauer spectroscopy and X-ray powder diffraction (XRPD) study revealed a series of $(Bi_{1-x}Fe_x)FeO_3$ perovskite-type structures, where the Bi-site is partially shared by Fe [4]. To further tune the multiferroic property either from electric or magnetic approaches, or simultaneously both directions, the present study reports two compositions $(Bi_{0.83}Fe_{0.17})FeO_3$ and $(Bi_{0.67}Fe_{0.33})FeO_3$. Samples were synthesized by a sol-gel route at 873 K. Starting from X-ray amorphous nano-crystallites heating time and temperature exclusively determine the iron occupancy on the perovskite *A*-site [4]. The materials were investigated by temperature-dependent neutron time-of-flight powder diffraction at POWGEN (SNS, Oak Ridge, USA) from 10 K to 650 K. A deep minimum has been observed at about 240 K for the change of *z*-coordinates of Fe-site and O-sites in the $(Bi_{0.83}Fe_{0.17})FeO_3$ system, which has been shifted to 260 K for $(Bi_{0.67}Fe_{0.33})FeO_3$. Any anomaly was not observed for the temperature-dependent atomic displacement parameters (ADPs) and thermal expansion. Fitting the ADPs using Debye approach, all three atom-sites showed usual harmonic behavior, leading to a Debye temperature of 288(10) K and 368(13) K for$(Bi_{0.83}Fe_{0.17})FeO_3$ and $(Bi_{0.67}Fe_{0.33})FeO_3$, respectively. Notably, the Bi/Fe and O-sites show higher static disorder in (Bi0.67Fe0.33)FeO3 than those in $(Bi_{0.83}Fe_{0.17})FeO_3$. The thermal expansion of the metric parameters were modeled using DEA approach [5,6. The intrinsic optical phonon anharmonicity has been extracted from the temperature-dependent Raman spectra. Both samples showed significantly a higher frequency shift of two selective modes than those of the stoichiometric $BiFeO_3$ perovskite. The damping factors obtained from the simplified Klemens phonon decay model distinguishes between the phonon and magnon contributions to the microscopic heat capacity of the modes.

References:
[1] F. Huang et al. Sci. Rep. 3 (2013) 2907
[2] A. Palewicz et al. Acta Cryst. B63 (2007) 537
[3] A. Jaiswal et al. J. Phys. Chem. 114 C (2010) 2108
[4] A. Kirsch et al. Z. Naturforsch. 71 b (2016) 447
[5] M.M. Murshed et al. Mater. Res. Bull. 84 (2016) 273
[6] M.M. Murshed et al., J. Solid State Chem. 229 (2015) 87

P4

Spin density of a cAAC-based stable radical from single-crystal X-ray diffraction

M. Flatken[1], B. Dittrich[1]
[1]Heinrich-Heine Universität Düsseldorf, Anorganische Chemie und Strukturchemie II, Düsseldorf, Germany

When in 1998 wavefunction fitting was introduced [1], it emerged that the evaluation of single crystal x-ray diffraction experiments permits to extract spin density. Procedures require fitting coefficients of a quantum chemical basis set. These coefficients are adjusted to best reproduce the election density of a crystallographic structural model, which is in turn obtained via Fourier transform of measured intensities. At the same time the resulting wavefunction is energy-minimized. In 2008 technical feasibility of the approach was significantly improved by introduction of Hirshfeld atom refinement (HAR) [2].

In HAR more accurate structural models can be refined by relying on aspherical scattering factors of the whole molecule. Such scattering factors are obtained by single-point energy computation of the structure under investigation, and Fourier transform of the quantum chemical electron density, partitioned using Hirshfeld's partitioning scheme [3]. The now atomic scattering factors are then least-squares adjusted to the experimental Bragg data. This gives a slightly different structure, and the process is repeated to convergence.

The combination of wavefunction fitting after successful HAR has been termed wavefunction refinement [4]. The ability to extract a minimal energy wavefunction that is in best agreement with crystallographic measurements gives access to many properties formerly only accessible by theory, like the spin density. We here present first results of a cAAC-based stable radical [5] re-measured to higher than conventional resolution at the Petra III synchrotron. Theoretical calculations have already been performed on the system and will serve as a benchmark for judging on the success of the experimental spin density determination.

References
[1] D. Jayatilaka, Phys. Rev. Lett. 1998, 80, 798–801.
[2] D. Jayatilaka, B. Dittrich, Acta Cryst. A 2008, 64, 383–393;
S. C. Capelli, H. B. Bürgi, B. Dittrich, S. Grabowsky, D. Jayatilaka,
IUCrJ. 2014, 1, 361–379.
[3] F. L. Hirshfeld, Theoret. Chim. Acta (Berl.) 1977, 44, 129-138.
[4] S. Grabowsky, P. Luger, J. Buschmann, T. Schneider,
T. Schirmeister, A. N. Sobolev, D. Jayatilaka,
Angew. Chem. Int. Ed. 2012, 51, 6776–6779.
[5] K. C. Mondal, H. W. Roesky, A. C. Stückl, F. Ehret, W. Kaim,
B. Dittrich, B. Maity, D. Koley, Angew. Chem. Int. Ed. 2013, 52, 11804–11807.

P6

Neutron radiography study on temperature and moisture effects on the tensile strength of bentonite moulding sand

G. Jordan[1], K. Schiebel[1], B. Schillinger[2], A. Kaestner[3], W. W. Schmahl[1]
[1]LMU, Geo- & Umweltwiss., München, Germany
[2]FRM II, Antares, Garching, Germany
[3]Paul-Scherrer-Institut, ICON, Villigen, Switzerland

Green sand casting is the most economical and environmentally friendly casting methods to pour multiple castings. However, reusability is limited. On average, the casting sand is recycled up to 20 times before disposal. Temperatures of 500-750°C cause chemical alterations of the smectite minerals in the bentonite. These alterations lead to a decrease of the binding properties of the smectites (Odom, 1984) and, therefore, of the mechanical properties of the sand mould. Aim of this study was to unravel the influence of temperature and remoistening on the binding properties of bentonite bonded moulding sand.

To investigate the binding properties, tensile testing was conducted with sand samples pre-treated at temperatures from 120°C to 750 °C before moistening. For tensile testing, the specimens were heated from above to 310 °C, in order to induce a temperature and moisture gradient into the samples simulating the casting conditions. Moisture transfer and tensile testing were imaged by neutron radiography. Evaluating the radiographs, moisture and location of the rupture plane were determined precisely (Fig.1). Temperature was measured with thermocouples within the specimen.

Moulding sands pre-exposed to $T > 400$ °C show a strongly reduced tensile strength. Moisture transfer rate increases and steam penetrates deeper into the sample (Fig.1). Samples pre-exposed to $T > 600$ °C show a modified location and morphology of the rupture plane (inlays of Fig. 1): no cementation of the sand occurred and the rupture was no longer located at the evaporation front, where the temperature just reaches 100 °C for the first time. Instead, rupture took place in the loose sand of the dry zone.

The results show, that alteration by temperature strongly affects both water transport and mechanical properties. Both effects were successfully observed and quantified, making a step further to understanding binding mechanism in clayey sands.

Fig. 1: Moisture evolution with time of moulding sands exposed to 120°C (left) and 750°C (right). The water content is colour coded. The data were measured along vertical profiles of the radiographies (see inlays). The inlays show the radiographs taken one second before (a) and one second after (b) rupture. Measured zone boundaries are marked by dots, which are fitted by a square root function (white lines).

Figure 1

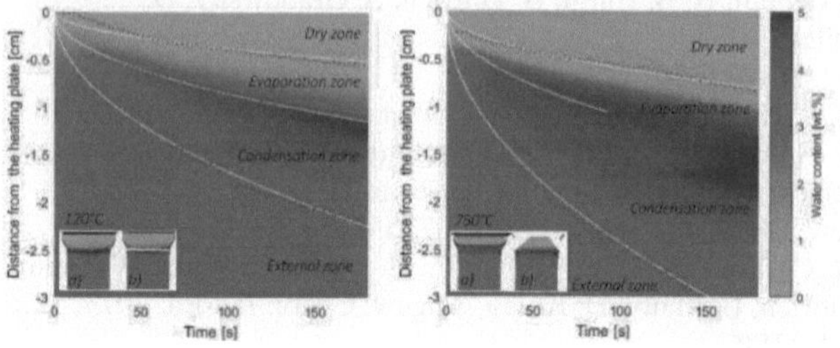

P7

Enhanced options at the BESSY II beamline KMC-2

D. Többens[1], G. Schuck[1], D. Wallacher[2], N. Grimm[2], S. Schorr[1]
[1]Helmholtz-Zentrum Berlin für Materialien und Energie (HZB), Struktur und Dynamik von Energiematerialien, Berlin, Germany
[2]Helmholtz-Zentrum Berlin für Materialien und Energie (HZB), Abteilung Probenumgebung, Berlin, Germany

At the Berlin synchrotron BESSY II, the KMC-2 beamline [1] serves two permanent endstations, "XANES" and "DIFFRACTION". A graded SiGe monochromator provides radiation in the energy range of 4 – 14 keV with energy resolution of 1/4000 and beam intensity stabilized to an accuracy of 0.3 %. XANES uses this high-quality radiation for EXAFS and X-ray fluorescence measurements. The six-circle goniometer of the DIFFRACTION endstation can be set up for a wide range of scattering and diffraction measurements, including grazing incidence diffraction (GID), reciprocal space mapping, diffuse scattering, anomalous diffraction, and powder diffraction.

A particular strong point of KMC-2 is the wide range of sample environments available for both stations. Besides various high-temperature furnaces and gas-loading devices, of particular interest for crystallographers are the newly developed low-temperature environments [2], which allow experiments down to 15 K.

The time needed to switch between XANES and DIFFRACTION is only two hours. This makes it possible to combine both stations in the same experiment, making KMC-2 a very versatile beamline especially suited for a wide range of non-standard experiments.

References:
[1] D. M. Többens, S. Zander, Journal of large-scale research facilities, 2, A49 (2016)
[2] https://www.helmholtz-berlin.de/user/experimental-infrastructures/sample-environment/se-at-bessy/temp-environment-kmc2/

P8

Magnetic structure determination of pyroxenes in the solid solution series $LiFeGe_2O_6$ – $NaFeGe_2O_6$

G. Redhammer[1], A. Senyshyn[2], G. Tippelt[1], G. Roth[3]
[1]University of Salzburg, Chemistry & Physics of Materials, Salzburg, Austria
[2]FRM-2, Garching, Germany
[3]RWTH Aachen, Institute of Crystallography, Aachen, Germany

Solid solution compounds along the $Li_{1-x}Na_xFeGe_2O_6$ clinopyroxene series have been prepared by solid state ceramic sintering and investigated by bulk magnetic and calorimetric methods, the Na – rich samples with x > 0.7 were also investigated by low temperature neutron diffraction experiments between 4 K and 20 K.

For samples with x(Na) > 0.76 the crystal structure adopts the *C2/c* symmetry at all measuring temperatures. Magnetic ordering is observed below 20K with a slight decrease of T_N with increasing Na+ content. The magnetic spin – structures change distinctly as a function of chemical composition: Up to x = 0.72 it can be described by a collinear commensurate arrangement of magnetic spins, in space group $P2_1'/c$ with propagation vector **k** = (½, 0 0), an antiferromagnetic (AFM) coupling within the $Fe^{3+}O_6$ octahedral zig-zag chains and an alternating AFM and ferromagnetic (FM) interaction, depending on the nature of the GeO_4 tetrahedral chains.

Close to the structural phase transition, x(Na) = 0.75, magnetic ordering is observed below 15K, however it becomes incommensurately modulated with k = (0.344, 0, 0.063). At 4 K the magnetic spin structure best can be described by a cycloidal arrangement within the M1 chains, the spins are within the **a** - **c** plane. Around 12 K the cycloidal structure transforms to a spin density wave (SDW) structure. For the *C2/c* structures, a coexistence of a simple collinear and an incommensurately modulated structure is observed down to lowest temperatures. For 0.78 < x < 0.82 a collinear structure with **k** = (0 1 0), space group *C2'/c* and an AFM spin structure within the M1 chains and an FM one between the spins is dominating, however, the modulated structure dominated the collinear one in the samples with x(Na) = 0.88. Here the magnetic propagation vector is **k** = (0.28, 1, 0.07) and the spin structure corresponds to a cycloidal structure within the M1 chains. As for the other samples, a transition from the cycloidal to a SDW structure is observed. Based on the neutron diffraction data, the appearance of two peaks in the heat capacity of Na rich samples can now be interpreted as a transition from a cycloidal magnetic structure to a spin density wave structure of the magnetically ordered phase for the Na – rich part of the solid solution series.

P9

An Approach to Forward Model Time-of-Flight Neutron Diffraction Data for Uniaxial Load Conditions Using the Example of Sandstone

S. Breuer[1], C. Scheffzük[1,2], B. Müller[1,3], F. Schilling[1]
[1]Karlsruhe Institute of Technology (KIT), Institute of Applied Geosciences, Div. Technical Petrophysics, Karlsruhe, Germany
[2]Joint Institute for Nuclear Research (JINR), Frank Laboratory of Neutron Physics, Dubna, Russian Federation
[3]State Research Center for Geothermal Energy, Karlsruhe, Germany

Mechanical properties of (sedimentary) rocks such as stress-strain relations are essential to understand processes within the Earths crust, *e.g.* the dynamic processes in oil and gas reservoirs, underground gas storages or in geothermal energy exploration.

Here, measurements of *in situ* lattice strains in bulk samples are realized by high-resolution neutron diffraction.

This study[1] deals with load-dependent deformation data of a cylindrical homogeneous sandstone specimen, examined by time-of-flight neutron diffraction method at the pulsed neutron source IBR-2M (JINR, Dubna, Russia). This approach focusses on a prediction of measured diffraction data at elevated uniaxial loads based on the load-free measurement and the known elastic properties of single crystal α-quartz. A spherical elastic-mechanical model is presented. It incorporates changes in pore space due to compaction of the sediment. This models behavior is taken to predict the shift of the diffraction pattern lattice *hkl* peaks based on the models proposed by Voigt[2] and Reuss[3] and combinations of these. Modeling results for distinct applied uniaxial stresses are compared to measured data.

It is shown that it is possible to forward model (*i.e.* predict) diffraction data at elevated uniaxial load-stages based on the load-free measurement by using an appropriate elastic-mechanical model describing the probed specimen. In case of the exemplarily treated sandstone, measured diffraction patterns of three uniaxial load-stages are in good agreement with model predictions using the Voigt model. The model predicts both well, the variation in full width at half maximum (FWHM) and the shift of the diffraction lines with increasing load. The diffraction pattern forward model quality is in the range of R_{wp} ~1-2 % (weighted profile *R*-factor).

References:
[1] S. Breuer (2015). Forward Modeling Time-of-Flight Neutron Diffraction Data for Uniaxial Load Conditions (Master thesis, publication submitted: Breuer S., Schilling F.R., Scheffzük Ch. & Müller B.; *J. Appl. Cryst.*). Karlsruhe Institute of Technology, Karlsruhe.
[2] W. Voigt (1928). *Lehrbuch der Kristallphysik*. Leipzig, B.G. Teubner.
[3] A. Reuss (1929). *Z. Angew. Math. Mech.*, **9**. DOI: 10.1002/zamm.19290090104.

P10
The automatized sample changing system at the High Resolution Powder Diffraction Beamline P02.1 at PETRA III, DESY.

M. Etter[1], J. Bednarcik[1], M. Wendt[1], S. Wenz[1]
[1]Deutsches Elektronen Synchrotron, PETRAIII, Hamburg, Germany

Synchrotron powder X-ray diffraction has become a standard tool in order to investigate the crystalline behavior of polycrystalline materials like minerals, pharmaceuticals and all kinds of organic and inorganic products. Compared to laboratory powder X-ray diffraction devices the usage of synchrotron radiation at dedicated beamlines is superior due to the much higher photon flux, which allows for faster data acquisition rates. These higher acquisition rates can be either used in order to perform time-resolved studies of dynamic processes in matter or they can be used in order to process a large number of samples. However, with respect to the processing of large sample numbers, the number of samples processed in a certain time frame is not limited by the acquisition rate of modern detectors but rather by the time which is required for a manual changing of these samples.

In order to overcome this drawback a new high-throughput automatized sample changing system will be installed at the high-energy high resolution powder X-ray diffraction beamline P02.1 at PETRAIII/DESY. This high-throughput automatized sample changing system consists of a commercial Stäubli robotic arm (see figure 1) modified with a special pneumatic gripper (see figure 2), new types of holders for capillaries and an interchangeable magazine system which has a capacity for up to 300 capillary samples. In combination with our heating / cooling devices (hot-air-blower / cryostreamer) it will be possible to measure automatized up to 300 different capillary samples in a temperature range between 90 K and 1200 K. Moreover, due to the high-energy of the beamline and the movable area detector system an easy switching between standard powder diffraction measurements and total scattering/pair distribution function measurements will be possible.

In the current presentation, the latest developments regarding the automatized sample changing system and the corresponding software at the PETRAIII beamline P02.1 is given.

Fig. 1: Stäubli robotic arm at beamline P02.1.

Fig. 2: Gripper for special capillary holders.

Figure 1

Figure 2

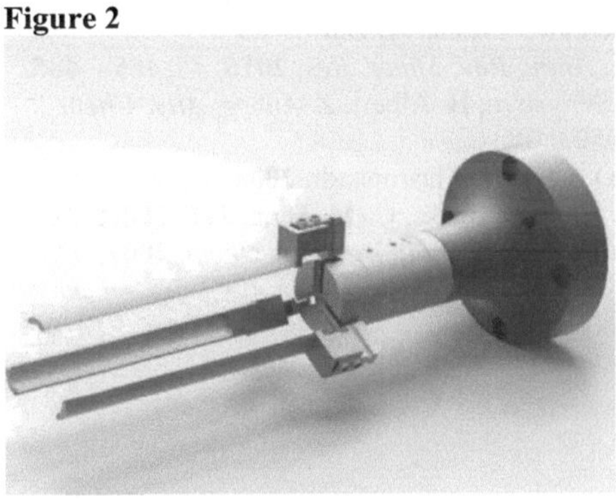

P11

Transition metal diborides: versatile materials with interesting crystal structures

K. Hofmann[1], M. Kayhan[1], B. Albert[1]
[1]Technische Universität Darmstadt, Eduard-Zintl- Institut für Anorganische und Physikalische Chemie, Darmstadt, Germany

Diborides of the transition metals exhibit favorable properties like superconductivity [1], hardness [2], high melting points, and magnetism [2]. MB_2 (M = Sc, Y, Ti, Zr, Hf, V, Nb, Ta, Cr, Mo, Mn) crystallizing in the AlB_2-type are well known, but diborides of the heavy metals with different crystal structures have also attracted a lot of interest lately, especially due to their extreme hardness [3].

On the basis of X-ray and neutron powder data we were able to confirm that the structures of the diborides of Re, Ru and Os contain puckered sheets of condensed B_6-rings comparable to the chair- and boat conformation of cyclohexane, and also described new variants like Os_2B_3 [4]. For molybdenum diboride two modifications are known, one crystallizing in the AlB_2-type with boron deficiency (MoB_{2-x}), and the second one in the so-called Mo_2B_4-type [5, 6]. Mo_2B_4 (s. g. *R* -3 *m*, no. 166) contains planar and corrugated boron atom layers. The crystal structure of W_2B_4 has been a topic of debate for years. On basis of X-ray data (single crystal and powder) a structure model (s. g. *P* $6_3/mmc$, no. 194) with planar and puckered boron sheets, very similar to Mo_2B_4, was derived. Neutron powder data on the other hand showed small extra reflections indicating a symmetry reduction and a larger unit cell. A new structure model (s. g. *P* $6_3/mcm$, no. 193) with partially occupied tungsten and boron sites is proposed.

Recent theoretical investigations show that the different structure types of the transition metal diborides can be attributed to a combination of the valence electron count, the size of the metal cation, and the bonding situation [7].

References:

[1] M. Kayhan, E. Hildebrand, M. Frotscher, A. Senyshyn, K. Hofmann, L. Alff, B. Albert, *Solid State Sci.* **2012**, *14*, 1656 – 1659.
[2] N. Kalyon, planned PhD thesis, TU Darmstadt
[3] R. B. Kaner et al., *Annu. Rev. Mater. Res.* **2016**, *46*, 465 – 485.
[4] M. Frotscher, A. Senyshyn, B. Albert, *Z. Anorg. Allg. Chem.*, **2012**, *638*, 2078 – 2080.
[5] M. Frotscher, PhD thesis, TU Darmstadt, **2008**
[6] M. Frotscher, W. Klein, J. Bauer, C.-M. Fang, J.-F. Halet, A. Senyshyn, C. Baehtz, B. Albert, *Z. Anorg. Allg. Chem.* **2007**, *633*, 2626-2630.
[7] J. Brgoch et al., to be published

P12

The neutron time-of-flight diffractometer EPSILON for residual and applied strain analysis on geological samples under uniaxial and triaxial load conditions

C. Scheffzük[1,2], B. I. R. Müller[1], S. Breuer[1], B. Altangerel[2,3], F. R. Schilling[1]
[1]Karlsruhe Institute of Technology, Institute for Applied Geosciences, Karlsruhe, Germany
[2]Joint Institute for Nuclear Research, Frank Laboratory of Neutron Physics, Dubna, Russian Federation
[3]Mongolian Academy of Sciences, Institute of Physics and Technology, Ulaanbaatar, Mongolia

The knowledge of the stress and strain state in rock samples is of importance in the understanding of the rock properties for the interpretation of geodynamic processes and for geotechnical applications such as mining and tunneling. Recently, the topic of induced seismicity has become a major issue in geoenergy production (gas production, geothermal). Thus, the role of tectonic stress for the initiation of rock failure is of high scientific but also economic interest. The investigation of deformation with conventional lab scale experiments allows the determination of macro-strain in the cm- to m-range. On the nano- to micro-scale, the application of diffraction methods offers the investigation of the strain which is localized in the crystal lattice, the so-called intracystalline strain. This localization of strain in a sample with dimensions in cm-range is achieved by strain scanning. Because of the low absorption properties of neutrons in matter, like metals and minerals, neutron diffraction is an excellent method for the investigation of strain in bulk samples, especially in multiphase samples, like rocks.

The neutron time-of-flight strain diffractometer EPSILON operated at the pulsed neutron source IBR-2M at the JINR Dubna is designed for the investigation of residual and applied intra-crystalline strain of bulk samples. Because of the large wavelength-range (λ = 7.8 Å) diffraction pattern with *d*-spacings up to *d* = 5.6 Å can be investigated. Using the long flight path of about 107 m a very good spectral resolution is achieved. That allows the investigation of multiphase rocks containing minerals with lower crystal symmetry. The diffractometer is equipped with a uniaxial pressure device, allowing load states up to 100 kN, *e.g.* 150 MPa with sample diameters of 30 mm and 60 mm length1. A triaxial pressure device with axial pressures of up to 150 MPa, confining pressures up to 70 MPa and additional pore pressure up to 70 MPa allows the *in situ* investigation of the pore pressure stress coupling.

To investigate the spatio-temporal evolution of failure for cyclic axial loading and unloading of sandstone rock samples, residual and applied strain investigations were carried out in samples of porous Lahr sandstone (Southern Germany). The investigations concentrated on 3 locations within the sample: in the center, at the rim and at a location in-between (half-radius). To avoid immediate rock failure a pre-experiment to determine the failure stress has been performed.

P13

Thermal stability of Fe-ion containing oxide materials, studied by *in situ* high-temperature measurements

O. Chimitova[1], A. Sarapulova[2], D. Mikhailova[2,3], M. Knapp[2], N. Bramnik[2], G. Balachandran[2], Y. Protz[4], A. Komarek[4], B. Bazarov[1], J. Bazarova[1], H. Ehrenberg[2]
[1]Baikal Institute of Nature Management, Siberian Branch, Russian Academy of Sciences, Ulan-Ude, Russian Federation
[2]Karlsruhe Institute of Technology, Institute for Applied Materials – Energy Storage Systems (IAM-ESS), Eggenstein-Leopoldshafen, Germany
[3]IFW Dresden, Institute for Complex Materials, Dresden, Germany
[4]Max Planck Institute for Chemical Physics of Solids, Dresden, Germany

The field of application for inorganic complex oxide materials has increased within the last decades. The special place is held by mix valence iron compounds, which can possess interesting magnetic and electrochemical properties. For application of the new synthesized structure the temperature stability region plays important role.

The single crystal of $RbFe_5(MoO_4)_7$ was obtained by flux method in the vacuumed ampule and structure was solved with the space group $P2_1/m$. It consists of separate FeO_6-octahedra, and zigzag Fe_4O_{18}-units of edge-sharing FeO_6-octahedra, which are connected via corners with MoO_4-tetrahedra. There is no direct contact between MoO_4-units. This structure is antiferromagnetic with magnetic moment of Fe-ion: μ=12.7(1)μ_B/f.u., θ= -12.8(8)K. The thermal stability of presented compound was investigated by *in situ* high temperature synchrotron measurements on the European Synchrotron Radiation Facility (λ=0.40073 Å, Grenoble, France). The structure decomposes at higher than 800°C with obtaining Fe(III) molybdate and different Rb polymolybdates. The melting point 804°C was confirmed by thermal analysis.

Fig. 1 Structure evolution during heating and cooling investigated by synchrotron measurements- $RbFe_5(MoO_4)_7$ and MoK_α X-ray diffraction - $Na_xFe_{1/2}Mn_{1/2}O_2$

As was known from literature [1], new electrode material, P2-$Na_{2/3}Fe_{1/2}Mn_{1/2}O_2$ delivers 190 mAhg^{-1} of reversible capacity in the sodium cells with the electrochemically active Fe^{3+}/Fe^{4+} redox. In general P2-layered compounds can be promising cathode materials for Na-ion battery. At the high point of charge (4,2 V) initial layered structure ($P6_3/mmc$) changes to the disordered structure OP4 [1] (P-6m2). Thermal stability of electrochemically desodiated phase was investigated using STOE STADI P diffractometer with tube furnace and temperature controller, MoK_α-radiation (Institute for Applied Materials, KIT). Sample was prepared in capillary filled with Ar. As a result of experiment the layered-to-rock-salt transformation (>400°C) was observed with formation of cubic $FeMn_2O_4$ phase and O_2 release.

The structure transformation during heating and cooling for both compounds we can observe stepwise on the Fig.1 due to *in situ* high-temperature facilities.

References:
[1] Naoaki Yabuuchi et al. Nature Materials, 11 (2012) 512-517

Figure 1

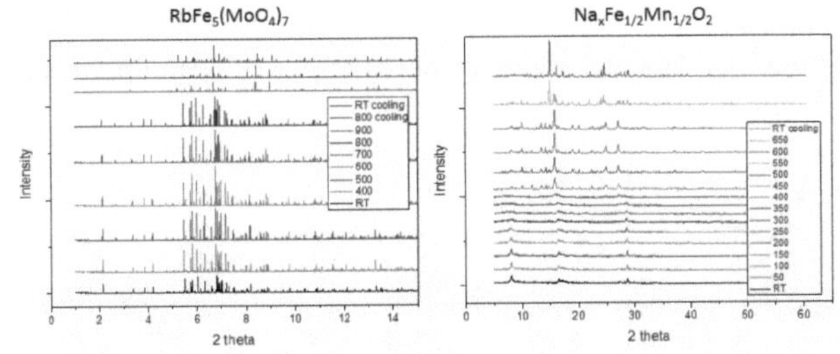

P14

Empirical asymmetry corrections in Rietveld refinements using the Split Pseudo-Voigt function (SPV) and evaluation of Rietveld plots using normalized difference curves

J. Birkenstock[1], T. Messner[1], R. X. Fischer[1]
[1]Universität Bremen, FB 5 - Geowissenschaften, Bremen, Germany

In para-focusing Bragg-Brentano type powder diffraction measurements the shapes of the reflections may be strongly asymmetric. At low angles the left half is usually less steep than the right one by curvature of the diffraction cones and axial divergence (1), at high angles the asymmetry has opposite sense by reverse curvature of the diffraction cones. At moderate angles significant left-halves asymmetry may arise from transmission effects in low-absorbing "infinitely thick" samples increasing with $\sin^2(2\theta)$ (2).

The well-known Caglioti equation (3) was modified and applied to the Split Pseudo Voigt (SPV) function to correct for all types of asymmetric peak shapes in BRASS2 (4). The parameters X_L(Ga), Y_L(Ga), X_L(Lo) and Y_L(Lo) as used in X·exp(Y·2θ) improve matching of low-angle asymmetric peaks (L = left half, Ga = Gaussian and Lo = Lorentzian) where X controls the slope with 2θ and Y the maximum 2θ angle up to which asymmetry correction is significant. At high angles asymmetry is accounted for by separate U, V and W parameters for the left and right halves of the SPV function. Transmission-related asymmetry is accounted for by the new $\sin^2(2\theta)$-related parameters Z_L(G) and Z_L(L). Significant improvements of profile fits are found, e.g., for diffraction data of zeolites.

We also suggest to replace the traditional difference plot $\Delta = (y_o - y_c)$ (with y_o = observed, y_c = calculated step intensities) in Rietveld diagrams by a normalized difference ratio plot with $\Delta_N = (y_o - y_c)/\sigma(y_o)$. The absolute standard uncertainties $\sigma(y_o)$ » $\sqrt{y_o}$ are largest for the largest step intensities and accordingly the difference curve usually displays the largest differences just where the reflections are. The new difference ratio plot Δ_N in BRASS2 also displays reference lines at $\pm\sigma$ and at $\pm3\sigma$, thus showing immediately where the differences are statistically significant or not. If all ΔN values lie within $\pm3\sigma$ (or are similar to those in pure background regions) the differences can be considered as not being significant.

References:
[1] LW Finger, DE Cox & AP Jephcoat (1994), J. Appl. Cryst., 27, 892-900
[2] HP Klug & LE Alexander (1974), Wiley & Sons, New York, 966 pp
[3] G Caglioti, A Paoletti & FP Ricci (1958), Nuclear Instruments, 3, 223-228
[4] J Birkenstock, T Messner & RX Fischer, Bremen Rietveld Analysis and Structure Suite, ver 2.2 – www.brass.uni-bremen.de or contact authors

P15

Characterisation of photonic crystal fibers using synchrotron radiation

M. Hagelstein[1], D. Batchelor[2], R. Simon[2], H. Hessdorfer[2], M. Frosz[3]
[1]KIT, IBPT, Eggenstein-Leopoldshafen, Germany
[2]KIT, IPS, Eggenstein-Leopoldshafen, Germany
[3]Max Planck Institute for the Science of Light, Erlangen, Germany

The transmission of X-rays of kagomé type Photonic Crystal Fibers (PCF) has been studied. The PCFs consist of a well defined and ordered structure of individual hollow channels of variable size and were produced at the Max-Planck Institute for the Science of Light.

The kagomé fibers are characterized by a large central channel of hexagonal shape forming the inner core, which is surrounded by a variable number of outer channels of a honeycomb structure. Typical dimensions range from a central core of ten micrometer diameter to sub-micrometer channel sizes in external areas of the fiber. The SiO_2 glass partitioning walls have a typical thickness of 250 nm. The fibers are used for single-mode guidance of ultra-violet radiation [1].

The characterization using monochromatic synchrotron radiation has been performed at the X-ray fluorescence facility at KIT [2]. An image of X-rays transmitted through a 15 mm long fiber is displayed in figure 1. Transmission of X-rays through hollow glass fibers is remarkably effective because many reflections take place with small losses. For a set of fibers the X-ray transmission characteristics and roughness of inner surfaces has been determined. For a photon energy of 8 keV the critical angle amounts to 0.22°. A best fit is obtained for a RMS roughness of 12 Å (figure 2).

Figure legends

Figure 1. X-ray transmission image of a kagomé PCF and an X-ray photon energy of 8 keV.

Figure 2. Measured X-ray transmission at a photon energy of 8 keV as a function of glancing angle (red), calculated reflectivity of an ideal flat surface (green), and calculated reflectivity of a surface with 12 Å RMS roughness (blue).

References
[1] F. Gebert, M. H. Frosz, T. Weiss, Y. Wan, A. Ermolov, N. Y. Joly, P. O. Schmidt, and P. St. J. Russell, "Damage-free single-mode transmission of deep-UV light in hollow-core PCF", Opt. Express 22, 15388-15396 (2014)
[2] R. Simon, G. Buth, M. Hagelstein, "The x-ray-fluorescence facility at ANKA, Karlsruhe: Minimum detection limits and micro probe capabilities", Nuclear Instruments & Methods in Physics Research, Section B 199, 554-8 (2003)

Figure 1

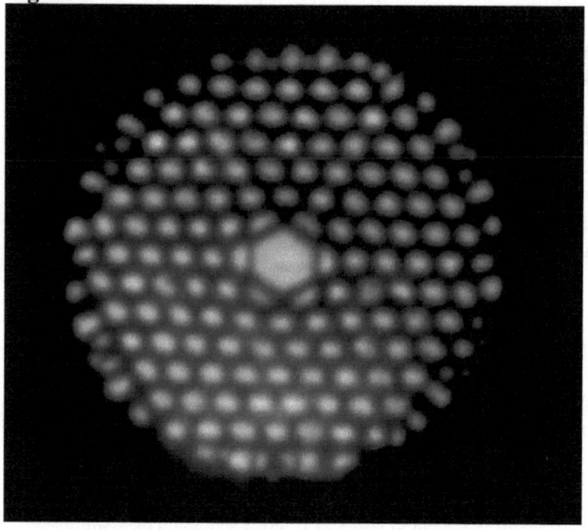

Figure 2

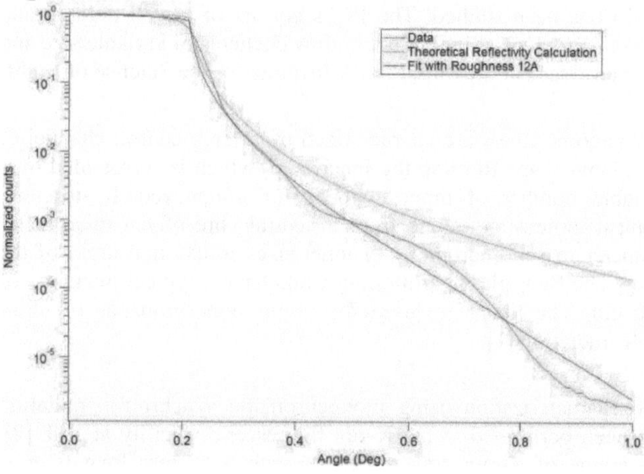

P16

In-situ synchrotron XRPD study of crystallization kinetics of $CaCO_3$ in liquid microjet

A. Mohammed[1], A. Cervellino[1], A. Testino[1], A. Carino[1]
[1]Paul Scherrer Institute (PSI), Villigen, Switzerland

Calcium carbonate, one of the most studied biominerals, has major applications across a broad spectrum of technologies. In order to investigate the crystallization kinetics, in-situ wide-angle (WAXS) and small-angle (SAXS) synchrotron X-ray scattering experiments are being performed at the X04SA-MS Beamline of the SLS synchrotron at the PSI, Villigen, Switzerland [1]. In particular, the SAXS signal is important for detecting aggregates and their size, independently of their atomic structure, while the WAXS will enable us to distinguish between amorphous clumps and crystalline NPs. The feasibility of the WAXS data collection has been established in recent tests, while a SAXS experiment has been performed in situ on a horizontal liquid microjet. This was generated using a capillary connected to a mixer. Four HPLC pumps were delivering solutions in order to obtain the desirable pH and saturation level of the system (Figure 1). The liquid was collected in a catcher where T and pH of the solution, under stirring, were monitored on line. After micro-jet optimization (pulsation damping, liquid jet diameter, solution composition, time delay between mixing point and liquid, X-ray beam focusing), measurements were carried out with stainless steel capillaries of 125 and 250 μm and with delay times of 0, 1, 60 s. The SAXS data, collected using a Mythen II detector [2], on $CaCO_3$ precipitation could reveal in some cases the presence of pre-nucleation amorphous clusters of 20-30 nm size. We are currently analyzing the data in order to verify the reproducibility and reliability of the measurements. The data are subtracted of the air and sodium carbonate solution background.

Refernces:
[1] P. R. Willmott et al., J. Synchrotron Rad. (2013) 20, 667-682
[2] A. Bergamaschi et al., J. Synchrotron Rad. (2010). 17, 653-668

Fig. 1: the experimental setup at the X04SA-MS beamline of the Swiss Light Source

Figure 1

Crystal physics

P17
Visualization of Modulation Functions in MoleCoolQt

C. Hübschle[1], V. Petříček[2], S. van Smaalen[1]
[1]Universität Bayreuth, Lehrstuhl für Kristallographie, Bayreuth, Germany
[2]Institute of Physics, Department of Structure Analysis, Prague, Czech Republic

In a recent extension of the molecular visualization program **MoleCoolQt**[1] modulation of aperiodic structures as described in output files of the program **Jana2006**[2] can be drawn. One of the possible visualizations is a so called *t*-movie. Three dimensional sections through (3+d) superspace are called *t*-sections. A *t*-movie shows these *t*-sections while varying *t* in each frame of the movie. *t*-movies can be saved into mp4 file format in HD or full HD resolution. For each visualized atom a graphical *t*-plot of the 3d displacements can be drawn.

References:
[1] C. B. Hübschle and B. Dittrich, *J. Appl. Cryst.* (2011). **44**, 238-240.
[2] V. Petricek, M. Dusek and L. Palatinus *Z. Kristallogr.* (2014). **229**(5), 345-352.

P18
Anisotropic displacement parameters from electronic-structure theory: how to perform and what to compare

J. George[1], V. L. Deringer[1,2], A. Wang[1], R. Wang[1], P. Müller[1], U. Englert[1], R. Dronskowski[1,3]
[1]Institut für Anorganische Chemie / RWTH Aachen University, Aachen, Germany
[2]Engineering Laboratory/ University of Cambridge, Cambridge, United Kingdom
[3]Jülich-Aachen Research Alliance (JARA-HPC) / RWTH Aachen University, Aachen, Germany

Anisotropic displacement parameters (ADPs) quantify thermal motion of individual atoms in crystals but they can also be used to identify problems in structure determination. In experimentally difficult cases (e.g., low ratio between observations and parameters), it would be desirable to predict ADPs reliably from theory.

Here, we compare calculated ADPs of several molecular crystals and metal carbonyls to experimental ones from neutron and X-ray diffraction [1,2,3]. This is done on the basis of first-principles harmonic and quasi-harmonic phonon calculations, as described in ref. 3 and ref. 4. Moreover, we use different a posteriori dispersion-corrections to the PBE-functional and the dispersion-corrected functional vdW-DF2 in the calculations. To assess the quality of the calculations, we also performed several customized diffraction experiments as benchmarks.

As exemplarily shown for the anisotropic displacement parameters of both pentachloropyridine and $Cr(CO)_6$, the harmonic phonon calculations at 100 K can result in nearly perfect ADPs (Figure 1).

Figure 1. Calculated ADPs compared to experimental ADPs derived by X-ray diffraction (left) for pentachloropyridine and (right) for $Cr(CO)_6$. Adapted from [1] and [2]. The first article is licensed under a Creative Commons Attribution-NonCommercial 3.0 Unported Licence, the second under a Creative Commons Attribution 3.0 Unported Licence

Moreover, we combine experiments and dispersion-corrected DFT to quantify lattice thermal expansion and ADPs in α-sulfur (S_8) [3]. Afterwards, ADPs are calculated in the harmonic approximation and also in the quasi-harmonic approximation (QHA) which includes the predicted lattice thermal expansion. Including thermal expansion improves the calculated ADPs for α-sulfur up to 200 K. Also, the anisotropic thermal expansion of α-sulfur is predicted in qualitatively correct agreement with experiment.

In the future, we would like to use such calculated ADPs in experimental charge-density analyses where only limited resolution of the diffraction data is available.

References:
[1] J. George, A. Wang, V. L. Deringer, R. Wang, R. Dronskowski, U. Englert, *CrystEngComm*, **2015**, *17*, 7414−7422. [2] V. L. Deringer, A. Wang, J. George, R. Dronskowski, U. Englert, *Dalton Trans.*, **2016**, *45*, 13680−13685. [3] J. George, V. L. Deringer, A. Wang, P. Müller, U. Englert and R. Dronskowski, *J. Chem. Phys.*, **2016**, *145*, 234512. [4] V. L. Deringer, R. P. Stoffel, A. Togo, B. Eck, M. Meven and R. Dronskowski, *CrystEngComm*, **2014**, *16*, 10907−10915.

Figure 1

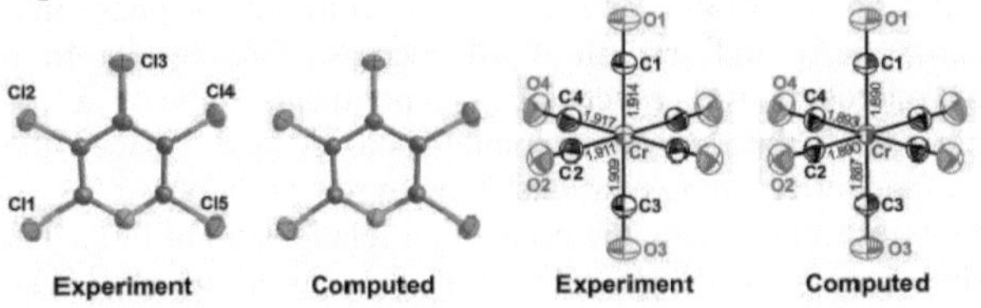

P19
Structure- property relations and structural instabilities in high-temperature piezoelectric materials of the oxoborate-family $RX_2Z_2O(BO_3)_3$

M. Münchhalfen[1], J. Schreuer[1], C. Reuther[2], J. Götze[2]
[1]Ruhr-Universität Bochum, Insitut für Geologie, Mineralogie und Geophysik, Bochum, Germany
[2]TU Bergakademie Freiberg, Institut für Mineralogie, Freiberg, Germany

The application of high-temperature piezoelectric materials has gathered importance, for example, in the fields of structural health monitoring and more efficient jet engines [1]. Particularly, members of the family of rare earth calcium oxoborates (RCOBs) crystallizing in the monoclinic space group Cm show high potential for high-temperature piezoelectric sensing applications, since they combine a high melting point at around 1770 K with high piezoelectric sensitivity and low electric resistivity [2]. Moreover, different possibilities of cation substitution allow for a large variability in chemical composition and tuning of certain physical properties.

$RCa_4O(BO_3)_3$ with R = Y, Gd single crystals were grown from melt using the Czochralski method and their electromechanical properties were studied from 100 K to 1500 K using resonant ultrasound spectroscopy, differential scanning calorimetry and dilatometry. The 13 and 10, respectively, independent components of the elastic and piezoelectric tensor of both compounds could be determined with high reliability. Above about 600 K more or less pronounced anomalies are observed in the temperature dependence of the investigated physical properties, which are probably related to order/disorder processes on certain cation sites.

Acknowledgements: The authors gratefully acknowledge the financial support of the DFG (PAK921/1, SCHR 761/4: Structure/property relationships and structural instabilities of high-temperature piezoelectrics of the oxoborate family.

References:
[1] Yu et al., IEEE Trans. Ultrason., Ferroelect., Freq. Control, 2014, 61, 8.
[2] Yu et al., Cryst., 2014, 4, 3.

Figure 1

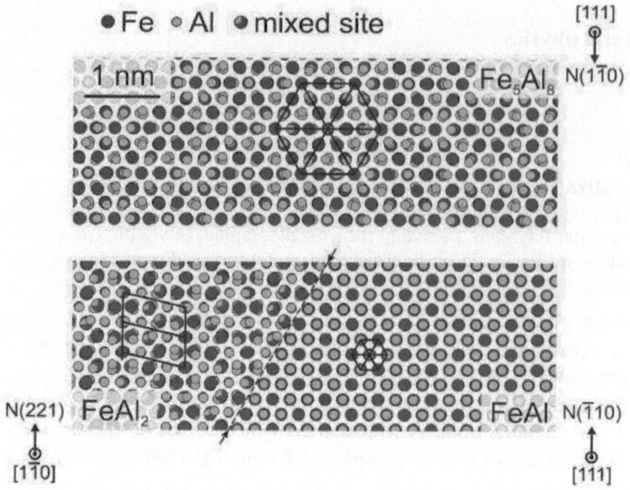

P20
Orientation relationship of eutectoid FeAl and $FeAl_2$

A. Schmitt[1], A. Kauffmann[1], S. Kauffmann-Weiss[2], T. Scherer[3], M. Heilmaier[1]
[1]Karlsruhe Institute of Technology (KIT), Institute for Applied Materials (IAM-WK), Karlsruhe, Germany
[2]Karlsruhe Institute of Technology (KIT), Institute for Technical Physics (ITEP), Karlsruhe, Germany
[3]Karlsruhe Institute of Technology, Institute of Nanotechnology (INT), Karlsruhe, Germany

In the aluminum range of 55–65 at.% in the Fe-Al system a eutectoid decomposition of the high temperature Fe_5Al_8 phase into B2-ordered FeAl and triclinic $FeAl_2$ occurs. This results in a lamellar microstructure. Since creep experiments indicate a co-deformation of the low temperature phases takes place, the interface and orientation relationship between both phases is of central interest. Previously, the orientation relationship of FeAl and $FeAl_2$ has been studied by Bastin et al. [1] and Hirata et al. [2]. However, both results are based on different crystallographic data regarding $FeAl_2$. Thus, the former results were re-evaluated on the basis of a recent re-determination of the $FeAl_2$ phase provided by Chumak, Richter and Ehrenberg [3]. The re-analyzed data reveals that both previously identified orientation relationships match subsequent to a rotation operation of 180° about a crystallographic axis of FeAl or by applying the inversion symmetry of the $FeAl_2$ crystal structure as it is suggested by the Chumak data set. Experimentally, the orientation relationship and interface were determined in as-cast fully lamellar material using orientation imaging microscopy and global texture measurements. The determined orientation relationships are in correspondence with the ones previously found. In addition, a preferential interface between FeAl and $FeAl_2$ was identified by means of trace analyzes using cross sectioning with focused ion beam. Based on these habit planes the orientation relation between the two phases can be described by $(114)_{FeAl2}$ || $(-101)_{FeAl}$ and $[1\text{-}10]_{FeAl2}$ || $[111]_{FeAl}$. Implications for the deformation behavior are stressed.

References:
[1] G.F. Bastin, F.J.J. van Loo, J.W.G.A. Vrolijk, L.R. Wolff, Crystallography of aligned Fe-Al eutectoid, J. Cryst. Growth. 43 (1978) 745–751. doi:10.1016/0022-0248(78)90155-0.
[2] A. Hirata, Y. Mori, M. Ishimaru, Y. Koyama, Role of the triclinic Al(2)Fe structure in the formation of the Al(5)Fe(2)-approximant, Philos. Mag. Lett. 88 (2008) 491–500. doi:10.1080/09500830802247151.
[3] I. Chumak, K.W. Richter, H. Ehrenberg, Redetermination of iron dialuminide, FeAl2, Acta Crystallogr. Sect. C. 66 (2010) i87–i88. doi:10.1107/S0108270110033202.

P21
Investigation of inversion domains in HTVPE GaN layers grown on sapphire substrates

M. Barchuk[1], M. Motylenko[1], G. Lukin[2], T. Schneider[2], O. Pätzold[2], D. Rafaja[1]
[1]TU Bergakademie Freiberg, Institute of Materials Science, Freiberg, Germany
[2]TU Bergakademie Freiberg, Institute of Nonferrous Metallurgy and Purest Materials, Freiberg, Germany

Gallium nitride (GaN) is one of the most promising direct wide bandgap semiconductor material for optical and optoelectronic devices. A typical heteroepitaxial growth of GaN yields a lot of microstructural defects due to a lattice mismatch between GaN and foreign substrates. Recently, we performed our first attempts to deposit 4 mm thick polar GaN layers using high-temperature vapor phase epitaxy (HTVPE) [1]. Besides threading dislocations (TDs) and basal plane stacking faults common in GaN, the HTVPE layers possess domains with inverse polarities, which density strongly depends on the applied growth conditions, in particular on the ratio between the partial pressures of nitrogen and gallium (V/III ratio) [2].

In this contribution, the presence of inversion domains in GaN grown by HTVPE is confirmed by transmission electron microscopy (TEM) and convergent beam electron diffraction (CBED). The TEM study determines the size of the domains with different polarity, while CBED indicates local lattice strains originating from the inversion domain boundaries (IDBs). Using experimental CBED data in combination with a JEMS simulation, we are able to define polarity of a particular domain.

The TEM and CBED studies were complemented by high-resolution X-ray diffraction (HRXRD). The measured reciprocal space maps (RSMs) recorded by means of HRXRD are simulated using a Monte Carlo based approach, which allows the density of threading dislocations (TDs) in GaN layers to be determined [3]. In our study, we demonstrate how the IDBs modify the shapes of the reciprocal space maps and, thereby, affect the accuracy of TDs density. Alternatively, the mosaicity model [4], which separates strain and size contributions to X-ray scattering, is applied for estimation of the TDs density and the size of inversion domains and these results are compared with data obtained from TEM.

References
[1] G. Lukin, C. Röder, M. Barchuk, G. Schreiber, O. Pätzold, J. Kortus, D. Rafaja, M. Stelter, *Phys. Stat. Solidi C* **11** (2014) 491-494.
[2] M. Barchuk, M. Motylenko, G. Lukin, O. Pätzold, D. Rafaja, *J. Appl. Cryst.*, submitted.
[3] M. Barchuk, V. Holý, B. Miljevic, B. Krause, T. Baumbach, J. Hertkorn, F. Scholz, *J. Appl. Phys.* **108** (2010) 43521.
[4] T. Metzger, R. Höpler, E. Born, O. Ambacher, M. Stutzmann, R. Stömmer, M. Schuster, H. Göbel, S. Christiansen, M. Albrecht, H.P. Strunk, *Phil. Mag. A* **77** (1998) 1013-1025.

P22

Thermoelectric Transport Properties in Magnetically Ordered Crystals

H. Grimmer[1]
[1]Paul Scherrer Institut, Research with Neutrons and Muons, Villigen PSI, Switzerland

The forms of the tensors describing thermoelectric transport properties in magnetically ordered crystals are given for frequently used orientations of the 122 space-time point groups up to second order in an applied magnetic field. It is shown which forms are interchanged for the point groups of the hexagonal crystal family by two different conventions for the connection between the Hermann-Mauguin symbol and the orientation of the Cartesian coordinate system. The forms are given in Nye notation, which conspicuously shows how the forms for different point groups are related. In the literature, effects occurring only in magnetically ordered crystals have been added to a related effect occurring in all point groups. It is shown that considering the two effects separately leads to simpler and stronger results. Errors in the literature are pointed out.

Figure: The thermoelectric transport properties under consideration

Figure 1

Even Tensors (invariant under $\bar{1}$ and 1') describing effects allowed in all 122 point groups		Magnetic Tensors (invariant under $\bar{1}$, change sign under 1') describing		Number of point groups allowing the effect
ρ^{0+}_{ij}	electric resistance (H=0)	ρ^{0-}_{ij}	spontaneous Hall effect (H=0)	31
ρ^{-}_{ijl}	(ordinary) Hall effect	ρ^{+}_{ijl}	linear magneto-resistance	66
ρ^{+}_{ijlm}	(quadratic) magneto-resistance	ρ^{-}_{ijlm}	(quadratic) anomalous Hall effect	66
k^{0+}_{ij}	heat conductivity (H=0)	k^{0-}_{ij}	spontaneous Righi-Leduc effect (H=0)	31
k^{-}_{ijl}	(ordinary) Righi-Leduc effect	k^{+}_{ijl}	linear magneto-heat-conductivity	66
k^{+}_{ijlm}	(quadratic) magneto-heat-conductivity	k^{-}_{ijlm}	(quadratic) anomalous Righi-Leduc effect	66
Σ'^{0+}_{ij}	Seebeck effect (H=0)	Σ'^{0-}_{ij}	spontaneous Nernst effect (H=0)	58
Σ'^{-}_{ijl}	(ordinary) Nernst effect	Σ'^{+}_{ijl}	linear magneto-Seebeck effect	69
Σ'^{+}_{ijlm}	(quadratic) magneto-Seebeck effect	Σ'^{-}_{ijlm}	(quadratic) anomalous Nernst effect	69
$T\Sigma'^{t0+}_{ij}$	Peltier effect (H=0)	$T\Sigma'^{t0-}_{ij}$	spontaneous Ettingshausen effect (H=0)	58
$T\Sigma'^{t-}_{ijl}$	(ordinary) Ettingshausen effect	$T\Sigma'^{t+}_{ijl}$	linear magneto-Peltier effect	69
$T\Sigma'^{t+}_{ijlm}$	(quadratic) magneto-Peltier effect	$T\Sigma'^{t-}_{ijlm}$	(quadratic) anomalous Ettingshausen effect	69

P23

Synthesis, Characterization of compounds of earth alkaline rare-earth aluminates

C. Otgonbayar[1], H. Pöllmann[1]
[1]Martin-Luther-Universität, Geowissenschaften und Geographie, Halle (Saale), Germany

Lasers based on $ABAlO_4$ systems (A= Ca, Sr, Ba and B=REE) are promising materials for optical, electronic and structural applications. Yb-doped $CaGdAlO_4$ is well known for its smoothest emission spectrum (from 990 to 1080 nm) and a relatively high thermal conductivity [1].

The abstract reports on the results concerning the sol-gel preparation of $SrAlEuO_4$, $SrAlYbO_4$ and $BaAlEuO_4$. The synthesised compounds are studied by powder X-ray diffraction analysis, IR, TG, REM and calorimetry.

Intermediate compounds are synthesised by Pechini method. The gels were prepared using stoichiometric amounts of analytical-grade $Sr(NO_3)_2$, $Ba(NO_3)_2$, $Al(NO_3)_3{\cdot}9H_2O$, Eu_2O_3 and $Yb(NO_3)_3{\cdot}5H_2O$. Additionally citric acid as a complexing agent and ethylene glycol as a dispersing agent was added to form the gel. The annealing temperature was between 1000 °C and 1500 °C.

$SrAlEuO_4$ belongs to the tetragonal space group I4/mmm. It is isostructural to $SrSmAlO_4$. $SrAlYbO_4$ could be synthesized by Pechini method. The synthesis results are given in Figure 1: an intermediate phase with composition $SrAlYbO_4$ is crystallized in the system. In contrast to strontium system there was no intermediate phase in barium system: $BaAl_2O_4$ and $BaEu_2O_4$ coexist in the system.

The present study demonstrates the versatility of the Pechini method to yield monophasic compounds earth alkaline rare-earth aluminates at low sintering temperature (1300 °C, 2 h) when compared to the temperature required for solid state synthesis (≥1400 – 1600 °C, over 48 h) [2]

References:
[1] Ricaud, Jaffres et al., Opt. Lett. **36**, 4134 - 4135 (2011)
[2] Chen, Xiao et al., J. Electroceram., **10**, 111 - 115 (2003)

Figure 1: XRDs of $SrAl_nYb_{x-n}O_4$(n=0; 0,2; 0,6; 1,0; 1,4; 1,8)

Figure 1

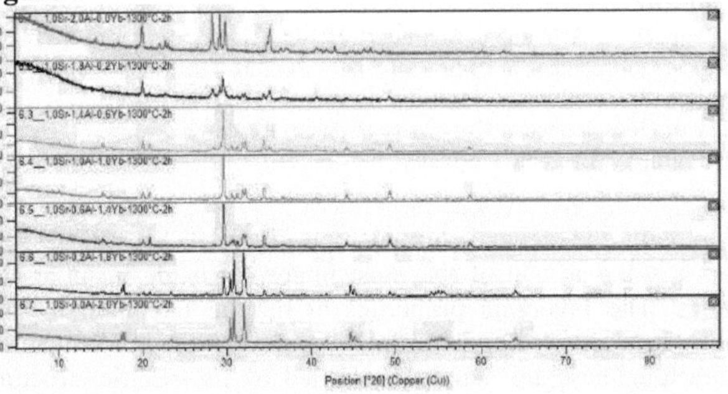

P24

Elastic properties and phase transitions of pure and Mn-doped åkermanite

M. Haiduk[1], J. Schreuer[1], M. Burianek[2], R. X. Fischer[2]
[1]Faculty of Geosciences at Ruhr University Bochum, Institute of Geology, Mineralogy, and Geophysics, Bochum, Germany
[2]Faculty of Geosciences at University Bremen, Research Group Crystallography, Bremen, Germany

Åkermanite, $Ca_2MgSi_2O_7$, belongs to the large family of melilite-type crystal species. The crystal structure is built-up of mixed tetrahedral sheets parallel (001) consisting of MgO_4 tetrahedra and Si_2O_7 groups. These layers are connected by Ca^{2+} ions with eight-fold coordination. At 358 K åkermanite undergoes a well-known phase transition from a normal crystal structure with space group P421m at higher temperatures to an incommensurately modulated one at lower temperatures [1, 2].

Our work focuses on how doping affects the phase transition of åkermanite at about 358 K. To this end pure and Mn-doped åkermanite single crystals were grown using the Czochralski technique. Structural features, heat capacity, thermal expansion and elastic properties of these crystals were studied in the range 100 K to 673 K employing X-ray diffraction techniques, differential scanning calorimetry, dilatometry, and resonant ultrasound spectroscopy, respectively.

The phase transition from normal to incommensurately modulated structure in åkermanite is clearly observable as discontinuities in both thermal expansion and elastic measurements. All eigenmodes exhibit a sharp softening at the transition temperature, whereas their quality factors remain initially nearly unaffected. However, below about 270 K strong ultrasound attenuation increases gradually.

References:
[1] Z. Li, S. K. Chan, S. Ghose, Elastic properties of the incommensurate phase of Akermanite, $Ca_2MgSi_2O_7$. Physics and Chemistry of Minerals **17(5)** (1990) 462-466.
[2] S. Haussühl, J. Liebertz. Elastic and thermoelastic properties of synthetic $Ca_2MgSi_2O_7$ (åkermanite) and $Ca_2ZnSi_2O_7$ (hardystonite). Physics and chemistry of minerals **31** (2004) 565-567.

P25

Structure-property relationships in mullite-type and related compounds

L. Hollenbeck[1], T. Rütten[1], J. Schreuer[1], H. Schneider[2]
[1]Faculty of Geosciences at Ruhr-University Bochum, Institute of Geology, Mineralogy & Geophysics , Bochum, Germany
[2]Faculty of Geosciences at University Bremen, Research Group Crystallography, Bremen, Germany

Mullite with the general composition $Al_2(Al_{2+2x}Si_{2-2x})O_{10-x}$ with $0.25 \leq x \leq 0.4$ is one of the most important oxide materials for ceramics. The favorable properties of mullite like high thermal stability, low density, low thermal expansion and high strength and fracture toughness are mainly controlled by its specific structural features [1]. In the crystal structure of mullite edge-sharing AlO_6 octahedra form octahedral chains which are cross-linked by TO_4 tetrahedra. The characteristic mullite-type backbone of parallel octahedral chains is found in many crystal species, although the linkage between the chains may vary significantly.

In our work we investigate how different types of linkage between mullite-type octahedral chains control the elastic properties of the corresponding crystal species. To this end we study the thermal expansion and the elastic properties of sillimanite ($x = 0$), 2/1-mullite ($x = 0.4$), sinhalite ($MgAlBO_4$), forsterite (Mg_2SiO_4) and rutile (TiO_2) in broad temperature ranges using dilatometry and resonant ultrasound spectroscopy, respectively.

The elastic properties of mullite and sillimanite are very similar. Both crystal species exhibit characteristic ultrasound dissipation phenomena close to room temperature and above about 1300 K [2]. First results on the elastic behavior of the structurally closely related sinhalite will be presented. On the other hand, the elastic anisotropy of forsterite and rutile differs significantly from the one of crystal species with more mullite-type linkage of the octahedral chains.

References:
[1] H. Schneider, R. X. Fischer, J. Schreuer, Mullite: Crystal structure and related properties. J. Am. Ceram. Soc. **98** (2015) 2948 – 2967.
[2] J. Schreuer, B. Hildmann, H. Schneider, Elastic properties of Mullite Single Crystals up to 1400°C. J. Am. Ceram. Soc. **89** (2006) 1642 – 1631.

P26

Influence of substrate potential shape on the diffusion of non-interacting Brownian particles

M. ouahmane[1], L. Arfa[1], L. EL arroum[1]
[1]universite Hasan II, Faculté des Sciences Ben M'Sik, Physique :Laboratoire de Physique de la matière condensée, casablanca, Morocco

In this work we present a study of diffusion mechanism of non-interacting Brownian particles submerged in a symmetric periodic Remoissenet-Peyrard potential which can represent the potential of a deformed substrate with deformability parameter s. Through this potential ions considered as Brownian particles can diffuse. This system is described by the Fokker-Planck Equation (FPE). We solve numerically this FPE using the Method of Continuous Fraction Matrix (MCFM) to calculate the dynamic structure factor S(q,ω). From S(q,ω) some relevant correlations functions are calculated. In particular, the half-line width λ(q) of the peak of the quasi-elastic dynamic structure factor S(q,ω) and the diffusion coefficient D(q). The results obtained in strong friction regime, and low temperature shows that the deformation of the substrate (s>0) involving potentials well-off leads to a diffusion mechanism described by a simple jump diffusion process with jump length close to lattice constant a. and the deformation of the substrate (s<0) involving potentials with wider wells leads to a dynamic of ions described by jump diffusion model with jump length close to constant a and also by liquid-like motion inside the unit cell. Our results chows that the shape of the substrate potential plays an important role on the diffusion process.

Figure 1

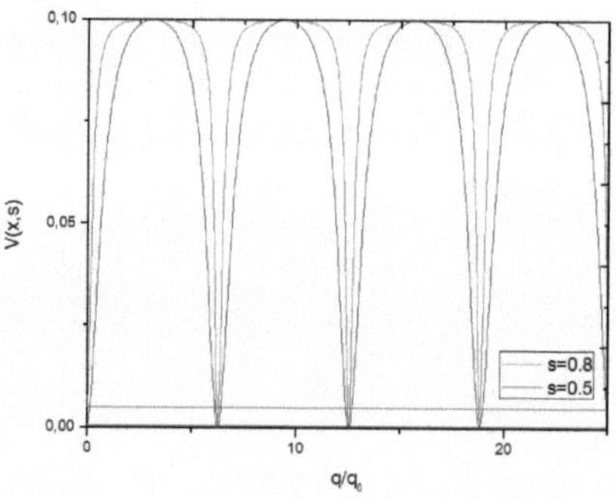

Figure 2

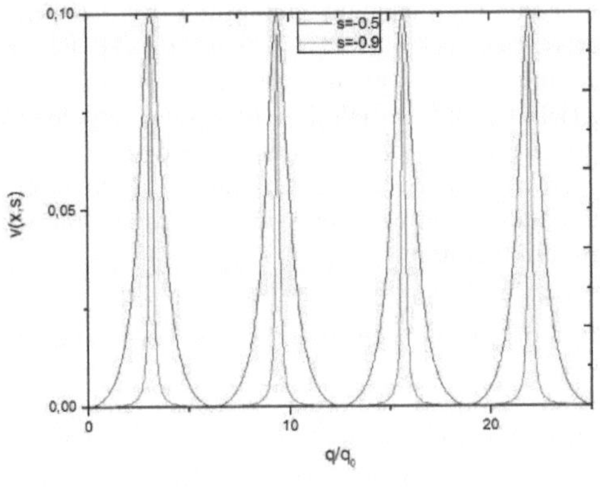

P27

X-ray times-d minus (Td-) diffraction of crystals

X. Gu[1]
[1]Central South University, Geosciences, Changsha, China

Exceptional diffraction peaks have been observed on a micro-diffractometer of Rigaku D/max Rapid IIR with longtime exposure for over 30 crystals in various symmetries including commonest crystals of quartz, silicon and beryl. The d-values (dT) of the exceptional peaks, ranging from 2.548 Å to 23.886 Å, are 2-, 3-, 4-times of the d-values (dN) of the normal (usually the strongest) diffraction peaks minus a small but inneglectable value, i.e., dT=ndN-Δ., where n=2, 3, 4 and Δ=0.003-0.099 Å (average 0.033Å), thus named as times-d minus (Td-) diffraction. The relative intensities of Td- diffraction lines are usually less than 1%, but can be 7.09-10.58% for single crystals of beryl, silicon and molybdenite (Fig. 1). The Td- diffraction lines are not allowed by the crystal structure according to the current crystallography theory and tend to indicate that diffraction may be caused by any crystals with interger times of the original unitcells.

Figure 1

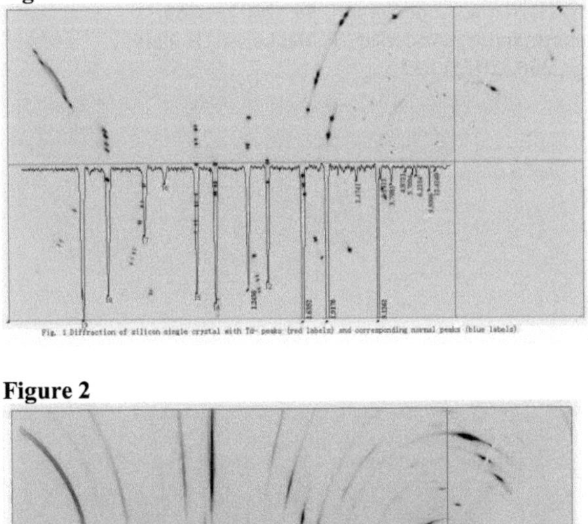

Fig. 1 Diffraction of silicon single crystal with Td- peaks (red labels) and corresponding normal peaks (blue labels)

Figure 2

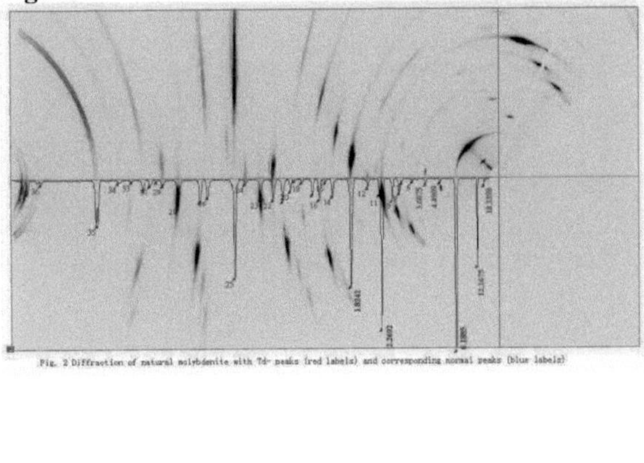

Fig. 2 Diffraction of natural molybdenite with Td- peaks (red labels) and corresponding normal peaks (blue labels)

P28

Curie temperature variations in synthetic Al- and Mg-substituted titanomagnetites constrained by thermal history

S. C. Lappe[1], J. Bowles[2], M. Jackson[3], E. Arenholz[4]
[1]RWTH-Aachen, Institut für Kristallographie, Aachen, Germany
[2]University of Wisconsin-Milwaukee, Department of Geosciences, Milwaukee, United States
[3]University of Minnesota, Institute for Rock Magnetism, Minneapolis, United States
[4]Lawrence Berkeley National Laboratory, Advanced Light Source, Berkeley, United States

One of the most important natural magnetic minerals is the titanomagnetite (*TM*) solid solution series ($Fe_{3-x}Ti_xO_4$, $0 \leq x \leq 1$). *TMs* are widely used in paleomagnetic studies. Therefore, it is essential to understand how they acquire and retain thermal remanent magnetisation (*TRM*). The main characteristic influencing *TRM* is the Curie temperature (T_C). So far it was assumed that T_C of *TMs* only depends on their composition. However, recently it has been discovered that some natural *TMs* with Al- and Mg-substitution show variations of T_C of >150°C, when annealed at moderate temperatures (300 - 425°C) for hours to months [1]. This means T_C also strongly depends on their thermal history. These time- and temperature dependent changes in T_C could have major consequences for many paleomagnetic studies, such as geomagnetic paleointensity estimates and paleomagnetic paleothermometry, as they may greatly affect the thermomagnetic behavior of *TMs* (e.g. [2]). One possible explanation for this phenomenon might be cation (re)ordering of the magnetic Fe^{2+} and Fe^{3+} cations between different sites within the crystal lattice. To test this theory, we synthesized *TMs* with different degrees of Al- and Mg- substitution and varying Ti-content ($0.25 \leq x \leq 60$). The magnetic susceptibility as function of temperature was measured to determine T_C. The specimens were then isothermally annealed for 10 to 1000 hours at 325 - 400°C. Subsequent T_C measurements showed that T_C on warming increases by up to 80°C with anneal time, whereas T_C on cooling is slightly lowered. The extent of variation and the maximum change of T_C depend on the anneal temperature and composition. X-ray magnetic circular dichroism (*XMCD*) of the Fe $L_{2,3}$-edge allows for determining site occupancy and valence state of the Fe cations. *XMCD* measurements of the samples indicate different Fe^{2+}/Fe^{3+} site occupancies for different compositions. However, there are no consistent differences between the *XMCD* spectra of specimens with same composition but different thermal histories. X-ray absorption spectra indicate no inter-site (re)ordering of the Ti^{4+}, Al^{3+} and Mg^{2+} cations for different anneal stages either. Therefore, it has to be assumed that while this is not the mechanism at work, there might be short-range cation (re)ordering within lattice sites or the process might be vacancy mediated.

References:

[1] Bowles, J. A., Jackson, M. J., Berquó, T. S., Solheid, P. A. and Gee, J. S. (2013), *Nature Communications, 4,* 1916.
[2] Yu, Y. and Dunlop, D. J. (2006), *J. Geophys. Res, 111,* B12S31.

P29

Antimony and arsenic tartrates of divalent cations: Crystal growth, crystal structures and vibrational spectroscopy

L. Bohatý[1], P. Becker[1], I. Matulková[2], I. Císarová[2], I. Nemec[2]
[1]Abteilung Kristallographie, Institut für Geologie und Mineralogie, Universität zu Köln, Köln, Germany
[2]Charles University in Prague, Department of Inorganic Chemistry, Prague, Czech Republic

The binuclear anionic antimony tartrate complex group $[Sb_2((+)C_4H_2O_6)_2]^{2-}$ forms a large variety of compounds, where the chirality of the complex group enforces non-centrosymmetry of the crystal structures. In the group of antimony tartrates of divalent cations of type $M^{II}[Sb_2((+)C_4H_2O_6)_2]\cdot xH_2O$ (abbreviated $M^{II}SbT\cdot xH_2O$) the majority of known crystals possesses even polar symmetry. The particular symmetry, together with the accessibility of large single crystals by crystal growth turns these compounds into attractive materials for the study of pyroelectricity, piezoelectricity and nonlinear optical properties (e.g. [1, 2]).

Since the availability of large single crystals is a necessary prerequisite for the investigation of all these properties the objectives of the contribution is on one hand an overview of crystal growth results of alkaline earth antimony tartrates. On the other hand, the question is pursued, whether structurally analogous arsenic tartrates exist.

Crystal growth of six different antimony tartrates, namely the hexagonal (s. g. $P6_4$) $CaSbT\cdot 2H_2O$, $SrSbT\cdot 2H_2O$, $PbSbT\cdot 2H_2O$, the tetragonal (s. g. $P4_1$) $BaSbT\cdot 3H_2O$ and the orthorhombic (s. g. $P2_12_12_1$) $CaSbT\cdot 9H_2O$ and $ZnSbT\cdot KNO_3\cdot 5H_2O$ was performed from aqueous solution by controlled evaporation of the solvent. Resulting crystals are of optical quality and possess (except of $PbSbT\cdot 2H_2O$) volumes of several cm^3. Synthesis of crystals of arsenic tartrates succeeded for $BaAsT\cdot 3H_2O$ (s. g. $P4_1$) and for two Sr-compounds, $SrAsT\cdot NaBr\cdot 8H_2O$ (s. g. $P2_1$) and $2SrAsT\cdot NaBr\cdot 14\ H_2O$ (s. g. $P2_12_12$). X-ray structure determination of all three crystals confirmed the existence of the binuclear complex $[As_2((+)C_4H_2O_6)_2]^{2-}$ in these structures and showed isomorphism of $BaAsT\cdot 3H_2O$ and $BaSbT\cdot 3H_2O$. A detailed comparison of $BaSbT\cdot 3H_2O$ and $BaAsT\cdot 3H_2O$ by means of structural and vibrational spectroscopy (Raman and IR) data is presented.

The results of the crystal chemical study of arsenic tartrate compounds encourage efforts to meet the challenge of growth of large single crystals of the new arsenic tartrates as the basis for their crystal physical investigation.

References:
[1] L. Bayarjargal et al., Cryst. Res. Technol. **43** (2008) 508.
[2] L. Bohatý et al., Cryst. Res. Technol. **50** (2015) 482.

P30

Linear optical properties, pyroelectricity and vibrational spectroscopy of polar guanidinium hydrogen phosphite and hydrogen selenite, $Gu(H_2PO_3)$ and $Gu(HSeO_3)$

I. Nemec[1], I. Matulková[1], I. Císarová[1], W. Krumbe[2], L. Anderson[3], P. Becker[4], L. Bohatý[4]
[1]Charles University in Prague, Department of Inorganic Chemistry, Faculty of Science, Prague, Czech Republic
[2]Leichlingen, Leichlingen, Germany
[3]II. Physikalisches Institut, Universität zu Köln, Köln, Germany
[4]Abteilung Kristallographie, Institut für Geologie und Mineralogie, Universität zu Köln, Köln, Germany

In the family of non-centrosymmetric guanidinium compounds attractive nonlinear optical crystals, ferroelectric crystals as well as non-ferroelectric polar crystals with remarkable pyroelectric and piezoelectric properties occur (e.g. [1-3]). In [4] the polar symmetry $P2_1$ and lattice parameters of the guanidinium compounds $Gu(H_2PO_3)$ and $Gu(HSeO_3)$ are given, together with general brief information about their crystal growth. The crystal structure of the former compound is described in [5]. However, up to the present no data of physical properties of the two compounds have been reported in literature.

The objectives of this contribution is a fundamental characterization of crystals of $Gu(H_2PO_3)$ and $Gu(HSeO_3)$ as optical materials, including X-ray crystal structure analysis and a vibrational spectroscopy (Raman and IR) study. In addition, for both polar crystals the pyroelectric coefficients and their temperature dependence were evaluated.

Using large single crystals grown from aqueous solution, refractive indices and their dispersion in the wavelength range 365 - 1083 nm were determined by the prism method. From the data phase matching conditions for collinear second harmonic generation (SHG) were analyzed in detail. Both crystals allow the realization of phase matching for SHG of both, type I and type II processes. The measurements of the pyroelectric coefficients in the temperature range 100 - 300 K, using a homemade measurement setup [6], show pronounced pyroelectricity but no indication for ferroelectric behavior of the two crystals.

The obtained promising results of our investigation encourage, as a next step, to analyze the nonlinear optical properties of the crystals of $Gu(H_2PO_3)$ and $Gu(HSeO_3)$ in detail.

References:
[1] L. Bohatý, P. Becker, E. Haussühl, I. Němec, O. Lux, H.J. Eichler, H. Yoneda, A. Shirakawa, A.A. Kaminskii, Z. Kristallogr. **230** (2015) 639.
[2] M. Szafrański, J. Phys. Chem. B **115** (2011) 8755.
[3] S. Haussühl, L. Bohatý, Z. Kristallogr. **167** (1984) 311.
[4] W. Krumbe, S. Haussühl, Z. Kristallogr. **178** (1987) 132.
[5] W.T.A. Harrison, Acta Cryst. E **59** (2003) o769.
[6] M. Ackermann, L. Andersen, P. Becker, L. Bohatý, Z. Kristallogr. **230** (2015) 633.

P31

Anelastic relaxation effects in sillenites

E. Haussühl[1], J. Schreuer[2]
[1]Goethe Universität Frankfurt, Institut für Geowissenschaften, Kristallographie, Frankfurt, Germany
[2] Ruhr-Universität Bochum, Institut für Geologie, Mineralogie und Geophysik, Bochum, Germany

Bi12*M*O20 with *M* = Si, Ti, and Ge (abbreviated as BSO, BTO, and BGO, respectively) are members of the sillenite group. The sillenites are relevant for industrial applications due to their prominent optic, elasto-optic, electro-optic and piezoelectric applications [1]. Sillenite crystals in general exhibit low acoustic wave velocities. Therefore, they are ideally suited for delay lines in electronic circuits. Single crystals of sillenites have so far been investigated in respect to their elastic behavior in the temperature range 250 K-300 K [e.g. 2, 3] and 300 K-773 K for BSO and BTO [4].

In this study the elastic properties of sillenites were investigated with the aid of resonant ultrasound spectroscopy (RUS) between ambient conditions and about 1100 K. We observed surprisingly strong and reversible acoustic dissipation effects, which are characterized by attenuation peaks at about 870 K for BSO, at 960 K for BGO and at 550 K for BTO, respectively. Those peaks resemble Snoek-effect like anelastic relaxation peaks. However, modes with small contributions of the elastic shear stiffness *c*44 and the piezoelectric coefficient *d*123 are almost unaffected by the dissipation. No discontinuities in the thermal expansion and the evolution of the elastic stiffnesses with temperature are observed.

The authors gratefully acknowledge Danil S. Pantsurkin (Nikolaev Insitute of Inorganic Chemistry SB RAS, Novosibirsk, Russia) for the provision of BGO-crystals and the Deutsche Forschungsgemeinschaft DFG for financial support of this investigation (HA 5137/3).

References:

[1] Mel'nikova, T.I., Kuz'micheva, G.M., Rybakov, V.B., Bolotina, N.B., and Dubovsky, A.B. (2011). Inorg. Chem., **50**, 2002-2009.
[2] Gospodinov, M., Haussühl, S., Sveshtarov, P., Tassev, V., and Petkov, N. (1988). Bulg. J. Phys., **15**, 140-143.
[3] Gospodinov, M., Haussühl, S., Sveshtarov, P., Dobreva, S., and Sampil, A. (1992). Mat. Res. Bull., **27**, 1415-1421.
[4] Shen, C., Zhang, H., Zhang, Y., Xu ,H., Yu, H., 1, Wang, J., and Zhang, S. (2014). Crystals, **4**, 141-151.

P32

Extension of the chemical composition of $RX_2Z_2O(BO_3)_3$ (R=Y, La-Lu, X=Ca, Sr, Z=Ca, Na, Y, La-Lu)

C. Reuther[1], J. U. Götze[1], D. C. Meyer[2], E. Mehner[2], J. Schreuer[3], M. Münchhalfen[3]
[1]TU Bergakademie Freiberg, Intitute for Mineralogy, Freiberg, Germany
[2]TU Bergakademie Freiberg, Institute of Experimental Physics, Freiberg, Germany
[3]Ruhr Universität Bochum, Institute for Geology, Mineralogy and Geophysics, Bochum, Germany

Oxoborates are known since 25 years [1]. Commonly the general composition of $RX_2Z_2O(BO_3)_3$ varies between R=Y, La, Pr-Tm and X/Z Ca. These substances are potential high temperature piezo- and pyroelectrics as well as nonlinear optical units due to the acentric space group Cm. Möckel et al.[2] demonstrated, that the $GdCa_4O(BO_3)_3$ shows a structural anomaly during heating at a temperature of approximately 850 °C confirming earlier observations by Mougel et al. [3] in 1998. Own experiments proved this anomaly also for compounds with other rare earth elements such as Y, La, Sm, Gd.

Results of own structural investigations and data from literature [e.g. 4] suggest a mixed occupancy of the R-/X Z-position depending on temperature and the used rare earth elements. Therefore, we assume an interdependent relation between the anomaly and the mixed occupancy.

One possible way to avoid the problem of mixed occupancy is to vary the chemical composition of R, X and Z-position. We will present results of the variation of R, X and Z with R=La, Gd, Y, X=Ca, Sr and Z=Ca or La, Y, Gd in combination with Na. The synthesis of powders was done by solid state reactions under different conditions. The experiments are aimed to synthesize isostructural oxoborates without a structural anomaly.

Acknowledgements: The authors gratefully acknowledge the financial support of the DFG (PAK921/1: Structure-property relationships and structural instabilities of high-temperature piezoelectrics of the oxoborate family).

References:
[1] Khamaganova, T., Trunov, V. & Dzhurinski, B. (1991): The crystal structure of calcium samarium oxide borates $Ca_8Sm_2O_2(BO_3)_6$, *Russian Journal of Inorganic Chemistry* 36, 484–485
[2] Möckel, R., Reuther, C. & Götze, J. (2013): REECOB: 20 years of rare earth element calcium oxoborates crystal growth research, *Journal of Crystal Growth*, 371, 70-76
[3] Mougel, F., Kahn-Harari, A., Aka, G. & Pelenc, D. (1998): Structural and thermal stability of Czochralski grown GdCOB oxoborate single crystals, *Journal of Materials Chemistry* 8, 1619–1623
[4] Ilyukhin, A. B. & Dzhurinskii, B. F. (1993): Crystal Structures of Binary Oxoborates $LnCa_4O(BO_3)_3$ (Ln = Gd, Tb, and Lu) and Eu2CaO(BO3)2, *Russian Journal of Inorganic Chemistry* 38, 847–850

P33

Crystallogaphy in my life

H. Preuß[1]
[1]Ruhestand, Hameln, Germany

The poster presents an overview to the crystallographic publications of the author 40 years after his postdoctoral qualification with a work on crystal physics of triclinic organic molecular crystals remaining as new as it was in the time of writing. The objects of investigation changed from polyethylene over complex salts with TCNQ to steel.

References:
[1] Heinz H. H. W. Preuss, Über das Auftreten von Kristallitgrenzen in neuartigen Bruchflächen von Polyäthylen, physica status solidi Vol. 3(1963) K 109 ff.
[2] CH. Hache, Heinz H. H. W. Preuss, Epitaktische Aufwachsungen von 7.7.7.8-Tetracyanodimethan auf NaCl-Spaltflächen, Kristall und Technik, 3 (1968) K 57 f.
[3] Heinz H. H. W. Preuss, On the Correction of Lattice Parameters for Two Isomorphous Compex Salts of 7.7.7.8 Tetracyanoquinodimethane (TCNQ). Kristall und Technik 4,4 (1969) K 7 - K 10
[4] Heinz H. H. W. Preuss, Über den direkten Nachweis der Existenz von Lamellen in massiven Polyäthylenproben durch mechanische Lamellentrennung, Plaste und Kautschuk, 22. Jg. Heft 12 (1975) S. 958
[5] Heinz H. H. W. Preuss, Trikline TCNQ-Komplexsalze als Modellkörper zur Untersuchung der Kristallplastizitä bei niederer Symmetrie, Dissertation B (Habilitationsschrift), Karl-Marx-Universität Leipzig, 1976, Abdruck in: Freiberger Forschungshefte B 204, 1979
[6] Heinz H. H. W. Preuss, Variable Kugelpackungsmodelle im fakultativen Unterricht Lehrgang Physik, Physik in der Schule, 24. Jg. (1986) 1 / 2, S. 16 – 26
[7] Heinz H. H. W. Preuss, Probability Analysis of the Azimuthal Distribution of Glide Traces on the Surface of Plastically Deformed Polycrystalline Metals, Cryst. Res. Technol. 21(1987) 2, 242 - 250
[8] Heinz H. H. W. Preuss, Freihand-Gruppenversuch mit Kandiskristallen, DPG-Tagung Bochum 2009, DD 28

P34

Determination of electrostatic magnitudes in a coumarin compound C17H13NO3

Y. Megrouss[1,2], N. BOUKABCHA[1,2], A. CHOUAIH[1,2], F. HAMZAOUI[1,2]
[1]University of mostaganem, Process Engineering, Belaasel, Algeria
[2]University of mostaganem, Process Engineering, BELAASEL, Algeria

During the last decade huge efforts have been focused in the development of new organic materials non-linear optical properties . These materials are known for their remarkable NLO properties that can be several orders of magnitude than those of known mineral compounds. These compounds have applications especially in the field of telecommunications and the generation of second harmonics.

We present a comparative study based on the exploitation of high resolution X-ray diffraction experimental data on the one hand, and theoretical ab initio calculations on the other. We performed a detailed structural and thermal analysis of the molecule and the calculation of the electrostatic properties of the molecular system such as the distribution of the charge density in the mean planes of the molecule and the electrostatic potential around the molecule .The experimental results have been obtained through the MOPRO software using the multipolar model Hansen-Coppens showing the distribution of the electron density in the form of spherical harmonics. Theoretical calculations were performed by the Gaussian 03 program using two methods of Hartree Fock and model of the DFT at the B3LYP / 6-31G.

The comparison of the results shows a good agreement between the experimental values thus testifying to the good quality of XRD data . The study revealed the nature of the intramolecular charge transfer existing within this molécule.Les detailed results relating to the calculation of the electrostatic potential and the distribution of the charge density in the crystal will be presented at the meeting.

References:
[1] J. Zyss, Molecular Nonlinear Optics: Materials and Devices, Academic Press, New York, 1994.
[2] K. Clays, B.J. Coe, Chem. Mater. 15, 642 (2003).
[3] Hamzaoui, F.; Drissi, M.; Chouaih, A.; Lagant , P.; Vergoten, G. Electron Charge DensityDistribution from X-ray diffraction Study of the M-Nitrophenol compound in the monoclinic.
[4] Jelsch, C., Guillot B., Lagoutte, A. & Lecomte C. J. Applied Crystallography 38, 38-54 (2005).Advances in protein and small-molecules charge-density refinement methods using MoPro.
[5] Hamzaoui, F.; Drissi, M.; Chouaih, A.; Lagant , P.; Vergoten, G. Electron Charge Density Distribution from X-ray diffraction Study . *Int. J. Mol. Sci.* **2006**, 7,103-115.

Figure 1

Deformation dynamic density map with a contour interval of 0.05 e Å^{-3}

Figure 2

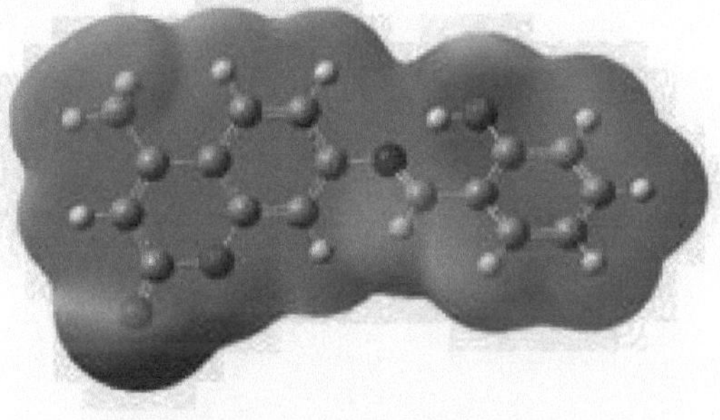

The electrostatic potential maps around the molecule

P35

Phase relationship around the Au-Si-Yb 1/1 quasicrystal approximants

T. Kurihara[1,2], T. Yamada[3], A. P. Tsai[3], Y. Prots[1], Y. Grin[1]
[1]Max-Planck-Institut für Chemische Physik fester Stoffe, Dresden, Germany
[2]Tohoku University, School of Engineering, Sendai, Japan
[3]Institute of Multidisciplinary Research for Advanced Materials, Sendai, Japan

Quasicrystalline (QC) and approximant (AP) phases in the systems of Yb and Au have attracted extensive attention in the field of aperiodic and heavy fermion systems since the discovery of quantum critical phenomena in $Yb_{15}Al_{34}Au_{51}$ QC and $Yb_{14}Al_{35}Au_{51}$ AP phases[1], and the observation of superconductivity in $Yb_{14}Ge_{22}Au_{64}$ and $Yb_{16}Ge_{20.5}Au_{63.5}$ AP phases[2]. In order to find new Yb and Au based QC and AP phases we have explored the ternary system Yb-Si-Au.

The samples $Yb_{14}Si_xAu_{86-x}$ and $Yb_{16}Si_yAu_{84-y}$ with various nominal compositions x=12～24 and y=16～24 were synthesized by arc melting of constituent elements with subsequent annealing at 670-720 °C for 1 week. The alloys were characterized by powder and single crystal X-ray diffraction.

The AC phase $Yb_{14}Si_xAu_{86-x}$ exists with wide homogeneity range of x=14~20, whereas for $Yb_{16}Si_yAu_{84-y}$ this range is rather small about y=20s. The lattice parameters of the AP phase (space group: Im-3) are different, e.g. 14.797 Å for $Yb_{14}Si_{20}Au_{66}$ and 14.446 Å for $Yb_{16}Si_{20}Au_{64}$, respectively. At the composition $Yb_{14}Si_{22}Au_{64}$, a coexistence of two AP phases is observed, similar to the $Yb_{15.3}Ge_{25.9}Au_{58.9}$ AP phases[3]. Evaluation of physical properties and temperature dependence of the coexistence of both AP phases are in progress.

[1] K. Deguchi, et al. *Nat. Mater.* **11**.12 (2012). [2] K. Deguchi, et al. *J. Phys. Soc. Jpn.* **84**.023705 (2015). [3] Q. Lin and J. Corbett, *Inorg. Chem.* **49**.10 (2010).

Figure 1

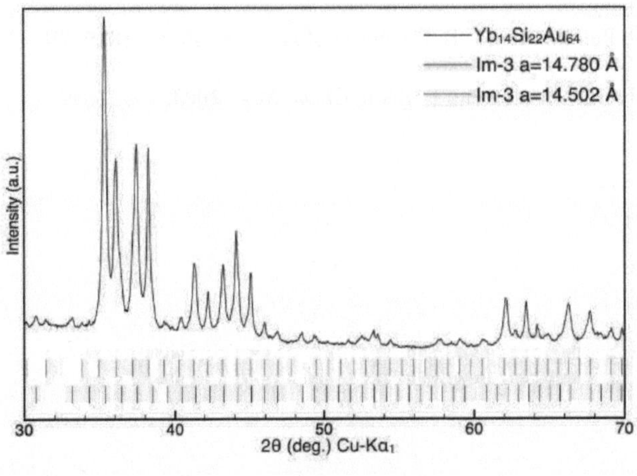

P36

Crystal Structure and Optical Properties of Triiodides and Dibromoiodides with Methylammonium, Formamindinium and Guanidinium Cations.

M. Daub[1], K. Eble[1], H. Hillebrecht[1]
[1]Albert-Ludwigs-Universität Freiburg, Institut für Anorganische und Analytische Chemie, Freiburg i. Br., Germany

Polyiodides are a well examined class of compounds that have been known about for a long time. First reports of polyiodides can be found in literature from about 200 years ago, where *Pelletier* und *Caventou* did experiments with iodine and strychnin.[1] Over the centuries, many different polyiodides have been characterized, of which the iodine content can vary greatly. Polyiodides are known for tri-, tera-, penta-... iodides up to anions like $[I_{29}]^{3-}$ or $[I_{26}]^{4-}$.[2]

In this contribution we present our systematic examinations of triiodides $[I_3]^-$ and dibromoiodides $[IBr_2]^-$ with methylammonium (MA, $[CH_3NH_3]^+$), formamidinium (FA, $[HC(NH_2)_2]^+$) and guanidinium (Gu, $[C(NH_2)_3]^+$) cations. Crystal structures of polyiodides and dibromoiodies are known with many different organic cations, however compounds with these organic cations have, to the best of our knowledge, not been described in literature thus far. The crystal structures of these compounds are based of different packings of the anions depending on the different organic cation and therefor depending on the NH···*X* hydrogen bonds that could be formed. For the dibromoiodides, we found orange-red compounds with all three cations ($GuIBr_2$: *I2/a*, a=8.711(3) Å, b=12.867(4) Å, c=22.963(5) Å, β=99.556(17)°; $FAIBr_2$: *C2/m*, a=9.0661(5) Å, b=7.6271(4) Å, c=6.0460(4) Å, β=112.472(4)°; $MAIBr_2$: *Pbca*, a=9.2971(3) Å, b=8.2842(3) Å, c=18.5619(6) Å). For the triodides a relatively stable, dark-brown compound was only formed with FA cations (FAI_3: *R3m*, a=7.5717(10) Å, c=13.3305(17) Å) With Gu cations a very unstable compound is formed (GuI_3: *R-3*, a=22.9631(13) Å, c=9.4799(5) Å), which easily decompose into $Gu_3(I_3)_2I$ (*C2/c*, a=42.0803(72) Å, b=13.0369(22) Å, c=23.2087(40) Å, β=111.884(8)°). The structural characterization will be supported by vibrational and optical spectroscopy.

References:
[1] P. Pelletier and J. B. Caventou, *Ann. Chim. Phys.*, **1819**, *10*, 164.
[2] P. H. Svensson and L. Kloo, *Chem. Rev.*, **2003**, *103*, 1649.

Fragment based ligand screening

P37
The expert system XDSAPP for data processing and fragment-screening
K. Röwer[1], S. Johannes[2], F. Huschmann[1,2], A. Heine[2], G. Klebe[2], M. S. Weiss[1]
[1]Helmholtz-Zentrum Berlin, Makromolekulare Kristallographie, Berlin, Germany
[2]University of Marburg, Department of Pharmaceutical Chemistry, Marburg, Germany

In the last decade, macromolecular crystallography (MX) has experienced constant improvements of beamline efficiency at synchrotron sources around the world, leading to exposure times below one second per image and complete data collection within a few minutes.

We have developed the expert software XDSAPP [1] to support users in the analysis of their diffraction data during measurements at the beamline with minimal effort and time. It mainly uses the diffraction data processing program XDS [2], along with additional software like POINTLESS from the CCP4 suite [3], XDSSTAT [4], SFCHECK [5] and PHENIX.XTRIAGE [6] for automatic decision making.

In addition, an independent refinement pipeline for automated ligand search based on PHENIX and COOT [7] is under development for use after processing with XDSAPP at the beamline.

References:
[1] Krug, M. et al. (2012). J. Appl. Cryst. 45, 568-572.
[2] Kabsch, W. (2012). Acta Cryst. D66, 125-132.
[3] Collaborative Computational Project (1994). Acta Cryst. D50, 760-763.
[4] Diederichs, K. (2006). Acta Cryst. D62, 96-101.
[5] Vaguine, A. et al. (1999). Acta Cryst. D55, 191-205.
[6] Adams, P.D. et al. (2010). Acta Cryst. D66, 213-221.
[7] Emsley, P. et al. (2010). Acta Cryst. D66, 486–501.

P38
Highly efficient fragment based lead discovery in the home lab
M. Adam[1], V. Smith[1], A. Heine[2]
[1]Bruker AXS GmbH, Karlsruhe, Germany
[2]Philipps-Universitaet Marburg, Inst. Pharm. Chemie, Marburg, Germany

Crystallography is frequently used as a secondary method to validate fragment-binding hits identified by biophysical screening. The capacity of crystallography to directly screen fragment libraries is often dismissed due to its supposed low-throughput and need for high-quality crystals.

Synchrotron beamlines are commonly used for solving fragment-protein structures by X-ray crystallography due to the perceived need for high brilliance beams to collect quality data from lower-quality crystals. Perceived high-throughput is achieved with extremely bright beams, fast detectors and automated sample handling. However, the need to transport crystals to beamlines when they are available typically disrupts structural feedback to the lead development chemists which is why crystallographic structures are not used as the primary screening method.

The D8 VENTURE is a home-lab system that combines immediate access with the METALJET high brilliance X-ray source and the fast PHOTON II CPAD detector. Such a system has the potential to reliably collect crystallographic data to provide ligand-protein structures almost immediately and provide near immediate structural feedback to the development chemists.

This work explores the capability of the D8 VENTURE system to rapidly collect high quality crystallographic data that would enable the use of a strategy relying on X-ray structure determination as the primary screening method. Using aspartic protease endothiapepsin as the target, we assess the capability of in-house X-ray structure determination to efficiently screen fragment soaks.

Figure 1

P39
X-ray Crystallographic Fragment Screening and Hit Optimization
N. Bertoletti[1], F. Braun[1], A. Metz[1], L. Zara[1], A. Heine[1], G. Klebe[1], S. Marchais-Oberwinkler[1]
[1]Philipps-Universität Marburg, Institut für Pharmazeutische Chemie, Marburg, Germany

Human 17β-hydroxysteroid dehydrogenase type 14 (*h*17β-HSD14) is the latest identified 17β-HSD member of the Short-chain Dehydrogenase-Reductase super family (SDR).[1] *In vitro*, *h*17β-HSD14 catalyses the NAD^+ dependent oxidation at position 17 of estradiol (E2) and 5-androstene-3β,17β-diol (5-diol) to estrone and dehydroepiandrosterone, respectively. Two variants of*h*17β-HSD14 are known. The first one was isolated from the retina and contains a serine at position 205. An allelic variant differs only by a threonine at this position, and was identified from melanotic melanoma cells. The *in vitro* turnover of both variants for E2 and 5-diol is equal.[2] Northern blot experiments revealed that *h*17β-HSD14 is predominantly expressed in brain, placenta and liver. As it is expressed in the brain, it may become a potential target for the treatment of neuronal diseases, which are estradiol level dependent.

Recently, we reported and characterized the apo (S205), holo (S205 and T205) and ternary complex crystal structures of *h*17β-HSD14 with estrone as well as with a nonsteroidal inhibitor class (T205).[2,3] As we started for the design of the ligands from an already existing library of 17β-HSD1/2 inhibitors, we affront the necessity to discover a new scaffold to overcome the selectivity issue toward HSDs.

We initiated a fragment-based lead discovery (FBLD) campaign by screening a 96 fragment library assembled considering a slightly extended Rule of 3 as a guideline. We believe that crystallographic fragment screening is a promising approach which may lead to the identification of more hits than other biophysical screening methods, especially for those ligands that have weak binding affinity,[4] while providing essential structural information about binding modes necessary for their further optimization. We will present the results of our fragment screen against *h*17β-HSD14 and we will disclose how we can optimize them in order to obtain more potent inhibitors.

Refernces:
[1] Lukacik et al. *Biochem. J.* **2007**, 402(3): 419–427
[2] Bertoletti et al. *J. Med. Chem.* **2016**, 59 (14), 6961–6967
[3] Braun et al. *J. Med. Chem***2016**, 59 (23), 10719–10737
[4] Schiebel et al. *ACS Chem Biol.* **2016**, 11(6):1693–1701

P40

Crystallographic fragment-screening (CFS) at the Helmholtz-Zentrum Berlin and why CFS should be the first choice

M. Weiss[1]
[1]HZB, Macromolecular Crystallography, Berlin, Germany

Fragment-screening constitutes a popular approach to identify small-molecule compounds, which are able to bind to macromolecular targets. Typically, the binding of fragments to their targets is identified by a cascade of biophysical methods and then further analyzed structurally by X-ray crystallography. Recently, however, this pre-screening cascade has been put on the spot and evidence is accumulating that X-ray crystallography should indeed be used as the primary screening method (Schiebel *et al.*, 2016). A large fragment-screening study (361 compounds *vs.* the protease endothiapepsin) has been conducted (Schiebel *et al.*, 2016) and will be presented, as well as some results using a new small, affordable and versatile compound library, which was assembled at the HZB (Huschmann *et al.*, 2016) and a new X-ray crystallography beam line dedicated for fragment screening experiments, which is currently being completed at the BESSY II synchrotron (Mueller *et al.*, 2015). To facilitate high-throughput crystallography we also give an example for efficient identification of fragment hits (Schiebel *et al.*, 2016).

References:
[1] J. Schiebel, N. Radeva, S. G. Krimmer, X. Wang, M. Stieler, F. R. Ehrmann, K. Fu, A. Metz, F. U. Huschmann, M. S. Weiss, U. Mueller, A. Heine & G. Klebe (2016). Six Biophysical Screening Methods Miss a Large Proportion of Crystallographically Discovered Fragment Hits: A Case Study. ACS Chem. Biol. 11, 1693-1703.
[2] F. U. Huschmann, J. Linnik, K. Sparta, M. Ühlein, M. Krug, X. Wang, A. Metz, J. Schiebel, A. Heine, G. Klebe, M. S. Weiss & U. Mueller (2015). Structures of Endothiapepsin-Fragment Complexes by Crystallographic Fragment-Screening Using a Novel, Diverse and Affordable 96-Compound-Fragment Library. Acta Cryst. F72, 346-355.
[3] U. Mueller, R. Förster, M. Hellmig, F. Huschmann, A. Kastner, P. Malecki, S. Pühringer, M. Röwer, K. Sparta, M. Steffien, M. Ühlein, P. Wilk & M. S. Weiss (2015). The Macromolecular Crystallography Beamlines at BESSY II of the Helmholtz-Zentrum Berlin: Current Status and Perspectives. Eur. Phys. J. Plus 130, 141-150.
[4] J. Schiebel, S. G. Krimmer, K. Röwer, A. Knörlein, X. Wang, A. Y. Park, M. Stieler, F. R. Ehrmann, K. Fu, N. Radeva, M. Krug, F. U. Huschmann, S. Glöckner, M. S. Weiss, U. Mueller, G. Klebe & A. Heine (2016). High-Throughput Crystallography: Reliable and Efficient Identification of Fragment Hits. Structure 24, 1398-1409.

P41

A Hit-Validated 96-Compound Library for Efficient Coverage of Chemical Space in Fragment-Based Lead Discovery

S. Glöckner[1], A. Metz[1], A. Heine[1], G. Klebe[1]
[1]Philipps-University Marburg, Pharmaceutical Chemistry, Marburg, Germany

The initial step of a Fragment-Based Lead Discovery (FBLD) project is the selection of a fragment library that efficiently covers the largest possible part of chemical space with a manageable number of compounds. A robust and ready-to-use solution was aspired to address this issue, resulting in the combination of two thorougly tested fragment libraries. An in-house extended Rule of 3 library with 361 compounds as well as a 100-compounds selection from the PDBeChemdatabase were previously screened against the aspartic protease Endothiapepsin with hit rates of 20 % and 10 %, respectively.[1-7] A diverse subset of all crystallographic hits against Endothiapepsin was complemented with five natural product- like derivatives from AnalytiCon to assemble a library of 96 fragment-like molecules that is available in 96-well plate format. This hit-validated library already led to hits against all seven targets under investigation in our group, including an impressive hit rate of almost 30 % against Protein Kinase A. Furthermore, selected fragments were tested against six additional targets at the Helmholtz-Center Berlin. Highlights and curiosities that have been encountered so far will be addressed.

References:
[1] Köster *et al. J. Med. Chem.* **2011**, *54*, 7784.
[2] Huschmann *et al. Acta Cryst.* **2016**, *F72*, 346.
[3] Schiebel *et al. ChemMedChem* **2015**, *10*, 1511.
[4] Schiebel *et al. ACS Chem. Biol.* **2016**, *11*, 1693.
[5] Schiebel *et al. Structure* **2016**, *24*, 1398.
[6] Radeva *et al. J. Med. Chem.* **2016**, *59*, 7561.
[7] Radeva *et al. J. Med. Chem.* **2016**, *59*, 9743.

Hot structures

P42
Pushing back frontiers: Synchrotron Quality Data From the D8 VENTURE In-House Solution

M. Adam[1], V. Smith[1], A. Doré[2]
[1]Bruker AXS GmbH, Karlsruhe, Germany
[2]Heptares Therapeutics, Herts, AL7 3AX, United Kingdom

Today macromolecular crystallography benefits from third generation synchrotrons, which have become the key factor to successfully collect good quality data quickly from poorly diffracting crystals. In-house systems cannot offer the same high flux X-ray beams. However, recent tremendous progress in the development of in-house machines with respect to X-ray beam intensity, goniometer precision, detector size and sensitivity, as-well-as software ease-of-use and functionality shines new light on the previous statement. We will report on examples where synchrotron quality data sets are matched by in-house system data sets. In-house data sets are collect within a reasonable time.

Our findings open access to high-quality data sets for regions where synchrotron facilities are difficult to access or IP considerations prevent the use of external facilities. In many cases diffraction data collection on-site can provides fast turnaround times to the synthesists, enabling key decisions to be made quickly and efficiently in real time.

As a first example we will present results obtained using the D8 VENTURE equipped with METALJET and PHOTON II detector. We investigated the Human Orexin-1, a G protein-coupled receptor (GPCR). These members of a superfamily of protein receptors are notoriously difficult to crystallize due to their instability when removed from the cell membrane, so remaining intractable to most SBDD platforms. We yielded a 2.77Å dataset in less than 2.5 hours. To the best of our knowledge the structure represents the first atomic resolution GPCR structure to be determined without the use of synchrotron radiation.

More examples from structural biology and charge density investigations are in progress.

Figure 1

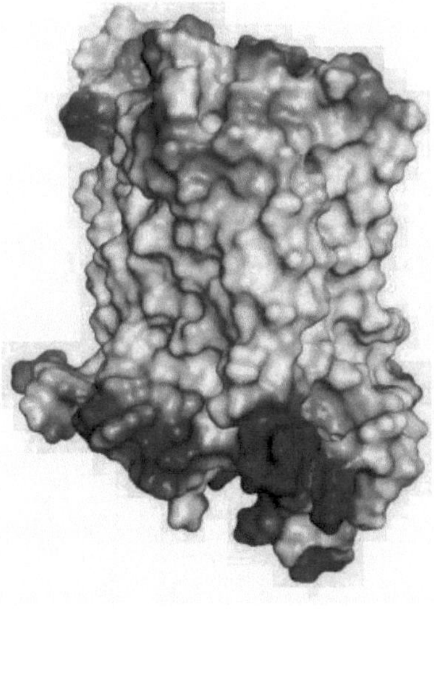

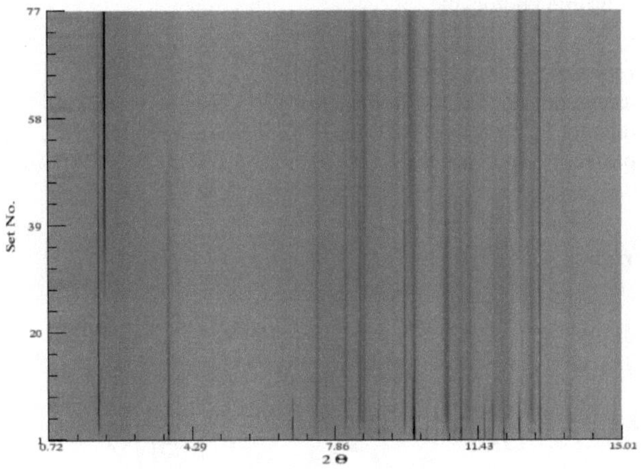

P44
Protein surface engineering enables a detailed view of protein cofactor interactions in a bacterial phytochrome

S. Nagano[1,2], P. Scheerer[3], K. Zubow[2], N. Michael[4], K. Inomata[5], T. Lamparter[6], N. Krauß[6,2]
[1]Justus-Liebig-Universität Gießen, Institut für Pflanzenphysiologie, Gießen, Germany
[2]Queen Mary University of London, School of Biological and Chemical Sciences, London, United Kingdom
[3]Charité - Universitätsmedizin Berlin, Institut für Medizinische Physik und Biophysik, Berlin, Germany
[4]Technische Universität Berlin, Institut für Chemie, Berlin, Germany
[5]Kanazawa University, Division of Material Sciences, Graduate School of Natural Science and Technology, Kanazawa, Japan
[6]Karlsruher Institut für Technologie, Botanisches Institut, Karlsruhe, Germany

Phytochromes are red light-absorbing photoreceptors that bind bilins as chromophores. We use the biliverdin-binding phytochrome Agp1 from the soil bacterium *Agrobacterium fabrum* which has a PAS-GAF-PHY tridomain as photochromic core module (PCM) as model system to study the molecular mechanisms of canonical phytochromes underlying intramolecular signal transduction from the PCM to the histidine kinase (HK) output module. Our attempts to determine the crystal structure of the PCM of Agp1 in the dark adapted Pr state resulted in crystals of the wild-type protein which maximally diffracted to 2.7 Å resolution. We tried the surface entropy reduction approach of protein surface engineering [1] in order to obtain crystals that belong to a different crystal form of improved diffraction quality. Of five mutant proteins that we obtained following this strategy only one belonged to a different crystal form and diffracted to a higher resolution (1.85 Å) than the wild-type crystals [2]. The most striking difference between the wild-type and mutant structures is that in the wild-type crystals the protein subunits are arranged as parallel dimers, whereas the subunits form antiparallel dimers in the mutant crystals. This suggests that the subunit interactions are rather weak in the PCM. Comparison of the subunits structures in the two crystal forms reveals the PHY domain to be in different positions relative to the PAS-GAF bidomains, indicating structural flexibility of the PCM in the Pr state. In the mutant structure the biliverdin (BV) chromophore can only be modelled as a sterically strained structure, that we attribute to the to the tight interaction between ring A of BV and Pro461 of the conserved PRxSF motif in the tongue of the PHY domain (Fig. 1). This structural detail was only detectable due to the relatively high resolution of the mutant crystal structure. We propose that the steric strain at ring A of BV and its release during photoconversion to the Pfr state contribute to the propagation of conformational changes during the phytochrome photocycle.

References:
[1] Derewenda, Z. S. (2004) *Structure* **12**, 529-535.
[2] Nagano, S., Scheerer, P., Zubow, K., Michael, N., Inomata, K., Lamparter, T., and Krauß, N. (2016) *J. Biol. Chem.* **291**, 20674-20691.

Figure 1

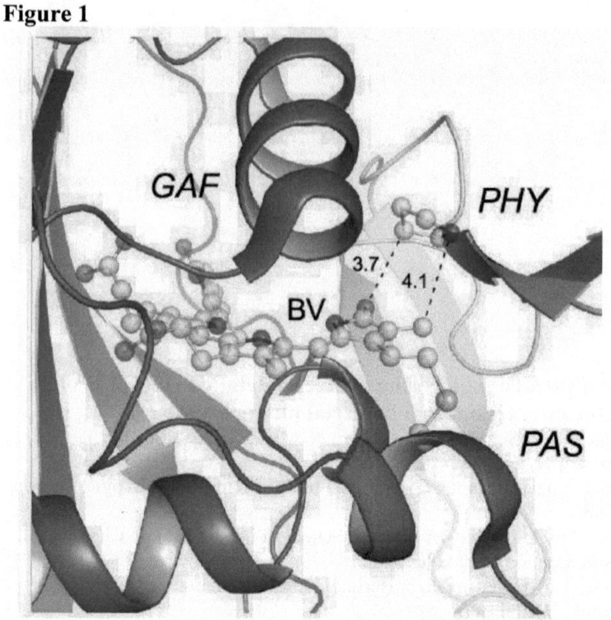

in situ/in operando studies of energy materials

P45

Accessing the pyroelectric coefficient – X-ray diffraction, density functional calculation, and electronic measurement

T. Leisegang[1,2], T. Weigel[1], M. Zschornak[1], T. Behm[1], C. Funke[1], S. Jachalke[1], E. Mehner[1], H. Stöcker[1], D. C. Meyer[1]
[1]TU Bergakademie Freiberg, Institut für Experimentelle Physik, Freiberg, Germany
[2]Samara National Research University, Samara Centre for Theoretical Materials Science, Samara, Russian Federation

The imminent threat of climate change, the shortage of oil as the main energy carrier and global population growth are current challenges for society which will further rise in importance in the coming decades. This, together with German, European and international policies to tackle these challenges and to work on an energy transition are important drivers for high technologies for efficient energy and material conversion processes. The increase in energy efficiency thus is becoming particularly important for further economic dynamics, in terms of innovative materials, corresponding devices, and decreasing production costs. The latter offers a high potential, especially given the fact that today, according to estimates more than 72 % of the converted energy worldwide gets lost in the form of waste heat. A large amount thereof, 63 %, is low-temperature waste heat for which new highly efficient materials are needed for conversion into, e. g., versatile electric energy. To make this waste heat usable, materials exhibiting the pyroelectric effect and high pyroelectric coefficients can be utilized. Pyroelectrics are characterised by a temperature dependent spontaneous polarisation.

For material and technology optimisation the access of the pyroelectric coefficient is necessary. For this purpose single crystal X-ray diffraction as well as density functional theory (DFT) can be utilised. From the crystal structures and electron densities the spontaneous polarisations and thus the pyroelectric coefficients can be deduced. Combining both methods is advantageous and comparable with macroscopic, electric measurements which will be shown.

On base of a standard pyroelectric material, lithium niobate, the ideal, microscopic, and macroscopic values of spontaneous polarisations and pyroelectric coefficients, respectively, were evaluated. Hence, a well-defined single crystal, prepared by focused ion beam technique, was used for determining the crystal structure and electron density in the temperature range of 80 K to 400 K. Based on the experimental as well as theoretical data the spontaneous polarisations and the pyroelectric coefficients were determined to 0.7791 Cm^{-2} and −119(23) $Cm^{-2}K^{-1}$ respectively; DFT derived values are 0,4048 Cm^{-2} and −73(20) $Cm^{-2}K^{-1}$.

This work was funded by the German BMBF (CryPhysConcept: 03EK3029A; R2R Battery: 03SF0542A) and SAB (PyroConvert: 100109976).

P46

Infrared Spectroscopic Study of Vibrational Modes Across the Orthorhombic-Tetragonal Phase Transition in Methylammonium Lead Halide Single Crystals

G. Schuck[1], D. M. Többens[1], S. Schorr[1], M. Koch-Müller[2], I. Efthymiopoulos[2]
[1]Helmholtz-Zentrum Berlin für Materialien und Energie GmbH, 14109 Berlin, Germany
[2]Deutsches GeoForschungsZentrum - GFZ, Telegrafenberg, 14473 Potsdam, Germany

In recent years interest focused on organic-inorganic perovskite as a highly efficient and low-cost material for future optoelectronic applications. One particular application could potentially boost the efficiency of commercial silicon photovoltaic modules from ~20 toward 30 % when used in tandem architectures [1].

Hybrid perovskites with ABX3 structure consist of an organic molecule i.e $CH_3NH_3^+$ (methylammonium abbreviated as MA) on A-position, Pb on B-position and halides (I, Cl, Br) on X-position. Recent studies revealed that especially the organic molecule influences the structural properties to a critical point. Three temperature dependent structural modifications in $MAPbI_3$ are known. The high temperature phase with space group *Pm-3m* at 327 K undergoes a phase transition to *I4/mcm* in which the number of disordered states of MA is lowered to 8 possible orientations [2]. In the orthorhombic (*Pmc*2$_1$) low temperature modification (T_c = 162 K) the orientation of the organic cation is fixed.

Our temperature-dependent infrared absorption studies of MA lead halide single crystals are focused on the analysis of the CH/NH rocking, C-N stretching, and CH/NH bending modes of the MA molecule in the 800-2000 cm^{-1} frequency range. By careful analysis of the spectral features of these IR-active modes (peak line width, position, and integral intensity), it was possible to identify drastic changes near the orthorhombic-tetragonal phase transition. These changes take place in a quite narrow temperature range ($\Delta T < 1K$), implying that they can be easily overlooked due to inadequate experimental conditions. Our observations indicate a high sensitivity of MA vibrational bands to structural changes. The experimental features of the vibrational modes are not described properly in current literature. Interconnect DFT simulations, detailed crystallographic investigations, and in-depth analysis of the vibrational modes are necessary in order to understand the dynamics of the structural changes taking place when the orientation of the MA cation freezes in the orthorhombic phase.

References:
[1] McMeekin et al., Science, 351 (2016) 151
[2] A. Franz, D.M. Többens and S. Schorr, Cryst. Res. Technol., 51 (2016) 534

P47

***In situ* Studies of NASICON-type $Ca_{0.5}Ti_2(PO_4)_3$@carbon for Sodium-Ion Batteries**

Q. Fu[1], A. Sarapulova[1], N. Bramnik[1], Z. Wei[2], F. Du[2], M. Knapp[1], H. Ehrenberg[1]
[1]KIT, IAM-ESS, Eggenstein-Leopoldshafen, Germany
[2]Jilin University, College of Physics, Changchun, China

Lithium-Ion Batteries (LIBs) have been successfully used in various applications such as portable electronic devices, vehicles and large-scale energy equipment. However, the high cost and uneven distribution of lithium sources hinder further large-scale application of LIBs in long-term aspects. Due to the abundance of sodium and the similar operation principle comparing to LIBs, Sodium-Ion Batteries (SIBs) have been paid more attention last years. SIBs are now considered as promising candidates to meet the demand for large-scale energy storage systems. Currently, the polyanionic materials with NASICON-type structure have been actively investigated for sodium-ion batteries due to open three-dimensional framework enabling fast Na+ diffusion and small volume expansion during cycling. Recently, the electrochemical insertion of sodium into $Ca_{0.5}Ti_2(PO_4)_3$ (CTP, space group R-3) with NASICON-type compound has been studied [1]. However, the *in situ* X-ray diffraction studies of CTP is still lacking. In our work, we studied the CTP as anode material for a Na-ion battery. The CTP has a similar formula with $Mg_{0.5}Ti_2(PO_4)_3$ studied by *in situ* diffraction(MTP, space group R-3c) [2], while it crystallizes in the different space group of R-3 [1]. In the space group R-3(CTP), the ions of Ca, Ti, P and O occupy 3a, 6c, 18f and 18f sites, respectively, while the ions of Mg, Ti, P and O occupy 6b, 12c, 18e and 36f sites, respectively. The material was synthesized by a sol-gel method with the starting materials $CaCO_3$, $(CH_3CH_3CHO)_4Ti$, and $NH_4H_2PO_4$. The galvanostatic cycling in sodium half-cells reveals high reversible capacity in the voltage range 0.01 V and 2.5 V vs Na^+/Na (see Fig.1). For the first discharge, the plateaus at around 2.0 V and 0.65V vs Na^+/Na were observed, which can be attributed to the Ti^{4+}/Ti^{3+} and Ti^{3+}/Ti^{2+} redox couples, respectively.

The *in situ* synchrotron diffraction (see Fig.2) reveals high reversibility of the structural changes occurring during cycling in the electrochemical cell. Two-phase reaction happened at the beginning of the discharge stages and the end of the charge process, corresponding to the plateau at high voltage of around 2 or 2.2 V, which is different from MTP in Li-ion cell [1]. Solid solution occurred in the region of second plateau (about 0.65 or 0.5 V). Below the second plateau, only some of the main reflections can be observed during the discharge process, while the others gradually disappeared, which might be due to a serious loss of crystallinity.

References:
[1] ACS Appl. Mater. Interfaces, 2016, 8, 35336
[2] JES, 2012, 159, A1716

Figure 1

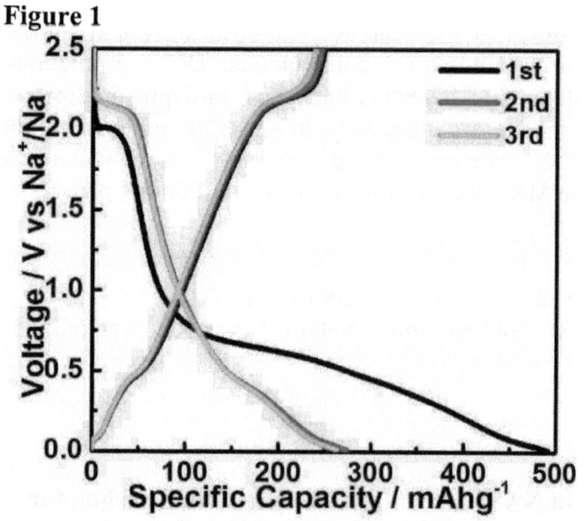

Figure 2

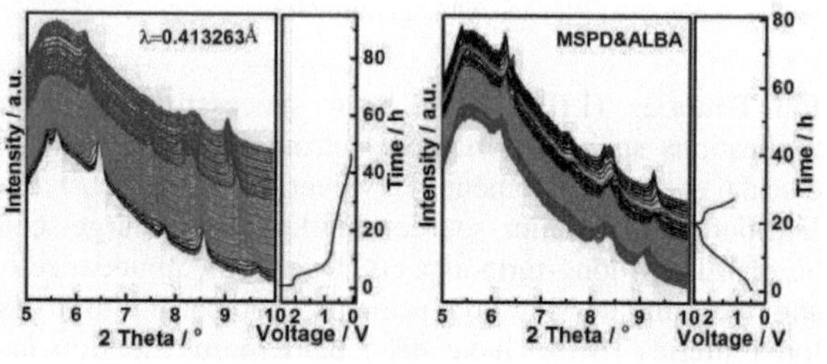

P48

Understanding of electrochemical mechanism of 4*d* layered cathode Li2RuO3: in situ synchrotron study combined with electrochemical kinetics

Y. Yao[1], X. Bie[2], B. Schwarz[3], L. Riekehr[3], C. Dräger[3], H. Ehrenberg[3], Y. Wei[1], F. Du[1]
[1]Jilin University, College of Physics, Changchun, China
[2]Kyoto University, Elements Strategy Initiative for Catalysts and Batteries (ESICB), Kyoto, Japan
[3]Karlsruhe Institute of Technology, IAM-ESS, Karlsruhe, Germany

Since the first demonstration of LiCoO2 as a cathode material for lithium storage, studies on such layered compounds have always been the mainstream in lithium-ion batteries. Recently, layered Li2RuO3 has attracted wide attention owing to its anion redox behavior, facilitating a high reversible capacity.1 Furthermore, the similarity in the crystal structure between Li2RuO3 and Li-rich materials, xLi2MnO3· (1−x) LiMO2 (M = Mn, Ni, Co), makes it more meaningful to further research on their structural evolution during the lithiation/delithiation processes.2 In this contribution, we report the high reversible capacity of about 307 mA h g-1 for Li2RuO3 between 2.0 and 4.5V at the current density of 20 mA g-1 and good rate capability with discharge capacity of 150 mAh g-1 at 10 C rate. Moreover, *in situ* synchrotron and *ex situ* HRTEM/SAED are employed to study the structural transitions during the initial charge-discharge processes. We prove that there are three phases involved in the charge compensation mechanism: two phases with C2/c space group and different cell parameters and another phase with R-3 symmetry play a role during the first cycle. In addition, we use electrochemical impedance spectroscopy (EIS) and galvanostatic intermittent titration technique (GITT) to study the kinetics and to determine the Li+-ion diffusion coefficient. Our studies will provide a significant insight to the electrochemical mechanism of Li2RuO3 and Li-rich composite compounds; it will also contribute to the design and elaboration of new cathode materials in future.

References:
[1] M. Sathiya, K. Ramesha, G. Rousse, D. Foix, D. Gonbeau, A. S. Prakash, M. L. Doublet, K. Hemalatha and J. M. Tarascon, Chem. Mater., 2013, 25, 1121–1131.
[2] Thackeray, M. M.; Kang, S.-H.; Johnson, C. S.; Vaughey, J. T.; Benedek, R.; Hackney, S. A. J. Mater. Chem. 2007, 17, 3112–3125.

Figure 1

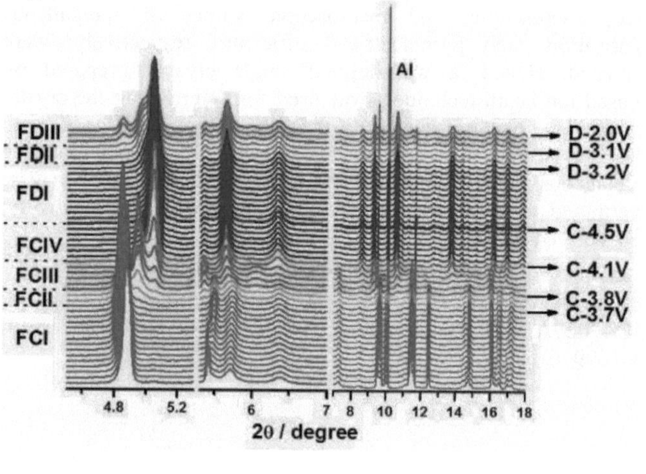

P49

Crystal chemistry of ternary Cu–Li–Sb phases and their structural relations to the Cu–Sb and Li–Sb compounds

A. Beutl[1], H. Effenberger[2], H. Flandorfer[1]
[1]Institute of Inorganic Chemistry – functional materials, Wien, Austria
[2]Universität Wien, Institut für Mineralogie und Kristallographie, Wien, Austria

Intermetallic electrodes have recently gained much attention as they have the potential to mandatorily increase the energy density of lithium ion batteries. One of the important systems is Cu–Li–Sb, as Cu-Sb anodes are considered as very promising showing high energy densities paired with good cyclability. The mechanism of the lithiation is still under discussion, especially due to the lack of phase diagrams and crystallographic data on ternary intermetallic compounds which were proposed only on the basis of theoretical calculations [1]. $CuLi_2Sb$ is the only one, for which the structure was already determined by means of powder X-ray diffraction. In the course of ongoing studies in the system Cu–Li–Sb new ternary compounds were found in samples synthesized by melting and long-term annealing. The crystal structures of $CuLi_{2-x}Sb$, $Cu_{2-x}Li_{1+x}Sb$ and $Cu_2Li_{1-x}Sb$ were studied by single-crystal X-ray investigations. Earlier similar investigations of the Cu–Li–Sn system revealed several new ternary compounds, see e.g. [2].

A general feature of the Cu–Li–Sb compounds is an ordered arrangement of the Sb atoms (ccp or hcp) but mixed and partial occupied M = (Cu, Li, □) sites. The extensive disorder makes them candidates for electrode materials to be used in Li-ion batteries.

$CuLi_{2-x}Sb$ exhibits the pseudo-symmetry *Fm*–3*m*; due to order of the Cu and Li atoms the acentric space group *F*–43*m* is verified, a = 6.248(2) Å, V = 243.9 Å³ = 4 * 61.0 Å³. Topologically related is *$Cu_{2-x}Li_{1+x}Sb$*. The average cell is cubic (*Fm*–3*m*, a = 6.169(2) Å, V = 234.8 Å³ = 4 * 58.7 Å³). The higher Cu:Li ratio is responsible for a distortion to a tetragonal *I* centered cell (a = 4.3832(1) Å, c = 6.0512(2) Å, V = 116.3 Å³ = 2 * 58.1 Å³). Even small crystal chips show adherent powder particles; broad reflection profiles as well as a splitting of the diffraction spots could not be resolved by single-crystal diffractometry. *$Cu_2Li_{1-x}Sb$* (*P*6₃/*mmc*, a = 4.3415(8) Å, c = 7.448(2) Å, V = 121.6 Å³ = 2 * 60.8 Å³) exhibits Cu and Li atoms at mixed or even partial occupied positions. An extensive displacement of the M atoms along [001] is evident.

The compounds mentioned above show close structural relations among each other as well as to the binaries Li_3Sb (*P*6_3/*mmc* and *Fm*–3*m* modifications), β-Cu_3Sb (*Fm*–3*m*), η-Cu_2Sb (*P*4/*nmm*) and δ-Cu_4Sb (*P*6_3/*mmc*).

References:
[1] M Nakayama, S Matsuno, J Shirakawa, M Wakihara: J Electrochem Soc **155** (2008) A505 (see also references therein).
[2] S Fürtauer, HS Effenberger, H Flandorfer: J Solid State Chem **220** (2014) 198; Z Kristallogr **230** (2015) 97; Z Kristallogr **231** (2016) 79.

P50

IN-SITU SYNTHESIS OF RUDDLESDEN-POPPER (RP) TYPE $La_2(Co, Ni)O_4$: PHASE CHANGES AND REDUCTION/OXIDATION MECHANISMS INVESTIGATED BY *IN-SITU* X-RAY POWDER DIFFRACTION

S. ORTATATLI[1], J. TERNIEDEN[1], C. WEIDENTHALER[1]
[1]Max-Planck-Institut für Kohlenforschung, Heterogeneous Catalysis, Mülheim an der Ruhr, Germany

Layered perovskite compounds with the general formula $A_{(n+1)}BO_{(3n+1)}$, known as Ruddlesden-Popper (RP) type, are studied due to their interesting physicochemical properties as well as their remarkable electrocatalytic performance as electrode materials in fuel cells[1]. The methods usually used for the synthesis of RP-type perovskites are not only complicated but also time consuming due to long calcination times at high temperatures and/or pressures[2, 3]. The unfavorable synthesis conditions also limit the applications of RP-type perovskites in large scale applications such as catalysis[4]. In this work, we evaluated synthesis conditions which allow a direct and easier preparation of RP-phases. The synthesis of layered $La_2CoO_{4\pm\delta}$ by heating $LaCoO_3$ under defined H_2 flow was monitored by *in-situ* X-ray powder diffraction (XRPD) experiments. When *in-situ* synthesized La_2CoO_4 is reheated under synthetic air flow, the conventional perovskite structure (in this case $LaCoO_3$) is formed. When the oxidized compound ($LaCoO_3$) is heated in H_2, reduction to $La_2CoO_{4\pm\delta}$ is achieved already at temperatures as low as 600 °C. *In-situ* synthesis of RP-type La_2NiO_4 has been achieved by heating as-synthesized $LaNiO_3$ under defined H_2 flow as well. With this study we show that *in-situ* XRPD provides a direct insight into the formation of RP-type $La_2(Co, Ni)O_4$ and allows control over the reversible phase change mechanism ($LaCoO_3 \leftrightarrow La_2CoO_4$). Results obtained from the *in-situ* XRPD studies can be transferred to design more practicable *ex-situ* synthesis routes for the production of RP-type $La_2(Co, Ni)O_4$.

References
[1] Kim, S. J., Kim, K. J., Dayaghi, A. M., Choi, G. M., *Int. J. Hydrogen Energ*, 41, **2016**, 14498.
[2] Skinner, S. J., *Solid State Sci.*, 5, **2003**, 419.
[3] Su, Y., Tsujimoto, Y., Matsushita, Y., Yuan, Y., He, J., Yamaura, K., *Inorg. Chem.*, 55, **2016**, 2627.
[4] Du, X., Zou, G., Wang, X., *Catal. Surv. Asia*, 19, **2015**, 17.

P51

P2-$Na_{0.66}Co_{0.95}Ti_{0.05}O_2$: a high performance cathode material for sodium ion batteries

N. Sabi[1,2], A. Sarapulova[3], S. Indris[3], H. Ehrenberg[3], J. Alami [2], I. Saadoune[1,2]
[1]University Cadi Ayyad, LCME, FST Marrakesh, Marrakesh, Morocco
[2]Mohammed VI Polytechnic University, Materials Science and Nano-engineering Department, Ben Guerir, Morocco
[3]Karlsruhe Institute of Technology, Institute for Applied Materials – Energy Storage Systems (IAM-ESS), Eggenstein-Leopoldshafen, Germany

Sodium-ion batteries have advantages compared to the current investigated technology thanks to the inexpensive cost of sodium. In a first initiative, parent NaxCoO2 was evaluated as an electrode material [1]. This material reveals many potential steps which some of them are assigned to structural changes during cycling. In our study, we choose to substitute small amount of cobalt (5%) with titanium by investigating a sodium based cathode material: P2-$Na_{0.66}Co_{0.95}Ti_{0.05}O_2$. This electrode material was prepared by the solid state method under 900°C for 12h. A highly crystalline pure phase, with space group $P6_3$/mmc, was obtained.

This layered material shows an initial discharge capacity of 116 mAh/g in the potential window of 2-4.2V with good capacity retention. Indeed, the electrochemistry of this material demonstrates a reduced number of potential steps, low polarization (0.04V), beside a reversible electrochemical phenomenon. The P2 structure was preserved till 4.2V, indicating a positive role of titanium substitution on the stability of the structure.

Fig. 1: Electrochemical behaviour and structural evolution with cycling of $Na_{0.66}Co_{0.95}Ti_{0.05}O_2$ electrode material, data from the MSPD beamline at ALBA.

Acknowledgements
This study was done under the ERANETMED project (Ref. ERANETMED-ENERG-11-151 STORENERGY). Ms Sabi is grateful to OCP and UM6P for the financial support.

References:

[1]Berthelot, R., Carlierand, D., Delmas, C. Electrochemical investigation of the P2–NaxCoO2 phase diagram, nature materials **10**,74–80 (2011)

Figure 1

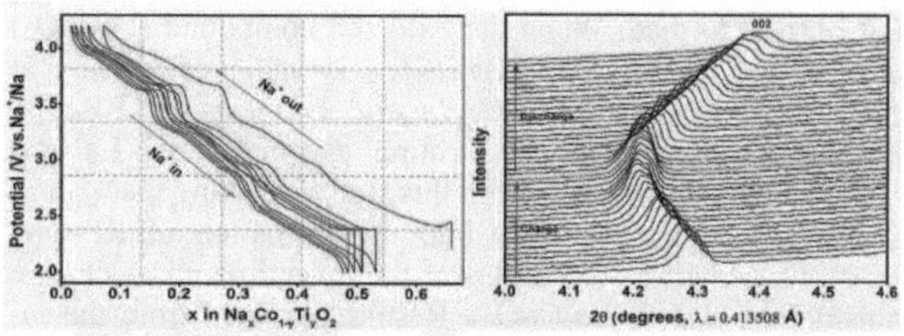

P52

In operando X-ray Total Scattering Experiments on Li-Oxygen Batteries

A. Fiedler[1], F. Scheiba[1], M. Yavuz[1], H. Ehrenberg[1]
[1]Karlsruhe Institute of Technology, IAM-ESS, Eggenstein Leopoldshafen, Germany

Rechargeable Li-O2 batteries are attracting more and more interest as their theoretical energy density is far higher than that of lithium ion batteries [1]. Typical nonaqueous air batteries consist of a lithium metal anode, a separator, a carbon electrode and an organic electrolyte.

In recent years, the pair distribution analysis (PDF) technique has been applied to amorphous, nanocrystalline, liquid and also disordered crystalline materials. It gives information about the local atomic arrangement in materials as well as the long range (average) structure. The PDF method was used to determine structural changes and to detect the occurrence of Li2O, LiO2, Li2O2 and Li2CO3 phases during the electrochemical reaction in the cell. Phase identification of these compounds by the Rietveld method is restricted due to the finite size of their crystals. We will show in this contribution that these restrictions can be overcome using PDF analysis.

Fig 1: The X-ray diffraction patterns Fig 2: PDFs of Li-air battery during cycling of Li-air battery during cycling

References:

[1] Peng, Z. Q.; Freunberger, S. A.; Hardwick, L. J.; Chen, Y. H.; Giordani, V.; Barde, F.; Novak, P.; Graham, D.; Tarascon, J. M.; Bruce, P. G., Oxygen Reactions in a Non-Aqueous Li+ Electrolyte , Angewandte Chemie Int. Ed. 2011, 50, 6351 (doi: 10.1002/anie.201100879).

Figure 1

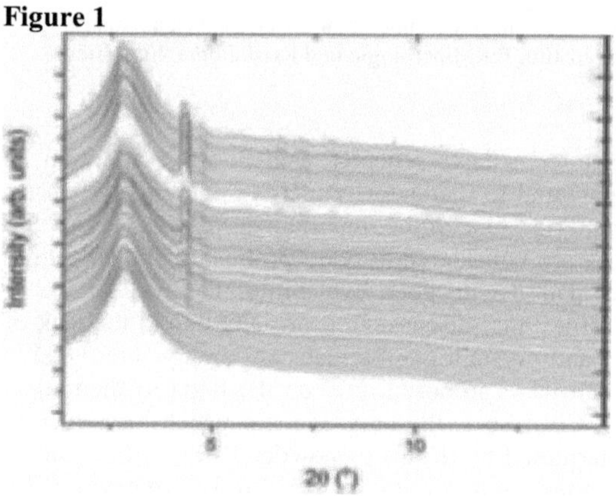

Figure 2

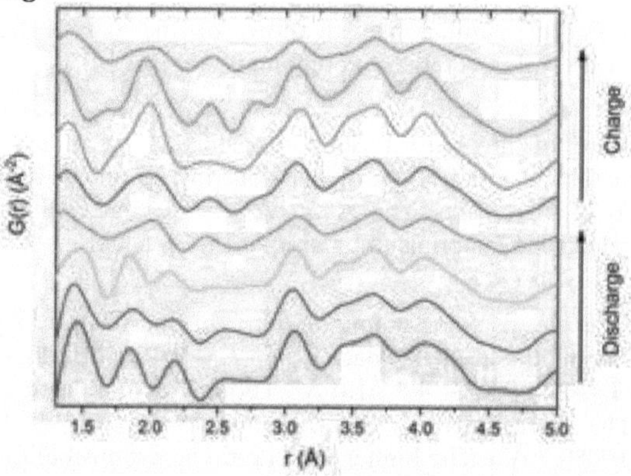

P53

Investigation of sodium intercalation mechanism in the layered oxide P2 structure by *in situ* synchrotron measurement

A. Sarapulova[1], S. Indris[1], N. Bramnik[1], M. Knapp[1], F. Fauth[2], H. Ehrenberg[1]
[1]Karlsruhe Institute of Technology, Institute for Applied Materials – Energy Storage Systems (IAM-ESS), Eggenstein-Leopoldshafen, Germany
[2]CELLS-ALBA Synchrotron, Barcelona, Spain

In situ synchrotron diffraction was used to study mechanism of Na insertion and extraction to the layered structure compound $Na_{2/3}Ni_{1/3}Mn_{2/3}O_2$ and its 3d-transition metal ion substituted derivatives $Na_{2/3}Ni_{1/6}Cu_{1/6}Mn_{2/3}O_2$ and $Na_{2/3}Ni_{1/3}Mn_{1/2}Ti_{1/6}O_2$. Structure and electrochemical mechanism of the synthesized materials are characterized by synchrotron diffraction (ALBA, Barcelona). All the synthesized materials are found to crystallize with the hexagonal P2 structure ($P6_3/mmc$ space group). As a result of experiment substituted samples demonstrated better electrochemical performance. The enhanced cycling stability of the substituted sample was attributed to the differences of the desodiation mechanism, specifically to the elimination phase transition to the O2 structure in high degree of charge. As know from literature [1] $Na_{2/3}Ni_{1/3}Mn_{2/3}O_2$ compound during electrochemical cycling undergo reversible phase transition P2→O2. Suppression of this transition by cation doping let us obtain P2 disordered structure with the some degree of stacking fault at the high degree of charge (Fig. 1).

Fig. 1 Stacking fault at the end of charge for the structure $Na_{2/3}Ni_{1/6}Cu_{1/6}Mn_{2/3}O_2$

Fig. 2 Different mechanisms during varied C-rate of charge and discharge

The mechanism of electrochemical processes in the layered structures of synthesized compounds $Na_{2/3}Ni_{1/3}Mn_{2/3}O_2$, $Na_{2/3}Ni_{1/6}Cu_{1/6}Mn_{2/3}O_2$ and $Na_{2/3}Ni_{1/3}Mn_{1/2}Ti_{1/6}O_2$ was investigated applying two regimes: 1) charge –discharge with the C-rate C/10, 2) slow charge C/40- fast discharge C/20 (Fig. 2). At the different C-rates we can observe different phase separations of the hexagonal phase. Thus, at the end of experiments the different mechanisms were obtained depending on the C-rate and the nature of substituting cations.

Acknowledgement
This study was done under the ERANETMED project (Ref. ERANETMED-ENERG-11-151 STORENERGY).

References:
[1] Zhonghua Lua, and J. R. Dahn. Journal of The Electrochemical Society, 148(11) A1225-A1229 (2001)4

Figure 1

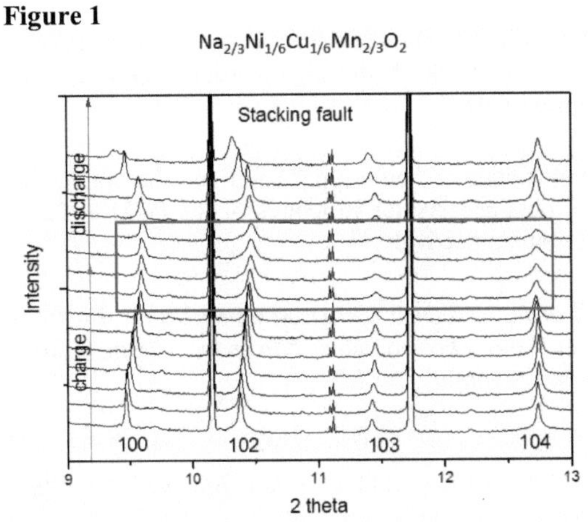

Figure 2

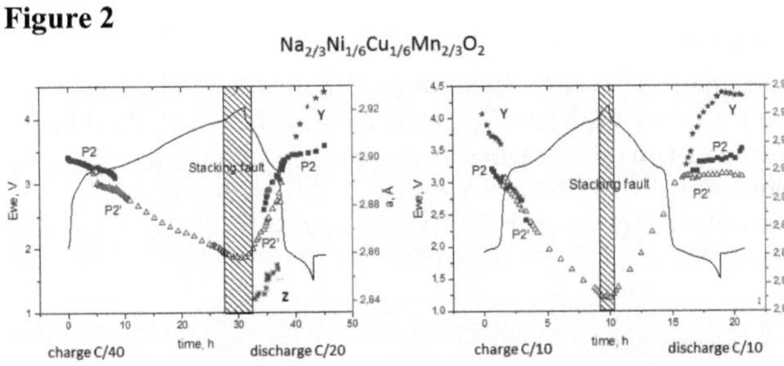

Macromolecular machines and switches

P54

Specificity of the rhodopsin transducin interaction

D. Kwiatkowski[1], N. Heyder[1], K. Lê Công[1], J. Tiemann[1], P. Hildebrand[1], M. Szczepek[1], P. Scheerer[1]
[1]Charité Berlin Universitätsmedizin, Institut für medizinische Physik und Biophysik, Berlin, Germany

G-Protein coupled receptors (GPCRs) transmit extracellular signals to activate a distinct set of intracellular heterotrimeric G-Proteins (Gαβγ) and/or arrestins. Upon activation GPCRs undergo major conformational changes, which allow the binding to G-proteins or arrestins [1-3]. There are 800 human GPCRs but puzzlingly only 16 human G-Proteins sub-types [2]. The crystal structures of rhodopsin, a prototypical class A GPCR, bound to a peptide derived from a C-terminal binding region (GαtCT) of the G-protein transducin (Gαt) [1] and of the adenosine and adrenergic receptor bound to the different G-protein Gαs [3] provided for the first time structural insights in the various G-Protein-Receptor interactions and the activation mechanism. However, a detailed understanding of GPCR G-Protein interaction is still elusive. Interestingly, G-Proteins and arrestins bind in the same cytoplasmic crevice, which opens upon receptor activation. In particular, the GαCT shares a common binding consensus motif with the finger loop region of all four arrestins [4]. Our recent biochemical and structural studies on the interaction of rhodopsin with various G-protein sub-types provide new insights into the nature of the common recognition mechanism of G-proteins by GPCRs and help to understand G-protein specificity.

References:

[1] Scheerer P, Park JH, Hildebrand PW, Kim YJ, Krauss N, Choe HW, Hofmann KP, Ernst OP. (2008) ***Nature* 455** (7212): 497-502.
[2] Hofmann KP, Scheerer P, Hildebrand PW, Choe HW, Park JH, Heck M, Ernst OP. (2009). ***Trends Biochem Sci.* 34** (11):540-52.
[3] Choe HW, Kim YJ, Park JH, Morizumi T, Pai EF, Krauss N, Hofmann KP, Scheerer P, Ernst OP. (2011). ***Nature* 471** (7340): 651-655.
[4] Anantharaman V1, Abhiman S, de Souza RF, Aravind L. (2011). ***Gene* 475** (2):63-78.
[5] Rasmussen SG, DeVree BT, Zou Y, Kruse AC, Chung KY, Kobilka TS, Thian FS, Chae PS, Pardon E, Calinski D, Mathiesen JM, Shah ST, Lyons JA, Caffrey M, Gellman SH, Steyaert J, Skiniotis G, Weis WI, Sunahara RK, Kobilka BK. (2011). ***Nature* 477** (7366):549-555.
[6] Szczepek M, Beyrière F, Hofmann KP, Elgeti M, Kazmin R, Rose A, Bartl FJ, von Stetten D, Heck M, Sommer ME, Hildebrand PW, Scheerer P. (2014). ***Nature Communications* 5**:4801.

P55

Structural Characterization of Fluorescence Optimized Bacterial Phytochromes as Optogenetic Tool

L. Sauthof[1], A. Schmidt[1], M. Szczepek[1], B. Qureshi[1], T. Stevens[1], M. Fernandez Lopez[2], F. Velazquez Escobar[2], N. Michael[2], N. Krauß[3], T. Lamparter[3], P. Hildebrandt[2], P. Scheerer[1]
[1]Institute of Medical Physics and Biophysics/Charité - Universitätsmedizin Berlin, Berlin, Germany
[2]Institute of Chemistry/Technische Universität Berlin, Berlin, Germany
[3]Allg. Botanik, Photobiologie/Karlsruhe Institute for Technology, Karlsruhe, Germany

Phytochromes are photoreceptors which were found in plants, bacteria, cyanobacteria and fungi. They can be divided into two classes: the red-light (Pr) absorbing prototypical and the far red-light absorbing (Pfr) bathy phytochromes. The structure can be described as a highly conserved photosensory core module (PCM) consisting of PAS, GAF and PHY and a variable output module which is mostly a histidine-kinase [1, 2]. Bacterial phytochromes are used as a template in optogenetic engineering. In comparison to other fluorescence proteins, they allow *in vivo* deep tissue imaging because of their advantageous absorption maxima. Due to low quantum yield fluorescence optimized variants were engineered to enable these photoreceptors as optogenetic tool. We studied the photoactivatable infrared fluorescence protein (Agp2-PAiR2, figure 1) which based on the bathy phytochrome *Agrobacterium tumefaciens* (Agp2) [3, 4].

For the Pfr (parent) states of wild-type Agp2 and the engineered Agp2-PAiR2 we obtained crystal structures of the PCM at 2.5 Å and 2.0 Å resolution, respectively.

Figure 1: *PCM of fluorescence optimized mutant Agp2-PAiR2 in the Pfr-state and the chromophore biliverdin within 2Fo-Fc electron density map contoured at 1.0 σ.*

References:

[1] Auldridge M E, Forest K (2011), *Crit. Rev. Biochem. Mol. Bio.* **46**:67-88.
[2] Scheerer P, Michael N, Park JH, Nagano S, Choe HW, Inomata K, Borucki B, Krauss N, Lamparter T. (2010), *Chemphyschem.* **11** (6):1090-105.
[3] Velazquez Escobar F, Piwowarski P, Salewski J, Michael N, Fernandez Lopez M, Rupp A, Qureshi BM, Scheerer P, Bartl F, Frankenberg-Dinkel N, Siebert F, Andrea Mroginski M, Hildebrandt P (2015), *Nature Chemistry.* **7**(5):423-430.
[4] Piatkevich K D, Subach F V, Verkhusha V V (2013), *Commun.* **4**:2153.

Acknowledgements
We are grateful to Uwe Müller, Manfred Weiss and the scientific staff of the BESSY-MX/Helmholtz Zentrum Berlin für Materialien und Energie at beamlines BL14.1, BL14.2 and BL14.3 operated by the Joint Berlin MX-Laboratory at the BESSY II electron storage ring (Berlin-Adlershof, Germany) and the scientific staff of the European Synchrotron Radiation Facility (ESRF, Grenoble) at beamlines ID29, ID23-2, ID30A-1-3, ID30B and ID23-1 for continuous support. This work was financially supported by the Deutsche Forschungsgemeinschaft – DFG – SFB 1078-B6.

Figure 1

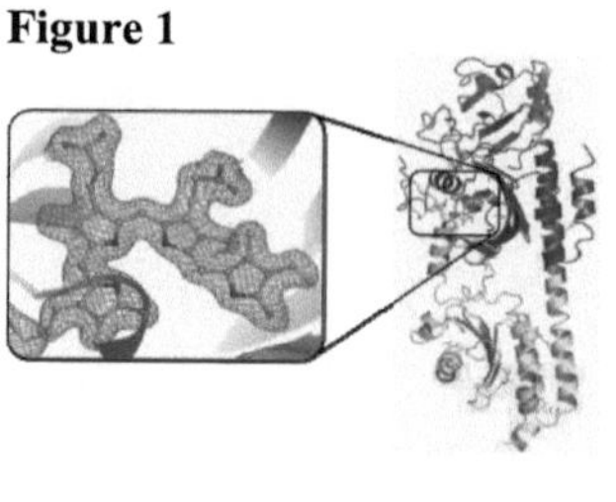

Materials for energy storage and conversion

P56

Structural and spectroscopic characterizations of layered tungstate hosting as hydrogen storage and conduction material

N. Lefeld[1,2], M. M. Murshed[1,2], P. Bottke[3], T. M. Gesing[1,2]
[1]Universität Bremen, Institut für Anorganische Chemie und Kristallographie, Bremen, Germany
[2]Universität Bremen, MAPEX Center for Materials and Processes, Bremen, Germany
[3]Universität Oldenburg, Technische Chemie, Oldenburg, Germany

The 3D framework-type and 2D layered-type tungstates are known for their capability to intercalate cations into the interstitial sites and between the weakly bonded layers, respectively [1-3]. While electron injection into the WO_3 host matrix is very rapid, cation diffusion is kinetically slow and often requires high activation enthalpy. However, if the configuration of O-H···O linkage is linear, the proton-transfer barrier was reported to vanish at an O-O distance of less than 250 pm, and the dynamical hydrogen bonds activation enthalpy significantly reduces [4]. Here we report on the synthesis and characterization of a Ruddlesden-Popper-like tungstic acid with the nominal composition of $H_2W_2O_7 \cdot nH_2O$, composed of double-layers of corner-sharing WO_6 octahedra. Selective leaching of $Bi_2O_2^{2+}$ from a polycrystalline Aurivillius phase $Bi_2W_2O_9$ using HCl kept the WO_6 double-layer intact [1,5]. The vacant A-site within the perovskite-type WO_6 double-layer could give rise to an enhanced storing capacity compared to previously reported poor performing single-layer tungstic acid [3]. Despite the prospective usages for battery and energy materials the crystal structures of these hydrated tungstates harbors considerable debates due to either lack of suitable single crystals or powders with sufficient crystallinity. Especially the extent of expected regimes with distortion, defects and intergrowth introduced upon leaching remains unclear. Thermal analysis of the compound shows a two-step weight loss, which might be attributed to both coordinated and interlayer water. Fitting of ^1H MAS NMR spectrum reveals at least three different bonding environments for hydrogen. The vibrational frequency of the W-O stretching modes observed by FTIR and Raman spectroscopy indicate the double bond nature to the layer-edge terminal oxygens and high degree of distortion in the WO_6 octahedra. The spectroscopic features of the O-H···O linkage could help to set appropriate constraints for the accurate determination of the crystal structure. An indirect band-gap of 2.62(3) eV determined for the bright yellow compound via diffuse reflectance UV-Vis spectra results in an absorption of blue light ($\lambda = 474$ nm), opening the possibility for visible light catalytic activity.

References:

[1] L. M. Kuti et.al., Inorg. Chem. 48 (2009) 6804.
[2] Y. Li et.al., Solid State Ionics 134 (2000) 271.
[3] C. Guéry et.al., J. Solid State Electr. 1 (1997) 199–207.
[4] K.-D. Kreuer et.al., Chem. Rev. 104 (2004) 4637.
[5] R. E. Schaak et.al., Chem. Commun. (2002) 706.

P57

Crystallographic Investigations on $Li_7La_3Zr_2O_{12}$ Solid Electrolytes

H. Stöffler[1], M. Yavuz[1], A. Roters[2], M. Schneider[2], S. Indris[1], H. Ehrenberg[1]
[1]Karlsruhe Institute of Technology (KIT), Institute for Applied Materials – Energy Storage Systems (IAM-ESS), Eggenstein-Leopoldshafen, Germany
[2]SCHOTT AG, Mainz, Germany

All-solid-state batteries (ASSB) might provide a promising next generation battery technology with many advantages over conventional Li-Ion batteries such as safety, durability, and high energy density. The garnet-type Li-ion conducting $Li_7La_3Zr_2O_{12}$ (LLZO) is a promising solid electrolyte for Li-based batteries due to its high Li-ion conductivity and chemical stability against elemental lithium. [1]

LLZO can exist in two different modifications: a low-conductivity tetragonal phase and a high-conductivity cubic phase.

LLZO powders with two different average particle sizes (0.4 and 1 µm) were investigated. A detailed structural characterisation at room temperature is performed using neutron and synchrotron diffraction. Rietveld refinements of neutron and synchrotron diffractograms confirm the cubic LLZO structure with some additional impurities such as Li_2CO_3, $La_2Zr_2O_7$, and/or LiOH, depending on the particle size. High temperature X-Ray diffraction enables the observation of phase transitions at temperatures ranging from 25 °C to 800 °C.

References

[1] Murugan, R.; Thangadurai, V.; Weppner, W. Fast Lithium Ion Conduction in Garnet-Type $Li_7La_3Zr_2O_{12}$. Angew. Chem., Int. Ed. 2007, 46, 7778−7781.

P57

Investigation on the difference in storage behavior between Na-Ions and Li-Ions battery materials.

L. ASMAA[1], A. El Bouaril[1], O. Brahim[2,3]
[1]Laboratoire de Physico-chimie des matériaux appliqués LPCMA,, chemistry, Casablanca, Morocco
[2]CIC Energigune, Parque Tecnológico, C/ Albert Einstein, Miñano (Álava),, Spain
[3]CIC Energigune, Parque Tecnológico, C/ Albert Einstein, Miñano (Álava, Spain

Over the past decade the great challenge is environmental pollution, climate warming, and energy crisis, so in order to overcome all this problems. A variety of renewable and clean energy sources, such as the wind, waves and sun are growing rapidly. However, the increasing use of renewable energy sources also brings other problems. To-date one of the most promising means energy storage technologies is the sodium and lithium ions batteries.

In this work we compare the difference in storage behavior between Na and Li in their analogous electrodes and summarize the sodium storage mechanisms in the available electrode materials. This study also includes some new results from our group on the synthesis, description of the structure, characterization using the Rietveld refinement, optical and magnetic studies of the both Nasicon-type phosphate Li3A(PO4)3 and Na3A(PO4)3.

Materials Science

P59

Influence of heat treatments on low oriented melt-spun poly(3-hydroxybutyrate) fibers

B. M. Liebeck[1], P. K. Manvi[2], T. Vad[2], G. Seide[3], T. Gries[2], G. Roth[4]
[1]RWTH Aachen University, Institut für Kristallographie, Aachen, Germany
[2]RWTH Aachen University, Institut für Textiltechnik, Aachen, Germany
[3]Maastricht University, Faculty of Humanities and Science - Biobased Materials, Geelen, Netherlands
[4]Institut für Kristallographie, RWTH Aachen University, Aachen, Germany

Introduction

Poly(3-hydroxybutyrate) (PHB) is a semi-crystalline, thermoplastic, and fully bio-degradable polymer produced by bacteria as carbon storage material. Considering the environmental problems caused by synthetic petroleum based plastics, biopolymers can be an ecologically worthwhile alternative. With a market share of 22% polymer fibers represent one of the largest fields of application for polymer materials. Therefore, fiber production from biopolymers is an important field of research.

Objective

Establishing efficient (melt-)spinning processes for biopolymers, such as PHB, is a challenging task. Therefore, knowledge about the crystallization behavior and the orientation distribution of the crystallites is of great importance.

Materials and methods
Melt-spun PHB fibers, produced on a micro-extruder with a maximum take-up speed of 200 m/min were characterized using thermal analysis and scanning electron microscopy (SEM). Furthermore X-ray diffraction (WAXD and SAXS) was applied to obtain information on how the selected spinning process influences the crystalline fraction of the polymer. Along with the as-spun fibers, annealed samples were analyzed to find out how additional heat treatment affects the crystal structure.

Results
The obtained results reveal that the annealing step significantly influences unit-cell parameters, crystallite size and crystallinity of the samples, while a moderate change in processing parameters shows no distinct effect on the PHB structure. A strong anisotropic peak broadening is found in the as spun fibers. This phenomenon is clearly reduced by annealing, indicating a stronger crystallite growth along certain directions. The orientation distribution of the crystallites is comparatively broad for fibers, and the resulting Hermans orientation factors are found to slightly increase with the take-up speed.

Conclusion
The crystalline structure of melt-spun PHB fibers can be significantly influenced through a subsequent annealing step, while the axial orientation of the crystallites is – due to the low take-up speeds – of low value which can be easily improved by increasing the take-up speed and/or by subsequent mechanical drawing of the fibers.

Figure 1

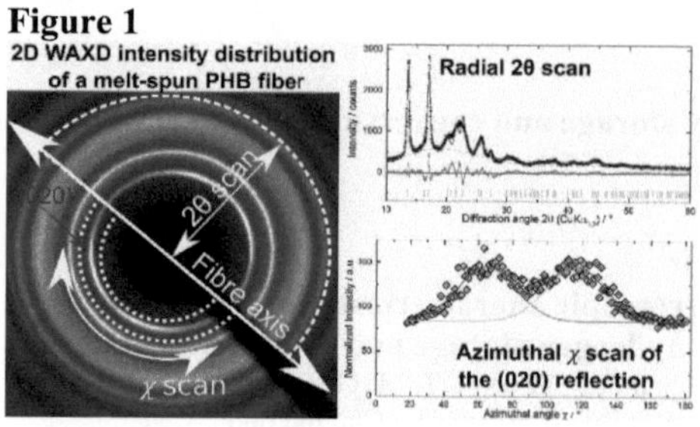

P60

Cation distribution and point defect concentration in off-stoichiometric $Cu_2ZnGeSe_4$ compound semiconductors

R. Gunder[1,2], A. Franz[1], S. Schorr[1,2]
[1]Helmholtz-Zentrum Berlin, Berlin, Germany
[2]Freie Universität Berlin, Institut für Geologische Wissenschaften, Berlin, Germany

The composition of kesterite-type compound semiconductors can be chosen such that it facilitates the utilization of abundant and non-toxic elements. Moreover, cationic substitutions like Sn ↔ Ge (i.e. CZGSe), as done in this study, increases the flexibility in terms of band gap engineering, resulting in a band gap range being suitable for solar energy conversion both to electrical and chemical energy.

This study aims to synthesize off-stoichiometric, single phase CZGSe kesterite-type powder samples by solid state reaction according to the different off-stoichiometry types proposed in literature [1, 2]. The analytical emphasis is put on the structural response with respect to the extent of off-stoichiometry, particularly cation distribution and point defect concentration.

The weighed samples, altogether, comprise 30 off-stoichiometric Cu-poor/Zn-rich A-type, Cu-poor/Zn-rich/Ge-poor B-type, Cu-rich/Zn-poor/Ge-rich C-type and Cu-rich/Zn-poor D-type kesterites. In dependence of the route of off-stoichiometry the kesterite phase accommodates certain point defect types to be analyzed by refining neutron diffraction data using the Rietveld method [3, 4]. Basic structural information as well as quantitative phase content is retrieved by Rietveld analysis of X-ray diffraction data. The compositional characterization is done by quantitative electron microprobe analysis using WDX spectroscopy.

WDX spectroscopy confirmed the formation both of secondary phases in most of the samples as well as type mixtures mainly due to systematic loss in GeSe. The cation distribution exhibits a clear dependency on Cu/(Zn+Ge) ratio and type fractions. The lowest point defect concentration was found in a slightly Cu-poor sample and, eventually, increases significantly both towards Cu-poor and Cu-rich regime.

The results demonstrate that the kesterite-type structure can tolerate considerably deviations from stoichiometry, and that the limit has not been reached yet. However, the main problem remaining is the loss in germanium upon synthesis.

References:
[1] Lafond, A., et al., Z. anorg. Allg. Chem., 2012. **638**(15): p. 2571-2577.
[2] Rios, L.E.V., et al., J. Alloy Compd., 2015.
[3] Rietveld, H.M., J. Appl. Cryst., 1969. **2**(2): p. 65-71.
[4] Rietveld, H.M., Acta Cryst., 1967. **22**(1): p. 151-152.

P62

Study of exfoliated molybdenum disulfide (MoS2)

G. Munkhbayar1, S. Palleschi2, F. Perrozzi2, J. Davaasambuu1 and L. Ottaviano2, *1Department of Physics, Faculty of Arts and Science, National University of Mongolia 2Department of Physical and Chemical Sciences, University of L"Aquila, Italy*

M. Gombosuren[1,2], P. Stefano[3], P. Francesco[3], D. Jav[2], O. Luca[3]

[1]National University of Mongolia, Ulaanbaatar, Mongolia

[2]National University of Mongolia, Physics, Ulaanbaatar, Mongolia

[3]University, Physics, L'Aquala, Italy

In the last decade, single atomic layer of graphite1, great interest has grown on two dimensional materials. Two dimensional materials are attractive for use in next-generation nanoelectronic devices because, compared to one-dimensional materials, it is relatively easy to fabricate complex structures from them. MoS2 is one of two dimensional materials and MoS2 layers are bound together by weak van der Waals interactions and laminated 0.7 nm one to another. Micromechanical exfoliated technique2 (similar to graphite) is the simplest and cheapest way to obtain exfoliated molybdenite with planar dimensions of the order of µm.

Differently from graphene, MoS2 is semiconductor with an indirect band gap of 1.2 eV3 in bulk to direct band gap 1.8 eV for a monolayer MoS22,4, make MoS2 attractive for a variety of electronic and optoelectronic devices. We obtained mono and few layers MoS2 samples on the SiO2 (270nm)/Si substrate from bulk MoS2 crystal by micromechanical exfoliation technique.

Our results are combined characterization by quantitative optical microscopy, Atomic force microscopy (AFM), Raman and Photoluminescence spectroscopy.

P63

Thermal Behavior of ADN and ADN-Prills – Crystal and Micro Structure

M. Herrmann[1], P. B. Kempa[1], U. Förter-Barth[1], T. Heintz[1]

[1]Fraunhofer ICT, Pfinztal, Germany

ADN is a promising oxidator for solid propellants, e.g. for space applications or solid rocket propellants. The raw material is purchased and processed at Fraunhofer ICT in a pilot plant scale, in order to form so called ADN-prills, which are used in new propellant formulations. However, ADN exhibits mechanical and thermal instabilities and incompatibilities, e.g. when incorporated in binder matrices or mixed with other ingredients. Hence, a systematic investigation of the ADN and ADN-prills were started using temperature resolved X-ray diffraction for monitoring crystal and micro structures. The studies revealed details of partial eutectics, melting, but also temperature ranges of recrystallization and interactions with additives. Fig. 1 show a so-called waterfall-plot of a temperature resolved series, with a two-step phase transition into the melt and a partial recrystallization before the final melting step.

Fig. 1: Patterns of temperature resolved diffraction patterns of ADN during heating from 30°C (top) to 93°C (bottom).

Figure 1

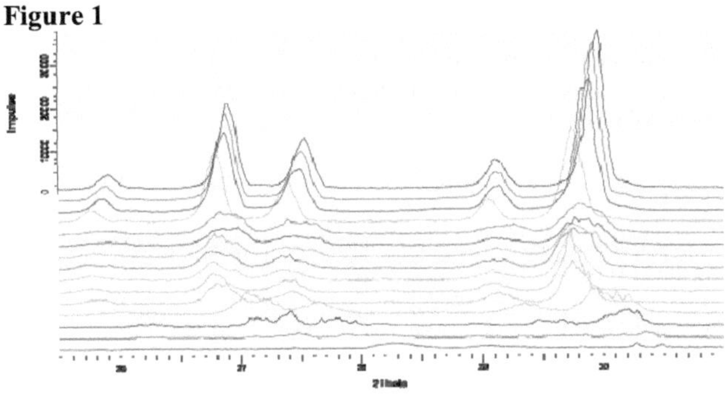

P64

Synthesis of Sphalerite – different ways to obtain chemically pure and doped Material

J. Heinrich[1], M. Schwarz[1], G. Heide[1]

[1]TU Bergakademie Freiberg , Institute of Mineralogy, Freiberg, Germany

Introduction

Sphalerite (sp-ZnS) is an important source of Zn and other technologically important elements like e.g. Fe, Cu, Cd [1] and trace elements like In and Ge [2]. A method to exploit sphalerite-containing deposits is biotic leaching. For a better understanding of the complex leaching behaviour, dissolution experiments require a reference material of high availability with a reproducible composition and homogeneity. Commercially offered ZnS is unsuitable as reference material and homogeneous natural ZnS is very rare [3]. The aim of this work is to apply different synthesis strategies to obtain homogeneous (un)doped sp-ZnS. Synthesis products are tested by powder XRD with respect to phase purity and microstructure (e.g. absence of wurtzitic stacking faults).

Methods

Precipitated, mainly sp-ZnS nanopowder used for the experiments is chemically pure but contains wurtzitic stacking faults. In2S3 and elemental Fe were used as dopants. Four different methods were used to produce a synthetic, doped sphalerite: state-of-the-art methods like 1) tempering in a furnace and 2) chemical transport reaction (CVD) or (in this context) new approaches like 3) spark plasma sintering and 4) high-pressure-high-temperature (HPHT) treatment using a toroid-type press [4]. A short summary of the experimental setups is given in table 1.

Results & Conclusions

Wurtzitic stacking faults, which are expected to be beneficial for leaching of ZnS did not disappear during treatments 1) to 3). Contrarily, with HPHT treatment the stacking faults were removed. Also a homogenous distribution of the Indium dopant was found by EPMA (electron probe micro analysis).

The comparison of different synthesis methods shows that only during HPHT treatment defect-free sphalerite was grown. This state was achieved by nucleation and growth after complete melting[5] of ZnS under HPHT. Thus, it is possible to dope commercial ZnS raw material and obtain homogeneously doped, defect-free sphalerite to be used as reference material for geochemical analysis and to calibrate microbiological leaching experiments. A synthesis time reduction from days (CVD) to minutes (HPHT) is also achieved.

Table 1: Overview of the different synthesis conditions

References:
[1] H.J. Rösler, Lehrbuch der Mineralogie, Dt. Verlag f. Grundstoffindustrie 1991.
[2] Z. Johan, Mineral. Petrol. 39 (1988) 211.
[3] P. Onuk et al., Geostand Geoanal Res. (2016) 1.
[4] L. Khvostantsev et al., High Pressure Res. 24 (2004) 371.
[5] A. Addamiano, P.A. Dell, J. Phys. Chem. 61 (1957) 1021.

Figure 1

Method	Temperature	Pressure	Transport agent
Furnace	540 – 900 °C	~ 1 hPa	-
CVD	1000 °C (source) – 900 °C (sink)	~ 1 hPa	Iodine
SPS	~ 1400 °C	~ 51 Mpa	-
HPHT	~ 1600 °C	~ 8 GPa	-

P65

Materials inspired by nature: synthesis and characterisation of perryite-related compounds $Ni_{8-x}Fe_xSi_3$ with $0 \leq x \leq 8$

S. Gallus[1], M. Chikovani[2], J. Persson[2], J. Voigt[2], K. Friese[2], A. Senyshyn[3], A. Grzechnik[1]
[1]RWTH Aachen, Institute of Crystallography, Aachen, Germany
[2]Forschungszentrum Jülich GmbH, Jülich Centre for Neutron Science JCNS-2 and Peter Grünberg Institute PGI-4 , Jülich, Germany
[3]Heinz Maier-Leibnitz Zentrum, Technische Universität München, Garching, Germany

Magnetocalorics display reversible thermal changes due to changes in a magnetic field. The thermodynamic process is a decrease of entropy in the magnetic subsystem compensated by an increase of the lattice entropy, being equal to a rise of temperature [1].

We search for new phases that have a large magnetocaloric effect. Our idea is to find out whether the *novel* materials already exist as minerals. For instance, perryite $(Ni,Fe)_8(Si,P)_3$ (R3c, Z=12) occurring in highly reduced stony meteorites [2] is similar to the materials with a magnetocaloric effect because it has a large number of Wyckoff positions which the magnetic ions could occupy.

Samples with the perryite-like compositions $Ni_{8-x}Fe_xSi_3$ ($0 \leq x \leq 8$) have been synthesised and chemically analysed. They have been examined with powder and single-crystal x-ray diffraction as well as powder neutron diffraction.

There occur two structural types in the system $Ni_{8-x}Fe_xSi_3$. For $0 \leq x \leq 4$, the materials have the structure related to $Ni_{31}Si_{12}$ (P321, Z=1) [3], whose c lattice parameter is about one third of that of perryite. The perryite structure is a stacking variant of the $Ni_{31}Si_{12}$ type [2]. The analysis of the single-crystal x-ray data and Rietveld refinements of neutron powder patterns show, that the phases with $x=2$ and $x=3$ have a doubled trigonal unit cell along the c axis compared to that of $Ni_{31}Si_{12}$. The material Ni_7FeSi_3 has the structure of $Ni_{31}Si_{12}$ (P321, Z=1). Magnetisation measurements exhibit an increasing temperature of magnetic transitions for the Ni-rich range with increasing Fe content. The sample $x=3$ has a magnetic transition close to room temperature. Temperature-dependent measurements display continuous changes of lattice parameters and no structural phase transitions.

The compounds with $5 \leq x \leq 8$ have the Fe_3Si structure (Fm-3m, Z=4) [4]. From the single-crystal x-ray and powder neutron diffraction data for $x=6$ and $x=5$, it is possible to determine the preference of Fe and Ni to occupy the Wyckoff sites *8c* and *4a*, respectively.

Further measurements to evaluate the magnetocaloric effect in the system $Ni_{8-x}Fe_xSi_3$ are underway.

References:
[1] V. Franco et al. (2012), *Ann. Rev. Mater. Res.*, 42, 305.
[2] A. Okada et al. (1991), *Acta Crystallogr. C*, 47, 1358.
[3] K. Frank et al. (1971), *Acta Crystallogr.* B, 27, 916.
[4] M. Schütte et al. (2003), *Z. Anorg. Allg. Chem.*, 629, 1846.

Molecular Crystallography at the Limits

P66

The XtaLAB mini II: A benchtop diffractometer for undergraduate institutions

J. Ferrara[1], E. Reinheimer[1]
[1]Rigaku Americas Corp, The Woodlands, United States

The XtaLAB mini was introduced at the 2008 IUCr in Osaka as a research grade benchtop diffractometer. Since then the XtaLAB mini has found its way into the undergraduate teaching curriculum with the most notable example being the University of Southampton.[1] This year the next generation benchtop, the XtaLAB mini II, was introduced at the ACA meeting in Denver. While the XtaLAB mini II has a number of hardware improvements, it is the software controlling the instrument and processing the data that represents the most important change, CrysAlisPro. CrysAlisPro is the widely popular data collection and data reduction program from Rigaku Oxford Diffraction. CrysAlisPro allows for automated data collection and reduction or step-by-step analysis so students can learn the basics of single crystal X-ray diffraction or delve deep into the nuances of problem crystals.

In this presentation we will explore the range of diffraction experiments one can perform with XtaLAB mini II focusing on the possibility to bring single crystal X-ray diffraction facilities to undergraduate institutions for both teaching and research.

References:
[1] S. J. Coles and L. K. Mapp *J. Chem. Educ.*, 2016, *93* (1), pp 131–140.

P67

An unconventional view of the molecular structure of a medicinal drug. Charge Density study of a new solid form of the antiretroviral drug lamivudine (3TC).

J. C. Tenorio[1,2], J. Ellena[1], C. W. Lehmann[2]
[1]São Carlos Institute of Physics - University of São Paulo, São Carlos, Brazil
[2]Max-Planck-Institut für Kohlenforschung, Mülheim an der Ruhr, Germany

The HIV infection currently affects millions of people worldwide. At present, the only practicable treatment is Antiretroviral Therapy (ART), which controls the virus replication but does not eliminate the virus itself. As a result, ART must be continued lifelong in order to prevent the virus from disseminating. In order to overcome this disadvantage, it is important to gain a better understanding of the molecular structure of antiretroviral drugs used in ART. Here we report a crystal engineering and Charge Density (CD) study of a derivative of Lamivudine (3TC), which is one of the well marketed and successful Nucleoside Reverse Transcriptase Inhibitors (NRTI). The investigation was undertaken in order to assess the physicochemical stability and the supramolecular properties of 3TC, since this compound has stability problems in the manufacturing process.[1,2]

The nitric acid salt derivative of 3TC (anhydride Lamivudine nitrate, 3TCH-NO_3) was synthesized and crystallized. 3TCH-NO_3 crystallizes in the space group $P2_1$, with one ionic pair per asymmetric unit. The crystal lattice exhibits an interesting hydrogen bond network, in which most of interactions are of the classical N–H...O and O–H...O types. In addition, some weak interactions of anion...π and non-classical C–H...S type can be observed.

High resolution X-ray diffraction data were collected ($\sin\theta_{max}/\lambda_{Mo} \approx 1.2$) and the experimental CD was determined by multipole model (MM) refinement. The experimental results were compared with models obtained by DFT calculations with a M06-X2 functional and 6-311++g(d,p) basis set (Fig.1). Some important aspects of the molecular conformation, topological analysis by quantum theory of atoms in molecules (QTAIM), supramolecular behavior and physicochemical stability of this new solid form are assessed and correlated with the CD distribution. This way, we hope to shed light on the pharmaceutical properties of a new solid form of the 3TC drug by a combination of crystal engineering and CD studies, which is a new approach in the analysis of a pharmaceutical compound.

Figure 1. Relief map of the theoretical Electron Density around the 3TC *cytosinic* fragment.

References:
[1] Martins, F. T.; Bonfilio, R.; De Araújo, M. B.; Ellena, J. J. Pharm. Sci. 2012, 101 (6), 2143–2154.
[2] Perumalla, S. R.; Sun, C. C. *Cryst. Growth Des.* 2014, 14 (8), 3990-3995.

Figure 1

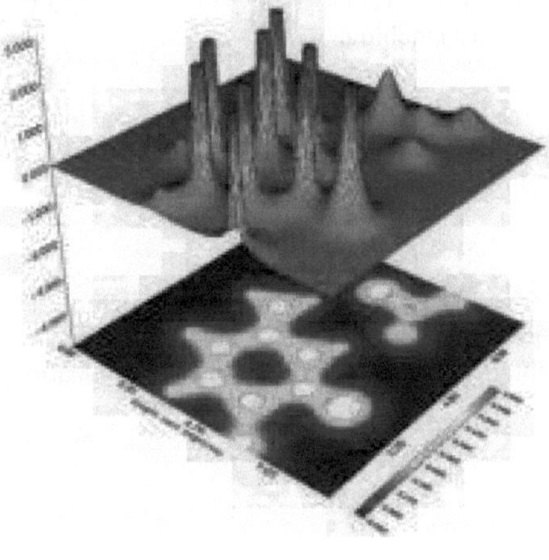

P68

The UV-triggered decomposition of 3-(4-pyridyl)-acetylacetone

K. N. Truong[1], M. Kremer[1], U. Englert[1]
[1]RWTH Aachen University, Institute of Inorganic Chemistry, Aachen, Germany

We discovered interesting structural changes in a crystalline complex after exposing the compound to direct sunlight.
During the investigation of the zinc-coordination chemistry of 3-(4-pyridyl)acetylacetone, we observed a decomposition of the material, depicted on the left hand side of the figure below, to a previously unknown structure featuring the new ligand (4-pyridyl)-acetone. Single crystal diffraction data were obtained for both structures, before and after the decomposition reaction. The structural change seems to be triggered by exposure to sunlight, while artificial light sources do not lead to the new structure. These observations suggest that UV-light might be the essential reason for this solid-state-reaction. Further studies in this direction will be performed during my master thesis at the RWTH Aachen University.

Figure 1

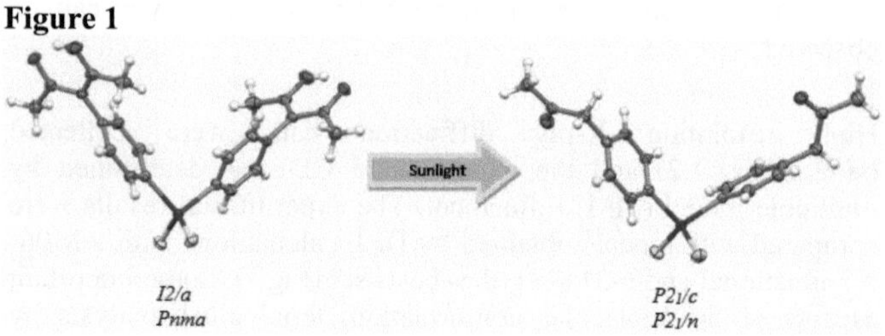

P69
Fostering what is needed tomorrow – facing problems in teaching crystallographic topics

S. Langenstück[1], U. Englert[1]
[1]RWTH Aachen, Inorganic Chemistry, Aachen, Germany

Newly enrolled students often encounter difficulties with the complexity of crystallographic topics. Problems like dealing with mathematics or the idea of 3D structures occur regularly and represent a challenge for successful teaching. Surveys among freshmen indicate that lack of interest is a popular reason for problems with these topis.

The field of didactics in chemistry tries to find new methods for supporting students in learning crystallography. A new interventional study fosters what kind of support is needed tomorrow and how inquiry-based teaching can be connected to the established ways of university teaching.

We expect to use the new RIASEC + N model [1] as a tool to increase the students' willingness to learn "difficult" topics such as crystallography and to eliminate potential obstacles on this way.

References:
[1] Dierks, P.O., Höffler, T.N., Blankenburg, J.S. Peters, H., Parchmann, I. **2016**: Interest in science: a RIASEC-based analysis of students interests**.** International Journal of Science Education, 38:2, S. 238-258

Nanoscience

P70
Thermal stability and properties of metastable $(Cr,Zr)_2O_3$

D. Rafaja[1], C. Wüstefeld[1], G. Abrasonis[2], S. Bräunig[1], C. Baehtz[2], M. Dopita[3], M. Krause[2], S. Gemming[2]
[1]TU Bergakademie Freiberg, Materials Science, Freiberg, Germany
[2]Helmholtz-Zentrum Dresden-Rossendorf, Rossendorf, Germany
[3]Charles University , Condensed Matter Physics, Prague, Czech Republic

This contribution illustrates the microstructure design in nanocomposites on the example of metastable oxides containing chromium and zirconium. Successive crystallization of amorphous Cr-Zr-O thin films, formation of the $(Cr,Zr)_2O_3/(Zr,Cr)O_2$ nanocomposites, thermal stability and thermally induced changes in the hexagonal crystal structure of metastable $(Cr,Zr)_2O_3$ were investigated by means of in situ high-temperature synchrotron diffraction experiments. The thin films of Cr-Zr-O were deposited at room temperature by using reactive ion beam sputtering from zonal Cr-Zr targets under oxygen flow. The resulting amorphous Cr-Zr-O solid solutions contained up to 15 at.% Zr. During the annealing in vacuum, the Cr-Zr-O solid solutions decomposed into two metastable phases, i.e., Cr-rich $(Cr,Zr)_2O_3$ and Zr-rich $(Zr,Cr)O_2$, which crystallized in hexagonal and tetragonal structure, respectively. With increasing Zr content in amorphous Cr-Zr-O, the start of the phase segregation and crystallization was shifted from 600°C at 3 at.% Zr to 1000°C at 15 at.% Zr. With the aid of the in situ high-temperature synchrotron powder diffraction experiments, it was found that the metastable $Cr_{2-2x}Zr_xO_{3-x}$ can accommodate up to approx. 3 at.% Zr. The zirconium atoms occupy partially the Wyckoff positions 6b in the corundum-like crystal structure of Cr_2O_3 that are empty in the stoichiometric chromium oxide. The incorporation of Zr into the crystal structure of Cr_2O_3 inflated the elementary cell and modified the thermal expansion of $Cr_{2-2x}Zr_xO_{3-x}$. The tetragonal structure of zirconia was stabilized by chromium. The phase segregation during the crystallization led to the formation of $(Cr,Zr)_2O_3/(Zr,Cr)O_2$ nanocomposites. The size of crystallites in these nanocomposites decreased with increasing Zr content from 60 nm to 30 nm and increased only slightly at the highest annealing temperatures.

P71
Crystallographic characterization of laser-ablated and polymer-stabilized silver-gold nanoalloys (4 nm)

O. Prymak[1], J. Jakobi[2], C. Rehbock[2], M. Epple[1], S. Barcikowski[2]
[1]University of Duisburg-Essen, Inorganic Chemistry, Essen, Germany
[2]University of Duisburg-Essen, Technical Chemistry, Essen, Germany

Alloyed silver-gold nanoparticles are promising candidates to combine and tune the physico-chemical and biological properties of both metals. Such multifunctional nanoparticles are of interest in heterogeneous catalysis, electrocatalysis, and biomedicine. We have synthesized PVP-coated bimetallic silver-gold and monometallic silver and gold nanoparticles by laser ablation of solid alloy targets in liquids. After size separation by ultracentrifugation, uniform spherical nanoparticles with a diameter of 4 nm were produced and characterized by colloidal (DCS) and microscopic (TEM) methods. The combination of UV-vis spectroscopy with energy-dispersive X-ray spectroscopy (EDX) as line scan along the nanoparticle showed a homogenous distribution of the elements inside the nanoparticles (Fig. 1, left). By extended crystallographic investigation using X-ray powder diffraction (XRD) and subsequent Rietveld refinement (Fig. 1, right) a precise determination of the lattice parameters and the crystallite size was performed. They showed that the nanoparticles in the whole composition range consisted of one domain and were single crystals. For comparison, the same analysis, extended by the determination of the microstrain, was carried out for the bulk target materials. A slight deviation from the Vegards rule was found for both nanoparticles and bulk target materials with a minimum at the composition Ag:Au=50:50. This effect was only weakly pronounced and did not change after the laser ablation, i.e. the crystallographic nature and composition of the prepared silver-gold nanoparticles were not significantly influenced by the laser ablation.

Figure 1: Representative TEM micrograph with an inset of EDX line scans (left) and Rietveld refinement (right) of the laser-ablated Ag50Au50 nanoparticles. The sharp diffraction peaks belong to LaB6 added as internal crystallinity standard.

Figure 1

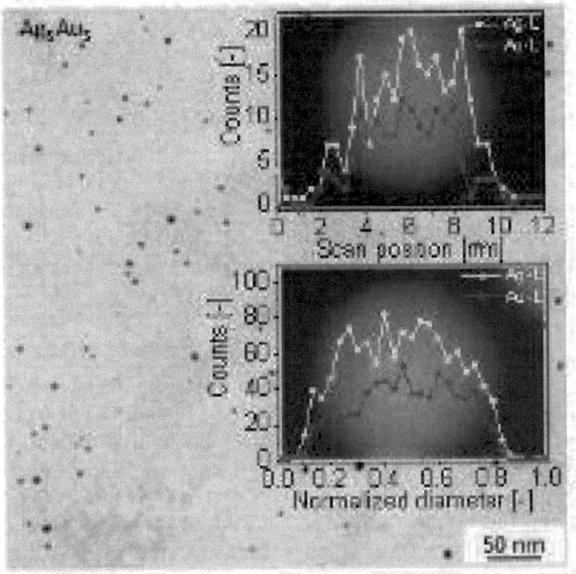

Figure 2

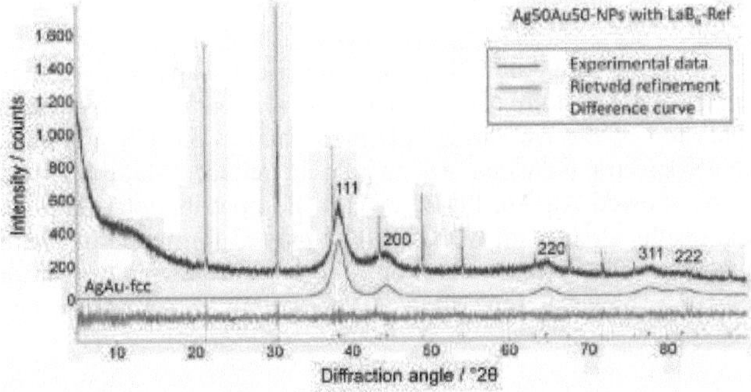

P72

Microfluidic Synthesis Of Ultrasmall AuPd Nanoalloys In Continuous Turbulent Flow

G. Tofighi[1], D. E. Doronkin[2,1], H. Lichtenberg[2,1], W. Wang[3], T. Sheppard[2,4], G. Rinke[5], J. D. Grunwaldt[2,4]
[1]Karlsruhe Institute of Technology, Institute for Chemical Technology and Polymer Chemistry, Karlsruhe, Germany
[2]Karlsruhe Institute of Technology, Institute of Catalysis Research and Technology, Karlsruhe, Germany
[3]Karlsruhe Institute of Technology, Institute of Nanotechnology, Karlsruhe, Germany
[4]Karlsruhe Institute of Technology, Institute for Chemical Technology and Polymer Chemistry, karlsruhe, Germany
[5]Karlsruhe Institute of Technology, Institute for Micro Process Engineering, Karlsruhe, Germany

Introduction

Due to their unique physical and chemical properties, ultrasmall metal nanoclusters (NCs) are attractive for numerous applications such as catalysis, nanoelectronics, sensing and biology. Introducing bimetallics with a variety of structures e.g. nanoalloys, core/shell and multi-shell nanoparticles provides new opportunities in science and technology. This requires synthesis techniques capable of controlling the intermediate processes during NC formation.

Objectives

Microfluidic synthesis of bimetallic NCs in continuous turbulent flow offers several benefits e.g. efficient and homogenous mixing, easier process handling as well as higher mass and heat transport within a wide range of operating conditions. The objective of this work was to explore the potential of microfluidics to produce AuPd nanoalloys.

Materials & Methods

A novel microfluidic setup that also allows *in situ* studies was used to produce AuPd NC colloids using $HAuCl_4$ and K_2PdCl_4 as precursor, $NaBH_4$ as reducing agent and polyvinylpyrrolidone as stabilizer. The setup consists of pressurized vessels that allow turbulent flow of reactants in a microreactor with cyclone micromixers for efficient mixing followed by a microchannel. AuPd NCs with different Au:Pd ratios were supported on TiO_2 and calcined at 380 °C for CO oxidation tests. For characterization, UV-Vis, XANES, EXAFS, TEM-EDX, ICP-OES, XRD were applied.

Results

TEM images of the AuPd NCs show spherical shape with average diameter of 1.1±0.4 nm (Fig. 1). This was also backed up by strong suppression of the Au surface plasmon band in the UV-Vis data. XANES spectra indicated Au and Pd in reduced state. EXAFS analysis showed Au-Au, Pd-Pd and Au-Pd contributions (Fig. 2). These results along with EDX and ICP proved formation of AuPd nanoalloys. The AuPd NCs on TiO_2 were aggregated after calcination, however they showed good catalytic activity in CO oxidation.

Conclusion

Nanoalloy AuPd NCs were produced in a microreactor under turbulent flow conditions via reduction reaction. The special micromixers integrated in the chip allow mixing times of the reactants below 2 ms to achieve an efficient arrangement of components in nanoalloys and *in situ* studies in future.

Figure 1: STEM image and size distribution of Au_5Pd_5 NCs.

Figure 2: Fourier transformed Pd K-edge EXAFS data.

Figure 1

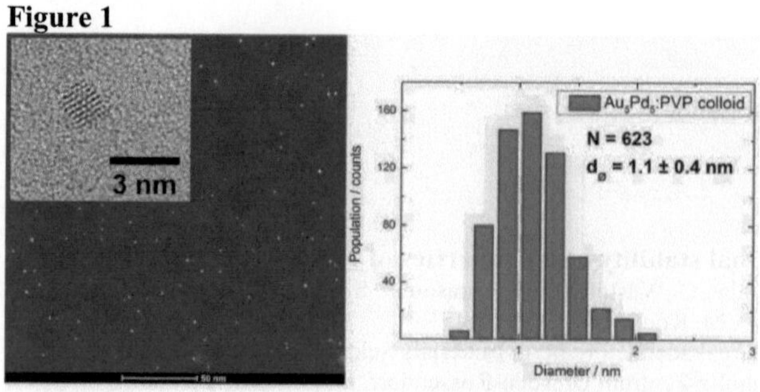

Figure 2

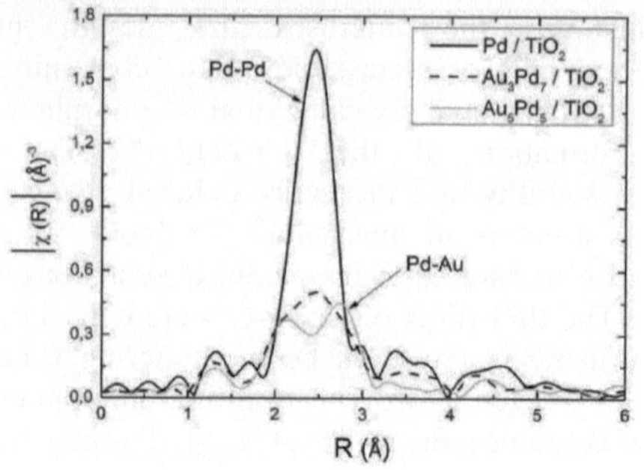

P74

Recent Developments on Incoatec"s Microfocus Source IμS for Material Science Applications

A. Stricker[1], B. Hasse[1], A. Kleine[1], J. Graf[1], C. Michaelsen[1]
[1]Incoatec GmbH, Geesthacht, Germany

Modern microfocus X-ray sources define the state-of-the-art for X-ray diffractometry and small-angle scattering. These sources are usually combined with multilayer X-ray mirrors which are excellent X-ray optical devices for beam shaping and preserving the source brightness. The optics provides a parallel or focused monochromatic X-ray beam magnified to a suitable size. Low power sealed microfocus sources, such as Incoatec"s IμS represent an attractive alternative to rotating anodes with a significant reduction in cost and maintenance. Power loads of a few kW/mm2 within anode spot sizes below 50μm deliver a compact brilliant beam.

In our presentation we will give an overview of representative experimental setups and results demonstrating the potential of our IμS in SAXS studies. These take advantage of the brilliance and outstanding beam quality of this low-maintenance microfocus source. We will show in more detail home-lab in-situ GISAXS measurements where we investigated how a multilayer grows during thin film deposition. This kind of experiments is typically done only at synchrotrons. With an IμS it is now also feasible in the home-lab.

We will further present results from our lately developed SCATEX pinholes, scatterless apertures which significantly reduce unwanted parasitic aperture scattering. Thus, these apertures enable system upgrades which increase the photon flux, the resolution and signal-to-noise ratio of the whole experimental setup. SCATEX pinholes are either made of Germanium for energies below 11.2 keV or of Tantalum for energies above 11.2 keV and are available with diameters ranging from 2 mm down to 20 μm and below. Therefore, these novel apertures are applicable to a wide range of different applications.

P75

In-situ time-resolved X-ray diffraction of a single GaAs nanowire during thermal annealing and at growth conditions

S. M. Mostafavi Kashani[1], D. Kriegner[2], P. Schroth[1,3,4], J. Vogel[1], L. Feigl[3,4], J. Jakob[3,4], A. Davtyan[1], D. Bahrami[1], T. Baumbach[3,4], U. Pietsch[1]
[1]Siegen University, Solid State Physics, Siegen, Germany
[2]Charles University, Department of Condensed Matter Physics, Faculty of Mathematics and Physics , Prague, Czech Republic
[3]Karlsruhe Institute of Technology (KIT), Institute for Photon Science and Synchrotron Radiation (IPS), Karlsruhe, Germany
[4]Karlsruhe Institute of Technology (KIT), Laboratory for Application of Synchrotron Radiation (LAS), Karlsruhe, Germany

We report on XRD experiments performed at synchrotron X-ray source PETRA III (DESY) at beamline P09 monitoring self-catalyzed GaAs NWs growth in-situ using a portable MBE[1] chamber mounted on a heavy load goniometer. A set of compound refractive beryllium lenses was used to focus the X-ray beam down to a size of about 2 x 6 μm^2 onto a single nanowire (NW) grown onto pre-patterned silicon substrate. The silicon oxide patterning parameters as well as the GaAs NW growth conditions were optimized for nucleation of NWs with NW spacing comparable with the X-ray beam size (see figure (a)).

Here, we report on time evolution of angular mis-orientation of neighbored GaAs NWs at a fixed substrate temperature of 630±25°C. The NWs were grown initially upright with respect to the Si(111) substrate and studied under different conditions namely UHV, arsenic background pressure, and GaAs growth conditions. The initially grown NWs were characterized by ex-situ SEM and the mis-orientation was probed by XRD.

After ex-situ SEM characterization, the specimen was reloaded into the vacuum chamber, degassed at 300°C in a separate load-lock, and transferred to the MBE chamber where the temperature was ramped to the initial growth temperature. A 2D cut through the reciprocal space map around the GaAs(111) Bragg peak was recorded with a time-resolution of ~2 seconds by means of a Pilatus 300K area detector (see figure (b)). The annealing procedure resulted in disappearance of the NW signal which is attributed to the NW tilting outside the detection plane (more than 7°). Figure (b) shows the XRD signal of two individual NWs where they disappear after few seconds. We interpret that by increasing NW tilting.

In another experiment, the same procedure was repeated under an arsenic pressure equivalent to ~0.2 ML/s 2D growth rate and similar tilting was observed. When the gallium shutter was opened mimicking the original growth conditions with a V/III ratio of 2, the NW signal re-appeared and returned to a nearly upright position. Within the experiment, no change of the phase composition was detected i.e. only tilting. In the presentation, we will discuss the results in detail.

References
[1] T. Slobodskyy, et al., Rev. Sci. Instrum. 83, 105112 (2012).

This project has been supported by German BMBF (05ES7CK and 05K13PS3)

Figure 1

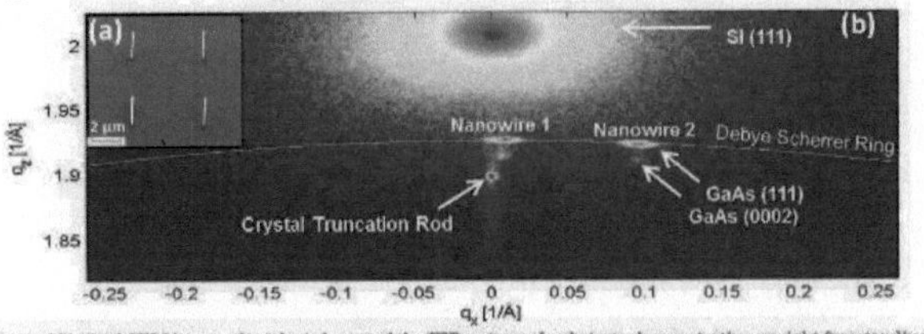

Fig (a) 30° tilted SEM image of a selected area of the FIB patterned substrate demonstrating upright freestanding NWs with 4 µm spacing. Using a focused x ray beam with a size of ~1.4×5.8 μm^2 we could identify single NWs separated by different tilt with respect to [111]. (b) shows. a 2D cut through a reciprocal space map around the GaAs (111) Bragg peak of 2 NWs. Additional features are visible as the Si(111) signal, NW signal with separated wurtzite (0002) and zinc-blende (111) segments, crystal truncation rod. The Debye Scherer ring is drawn for indicating possible tilts.

P76

Transrotational structure and its features revealed by TEM for 2 kinds of spherulites growing in amorphous films

V. Kolosov[1]
[1]Ural Federal University, Ekaterinburg, Russian Federation

Last 30 years growing interest is observed for synthesis and studies of unusual for condensed matter structures including quasicrystals, fullerenes, nanotubes. This paper presents some experimental results for the formation and microstructure on the nano-, meso-, and microscale of the other less-known **"transrotational"** microcrystals (discovered for diverse films of semiconductors, oxides and metals produced by different methods) focusing on the complicated crystallographic orientations for 2 type of spherulites. The main data have been obtained by diffraction transmission electron microscopy (TEM), primarily bend-contour method.

As initially shown earlier [1] the growth of crystals in nano-thin amorphous films (below 100 nm) is often accompanied by strong (up to 300 degrees per 1 µm) internal "crystal lattice bending" of non-dislocational nature. For such extraordinary microcrystals the new term "transrotational" crystals was introduced [2] since the translation of the crystal unit cell is accompanied by slight permanent rotation (up to ~ 0,1 degrees per unit cell) around the axis lying in the film plane (FP).

For spherulites growing in thin films the azimuthal misorientations are complicated by such "transrotation" and very complex textured crystalline aggregates are formed in result. Main features of crystallographic orientations are shown at the Figures in colors.

Fig. 1. Spherulite (hexagonal Se) with regular perturbation around 2 poles (where [001] is normal to FP) with nucleation center in between (where [001] is parallel to FP as is at the crystal periphery);

Fig.2. Spherulite (α-Fe_2O_3) with alternating circular single-crystalline ([001] normal to FP) odd zones and fine-grained ([001] parallel to FP) even zones.

References
[1] Bolotov I.E., Kolosov V.Y., Kozhyn A.V. *Phys. Stat. Sol.* 1982, **72a**, 645.
[2] Kolosov V.Yu., Tholen A.R. Acta Mater., 2000, **48**, 1829.
Partially supported by Ed. & Sci. RF Ministry, Program 211 (RF Government 02.A03.21.0006).

Figure 1

Figure 2

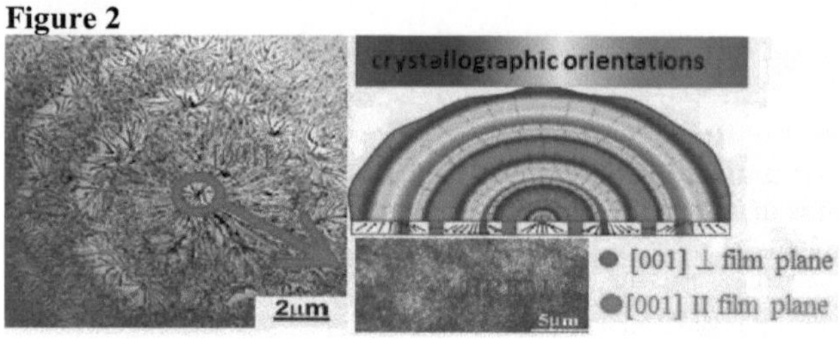

P77

Hydrothermal preparation of LFP particles with high crystalline for electrochemistry

G. Tian[1], F. Scheiba[1], H. Ehrenberg[1], K. Chakravadhanula[2]
[1]Karlsruhe Institut für Technologie, Institut für Angewandte Materialien, Eggenstein-Leopoldshafen, Germany
[2]Karlsruhe Institut für Technologie, Institute of Nanotechnology, Eggenstein-Leopoldshafen, Germany

To overcome a poor electronic conductivity, researchers are trying their best to obtain the ideal lithium iron phosphate (LFP) structure, including nano-scale particles with a uniform carbon coating and a large interface area with the electrolyte [1]. Here $LiFePO4$ hollow particles via hydrothermal synthesis are modified by a uniform carbon layer derived from polyelectrolytes (poly (diallyldimethylammonium chloride) and poly(sodium-4-styrenesulfonate)) After carbonized at 650 °C for 6 h, the residual product was denoted as LFP-C. As a reference, the uncoated LFP particles was denoted as LFP-P.

The phase purity and crystal structure for the as-prepared LFP-P/C were examined by XRD measurement with Rietveld refinement. In Fig. 1a & b, the XRD patterns indicate that all peaks of samples are well indexed to the orthorhombic LiFePO4 [2]. Fig. 1c present the TEM of LFP-C particles with typical hollow structure in the mono-dispersed state. Moreover, the HRTEM image in Fig. 1d exhibits a clear shell wall for hollow structure and an amorphous carbon layer on the surface of the particle shell, and the insert is the image for high crystal lattices of LFP, and the lattice fringe is 0.43 nm, related to the interplanar spacing of (010) facet exposed. 3 nm thick carbon layer is coated on LFP surface of particle. Ramen spectra proves in turn the polyelectrolytes and thermal treatment to result in the dense carbon layer for LFP-C. The LFP-C exhibits a high rate performance (159, 157, 154, 137 and 115 mAh/g at C/15, C/10, C/5, 1C and 5C, respectively) compared with LFP-P. In brief, LFP hollow cathode modified by self-assembled coating could improve electrical conductivity, which delivers superior specific capacity.

Fig. 1 Rietveld refinement of XRD (Cu Kα1) for LFP-P (a) and LFP-C (b); TEM image of LFP-C (c) and LFP-C (d).

References: :

[1] M.S. Whittingham, Lithium batteries and cathode materials, Chemical Reviews. 104 (2004) 4271–4301. doi:10.1021/cr020731c. [2] S. Yang, M. Hu, L. Xi, R. Ma, Y. Dong, C.Y. Chung, ZH-24 Solvothermal Synthesis of Monodisperse LiFePO 4 Micro Hollow Spheres as High Performance Cathode Material for Lithium Ion Batteries, ACS Applied Materials & Interfaces. 5 (2013) 8961–8967. doi:10.1021/am401990b.

Figure 1

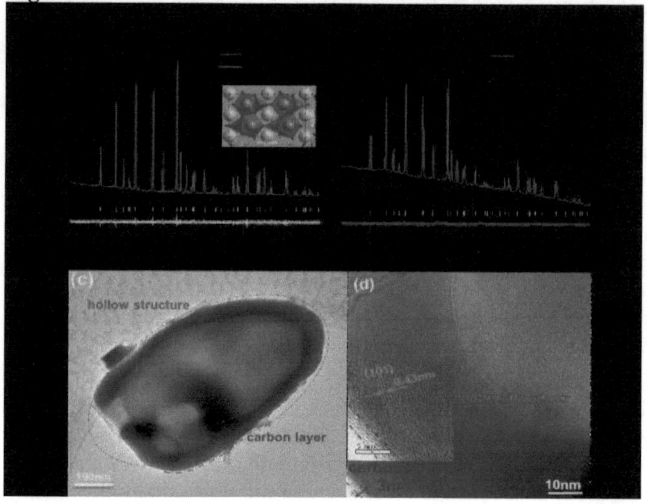

P78

Sulfur-Poly(acrylonitrile) as Cathode Material for Li/S Batteries – Correlating Electrochemical Performance and Chemical Structure

R. K. Zenn[1,2], M. Frey[1,3], S. Warneke[1,3], K. Müller[4], A. Hintennach[1], M. R. Buchmeiser[3,5], R. E. Dinnebier[2]
[1]Daimler AG, E-Drive, Böblingen, Germany
[2]Max-Planck-Institute for Solid State Research, X-Ray Diffraction, Stuttgart, Germany
[3]University of Stuttgart, Institute of Polymer Chemistry, Stuttgart, Germany
[4]Max-Planck-Institute for Solid State Research, Interface Analysis, Stuttgart, Germany
[5]Institute of Textile Chemistry and Chemical Fibers (ITCF), Denkendorf, Germany

Question

Li/S batteries are among the most promising next generation energy storage systems. However, Li/S batteries that use *cyclo*-S_8 as active material suffer from low rate capability and low cycling stability, etc. An alternative active material is the sulfur-containing composite sulfur-poly(acrylonitrile) (S-PAN), which overcomes some issues of *cyclo*-S_8.[1,2] Despite extensive research the structure-property relationships of S-PAN are still under debate.[3–5] In order to correlate chemical structure and electrochemical performance, the structure of S-PAN was determined and the Li/SPAN cells were analyzed *post-mortem.*

Methods

S-PAN was synthesized via thermal conversion of a PAN-based polymer at 550 °C with an excess of elemental sulfur. Non-covalently bound sulfur was removed via Soxhlet extraction. Elemental analysis, XRPD, XPS, ToF-SIMS, Raman, FTIR, ^{13}C-mCP-NMR spectroscopy were employed. For assigning vibrational bands DFT simulations were performed. Additionally, SPAN cathodes were analyzed after 1000 cycles *post-mortem* via XPS and XRPD.

Results
Complete dehydrocyclization and aromatization of PAN during synthesis led to condensed pyridine and benzene structures. Up to 46 wt.-% of sulfur was covalently bound to PAN in the form of thioethers, organic (pyridyl) di- and polysulfides, thiols, thioamides, thioketones.

In cycled, ***discharged*** cathodes solid Li_2S_x was identified. In cycled, ***charged*** cathodes the original sulfur functional groups were restored. Thus, in contrast to *cyclo*-S_8 all redox-reactions were reversible solid-solid reactions in Li/SPAN cells, which is one explanation for the good cycling stability.[6]

Conclusion
The structure determination confirmed and refined the structure proposed by Fanous *et al.*[5] and refuted other currently discussed structures.[3] *Post-mortem* analysis revealed that the SPAN-backbone binds sulfur reversibly via covalent bonds. This explains the high cycling stability of Li/SPAN cells compared to Li/S cells.

References:
[1] J. Wang *et al. Adv. Mater.* **2002**, *14,* 963.
[2] Z. Xu *et al. Angew. Chem.* **2016**, *128,* 10528.
[3] S. Zhang. *Energies* **2014**, *7,* 4588.
[4] S. Wei *et al. J. Am. Chem. Soc.* **2015**, *137,* 12143.
[5] J. Fanous *et al. Chem. Mater.* **2011**, *23,* 5024.
[6] R. K. Zenn, M. Frey *et al. ACS Energy Lett.* **2017**, *submitted for publication.*

Figure 1

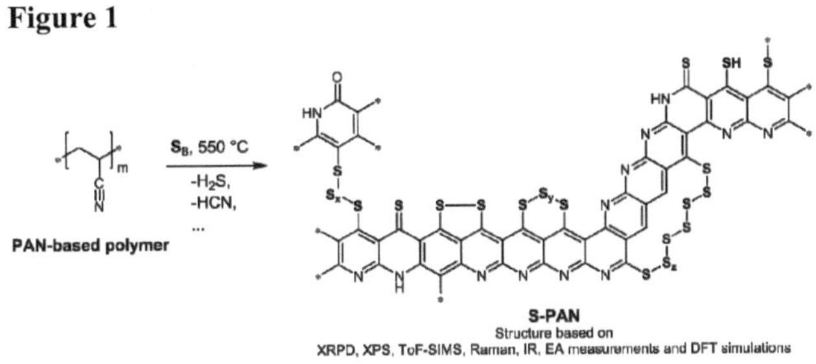

P79
EDTA EFFECT ON STRUCTURAL AND OPTICAL PROPERTIES OF Ag_2S NANOPARTICLES

A. SOUICI[1], N. DOUAS[2], W. MOUHAOU[1], S. OUHENIA[1], Z. CHAFI[1], N. KEGHOUCHE [2]
[1]University, Physics, Bejaia, Algeria
[2]University, Physics, Constantine, Algeria

Introduction
Materials at nanoscales dimensions possess specific physical properties which are different to those of the baulk. For their potential applications in nanotechnology and nanomedicine, semiconductors nanoparticles have been an increasing attention in the last two decades.[1] According to their negligible toxicity and appealing near-infrared photoluminescence, Ag_2S nanoparticles have a promises biomedical applications.[2,3]

Objectives
In this work we investigate the effect of EDTA molecule on the growth process, the optical and the structural properties of silver sulfide (Ag_2S) nanoparticles synthesized in colloidal solutions.

Materials & methods
Silver sulfate (Ag_2SO_4) and thioacetamide ($CH_3C(S)NH_2$) were used as precursors with ethylenediaminetetraacetic acid (EDTA). The latter plays the role of a stabilizing agent. The analysis of Ag_2S nanoparticles was followed by UV-visible spectrophotometry, X-ray diffraction and scanning electron microscopy (SEM).

Results
The UV-visible spectra and SEM observations show that the presence of EDTA molecules plays an essential role in determining the final size and the optical properties of Ag2S nanoparticles, figures 1 and 2. It was fond that during the formation of the first Ag_2S nanoparticles between 5 and 120s an isosbestic point is observed around 520 nm whereas the absorption band shift gradually towards the long wavelength with increasing size. The structural analysis by Rietveld refinement of the X-ray diffraction patterns confirm the monoclinic phase of Ag_2S nanoparticles with space group P21/n (n°14) and average size around 100 nm.

Conclusion
We have prepared Ag_2S nanoparticles by chemical way with the presence of EDTA as stabilized agent. The results illustrate that the concentration of EDTA molecule plays a crucial effect on the growth process, the final size and their optical properties.

Fig 1: Absorption spectra of Ag_2S nanoparticles

Fig 2: SEM images of Ag_2S nanoparticles

References:
[1] Sun CQ, Prog. Solid State Chem. 35, 1-159 (2007).
[2] Y. Zhang, G. Hong, Y. Zhang, G. Chen, F. Li, H. Dai, Q. Wang, ACS Nano (2012) 6, 3695–3702.
[3] C. Li, Y. Zhang, M. Wang, Y. Zhang, G. Chen, L. Li, D. Wu, Q. Wang, Biomaterials (2014) 35, 393–400.
[4] G Ouyang, CX Wang and GW Yang, Chem. Rev. 109, 4221-4247 (2009).

Figure 1

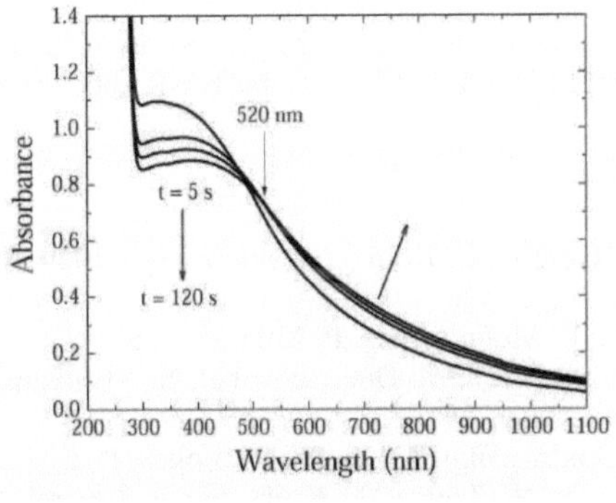

Figure 2

P80

Solving the Hydrogen and Lithium Substructure of Poly(triazine imide)/LiCl by a combination of electron diffraction tomography, nuclear magnetic resonance and X-ray total scattering

Y. Krysiak[1], M. B. Mesch[2], K. Bärwinkel[2], J. Senker[2], R. B. Neder[3], U. Kolb[1]
[1]Johannes Gutenberg University Mainz, Institute of Inorg. Chemistry and Analy. Chemistry, Mainz, Germany
[2]University of Bayreuth, Inorganic Chemistry III, Bayreuth, Germany
[3]Friedrich-Alexander-Universität Erlangen-Nürnberg, Chair for Crystallography and Structural Physics, Erlangen, Germany

Crystal structure solution using automated electron diffraction tomography[1] (ADT) is nowadays an accepted method for deriving full 3D structure information at atomic resolution from a nanocrystalline material. Crystals with special structural features like twinning[2], super-structural effects[3] or pseudo-symmetry could be solved ab-initio with ADT. In case of a reduced periodicity in the lattice due to electron radiation damage, disorder or stress/strain effects the intensities are more and more difficult or impossible to derive from diffraction patterns. Uncertainties in the intensity extraction allow often the detection of the major structure but cause problems for detailed structural insights being especially important for materials physical properties.

By combining nuclear magnetic resonance (NMR) measurements and the analysis of X-ray total scattering (PDF) providing short-range respectively long-range information with electron diffraction results, a full structural description can be derived. With this combination, we resolved the H/Li substructure of poly(triazine imide)/LiCl.[4] This sample with incorporated lithium chloride recently attracted substantial attention due to its photocatalytic activity for water splitting. In addition to the known major structure, electron diffraction tomography revealed the positions of the Li atoms and a symmetry reduction. NMR simulations added information about the protonation of the channels and the PDF analysis of X-ray powder diffraction data, simulated with DISCUS software package[5], proposed a long-range modulation of the layers.

Figure 1: Superposition of the structure solved in *C*2221 by electron diffraction (blue) and in *P*212121 by NMR spectroscopy (red).

References:
[1] E. Mugnaioli, T. Gorelik, U. Kolb, *Ultramicroscopy* **2009**, *109*, 758–765.
[2] S. Bhat, L. Wiehl, L. Molina-Luna, E. Mugnaioli, S. Lauterbach, S. Sicolo, P. Kroll, M. Duerrschnabel, N. Nishiyama, U. Kolb, et al., *Chem. Mater.* **2015**, *27*, 5907–5914.
[3] E. Mugnaioli, I. Andrusenko, T. Schüler, N. Loges, R. E. Dinnebier, M. Panthöfer, W. Tremel, U. Kolb, *Angew. Chem.* **2012**, *124*, 7148–7152.
[4] M. B. Mesch, K. Bärwinkel, Y. Krysiak, C. Martineau, F. Taulelle, R. B. Neder, U. Kolb, J. Senker, *Chem. – Eur. J.* **2016**, *22*, 16878–16890.
[5] T. Proffen, R. B. Neder, *J. Appl. Crystallogr.* **1997**, *30*, 171–175.

Figure 1

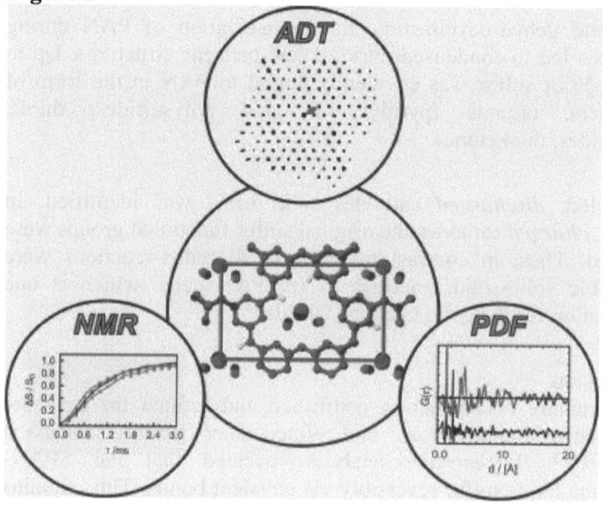

P81

A DEGAS Study of the Water and Hydrogen Release from Australian Opal

P. S. Thomas[1], K. Heide[2], G. Heide[3], K. Galonska[3]
[1]School of Mathematical and Physical Sciences / University of Technology Sydney, Sydney, Australia
[2]Institut für Geowissenschaften / Universität Jena, Jena, Germany
[3]Institut für Mineralogie / TU Bergakademie Freiberg, Freiberg, Germany

Two types of precious opal, opal-CT from Tintenbar and opal-AG from Coober Pedy, Australia have been investigated using DEGAS to investigate water and hydrogen release as a function of the temperature.

The directly-coupled-evolved-gas-analysis-system (DEGAS) comprises of a Netzsch STA 409 CD Skimmer coupled to a quadrupole mass spectrometer QMS 403/5 with the skimmer orifice removed to allow the transport of the evolved gases directly to the mass spectrometer, which are analysed in a multiple ion detection mode. Pieces of opal of the order of 5 mg were heated at a rate of 10K/min under vacuum (10^{-4} Pa) to 1300°C, while the release of volatiles, in particular, water (H_2O^+; m/z = 18 amu) and hydrogen (H_2^+; m/z = 2 amu) were monitored.

At low temperature a correlation between water and hydrogen release is observed indicating that the hydrogen signal results from H_2O-fragmentation in the mass spectrometer. At elevated temperature, however, the release of hydrogen from opal is detected independent from the release of water. The origin of hydrogen release between 600 and 1000°C can be explained through the inverse-micellar decomposition of silanol groups (Heide K. et al. 2008). DEGAS studies have yielded interesting results in the high temperature decomposition of strongly bound silanol functionalities through an inverse micellar decomposition process resulting in the formation of hydrogen as the evolved species (Heide K. and Földvari M., 2006; Heide K. et al, 2008; Thomas P. S. et al, 2014).

The two opal types showed significantly different hydrogen release profiles indicating a difference in the structure of the silicas of these opal types. Based on these differences, this study has demonstrated that DEGAS and hydrogen release can be applied to the characterisation of different morphologies of silica.

This technique should be applied to the structural characterisation of non-crystalline solids, e.g. opals.

References:
[1] Heide K., Földvari M., 2006. High temperature mass spectrometric gas-release studies of kaolinite $Al_2[Si_2O_5(OH)_4]$ decomposition Thermochim. Acta. 446, 106 – 112
[2] Heide K., Woermann E., Ulmer G., 2008. Volatiles in pillows of the Mid-Ocean-Ridge-Basalt (MORB) and vitreous basaltic rims, Chem. Erde. 68, 353 – 368
[3] Thomas P. S., Heide K., Földvari M., 2014. Water and hydrogen release from perlites and opal A study with a directly coupled evolved gas analyzing system (DEGAS). J. Therm. Anal. Cal. 115, 853 – 859

P82

Boehmite (γ-AlOOH) nanoparticles as model system for monitoring particle formation in water based reaction media within the second time scale

S. Diez[1]
[1]Friedrich Alexander Universität, Erlangen, Germany

Pair Distribution Function (PDF) modelling is applied to monitor the in-situ nucleation, growth and stabilisation of Boehmite (γ-AlOOH) nanoparticles. Boehmite, is the compound of interest for our study as it forms stacked, platelet-shaped nanoparticles from various sol-gel syntheses. Figure 1 depicts the general approach of the measurement at the ESRF and the corresponding data set of a common in-situ synthesis gathered over a reaction time of one hour.

Fig. 1: In-situ synthesis of pure boehmite precursor structures at room temperature. a) Schematic setup used for measurements at synchrotron facilities. b) Excerpt of a background corrected data set in which tiny particles and clusters result in amorphous appearing signals at the beginning of the synthesis (back). Slowly, peaks of flat boehmite layers increase in intensity (front).

The background corrected data sets (Fig. 1b) can be interpreted similarly to powder diffractograms in which nanoparticles usually (dependent on their sizes) give rise to broad peaks. In the present example, early exposures of the reaction exhibit a very broad signal which corresponds to amorphous and poorly crystallized particles. Measurements at a later point of time reveal, however, that broad peaks are emerging from the amorphous halo at positions characteristic for finely crystalline boehmite.

From these data sets PDFs are deduced[1] which represent a histogram of interatomic distances and reveal the local structure of these freshly grown nanoparticles. Modelling the pdf data[2] based on the known orthorhombic structure of boehmite provides evidence that crystalline domains of Boehmite nanoparticles grow stepwise towards a layered structure of about 3 nm size. Moreover, additional data will illustrate the influence of ligands inserted during the reaction on the outcome of the presented sol-gel synthesis. Current research of our institute focuses on modelling precursor phases at the start of the synthesis which eventually will transform into AlOOH sheets and conglomerate accordingly.

References:
[1] P. Juhas, T. Davis, C. L. Farrow, S. J. L. Billinge, *J Appl Crystallogr* **2013**, *46*, 560-566.
[2] T. H. P. R.B. Neder, *Diffuse Scattering and Defect Structure Simulations*, Oxford University Press **2008**.

Figure 1

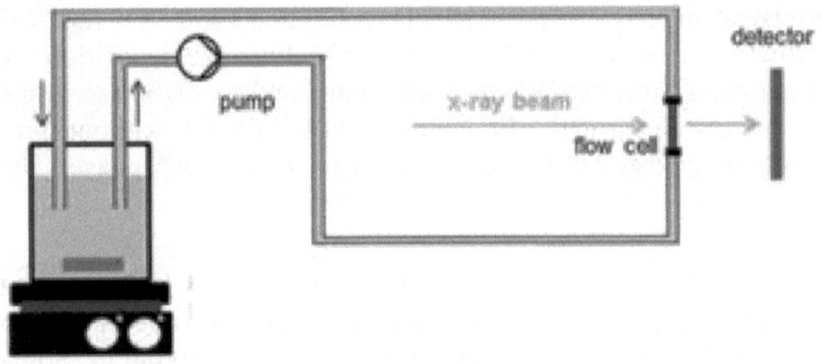

Figure 2

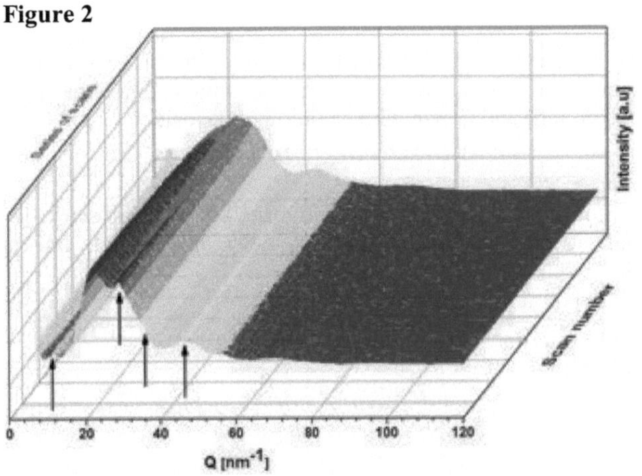

P83

In-situ time-resolved XRD and RHEED study of the polytypsim in GaAs nanowires

J. Jakob[1,2], P. Schroth[1,3], L. Feigl[1], S. M. Mostafavi Kashani[3], J. Vogel[3], U. Pietsch[3], T. Baumbach[1,2]
[1]Karlsruhe Institute of Technology, Institute for Photon Science and Synchrotron Radiation, Eggenstein-Leopoldshafen, Germany
[2]Karlsruhe Institute of Technology, Laboratory for Application of Synchrotron Radiation, Karlsruhe, Germany
[3]University of Siegen, Department of Physics, Siegen, Germany

The integration of III-V semiconductors on silicon is of particular interest to combine the standard semiconductor platform with direct band-gap materials. One approach to overcome the lattice mismatch between both material systems is the growth of III-V nanowires onto silicon substrates.

With a portable MBE system for in-situ X-ray investigations [1], we have studied the Zincblende (ZB) – Wurtzite (WZ) polytypism in gallium arsenide nanowires. The current setup allows for simultaneous time-resolved investigation of crystal structure evolution by means of X-ray diffraction (XRD) and by Reflecting High-Energy Electron Diffraction (RHEED) during the complete growth process.

The combination of these complementary methods provides detailed information on crystal structure and the distribution of WZ and ZB segments within the nanostructures. While XRD gives access to the shape and the crystal structure of the whole illuminated NW ensemble under inspection, RHEED as a surface sensitive technique offers the possibility to analyse the evolution of phases along the NW axis during the growth process.

In this contribution we demonstrate the feasibility of the approach at the example of the formation of crystal phases during selfcatalysed growth of GaAs nanowires onto Si(111). By analysing the relative intensity of the RHEED diffraction spots as a function of growth time it is found that in case of gallium pre-deposition the probability of forming WZ is higher compared to ZB phase in the early stages of growth. In contrast, NWs grown without pre-deposition quickly tend to a low fraction of the WZ phase. The results are verified by comparison with the XRD data.

We would like to acknowledge Hans Gräfe, Bärbel Krause and Svetoslav Stankov at the UHV laboratory of the Institute of Photon Science and Synchrotron Radiation, KIT and Jörg Strempfer, Sonia Francoual and David Reuther at P09, PETRA III, Desy. The work was supported by the BMBF projects (05ES7CK) and (05K13PS3).

Figure 1. Time-resolved RHEED intensities of phase selective reflections of GaAs nanowires grown without gallium pre-deposition (left) and grown with gallium pre-deposition (right).

References:

[1] T. Slobodskyy, P. Schroth, D. Grigoriev, A. A. Minkevich, D. Z. Hu, D. M. Schaadt, T. Baumbach, *Rev. Sci. Instrum.*, **83**, (2012), 105112.

Figure 1

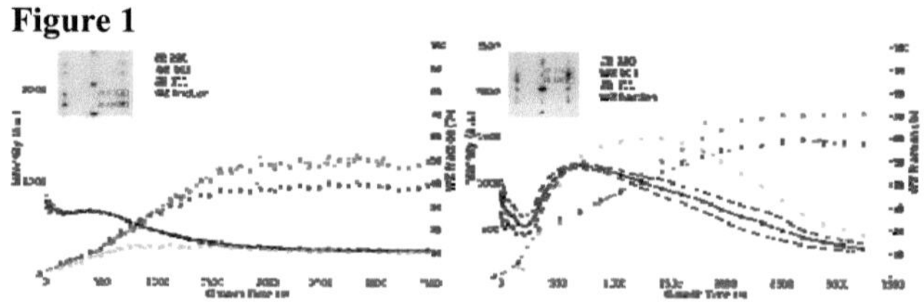

New approaches in structural biology

P84
Optimized Sample Centering for Your Best Experimental Result

M. Adam[1], J. Toorn[1]
[1]Bruker AXS GmbH, Karlsruhe, Germany

Recent improvements in X-ray instrumentation allow the investigation of ever smaller crystals.. Today"s samples are typically in a range of less than 100 µm in size, sometimes even less than 10 µm. These small crystals require mechanically stable goniometers with the best possible sphere of confusion. All our goniometers possess a sphere of confusion smaller than 7 µm. However, special attention must be paid on a proper alignment of the sample since a poorly aligned sample leads to not optimal data sets.

To accomplish the optical alignment video microscopes typically equipped with electronic crosshairs became the standard on modern single crystal diffraction systems. The center of the crosshairs needs perfectly to match with the center of the goniometer and the crystal needs to be perfectly centered in the crosshairs. The latter is often prevented by samples providing little optical contrast, resulting in a significant error. As a consequence the crystals often oscillates in the beam or—in the worst case—moves completely out of the beam during the data collection. While scaling programs, such as SADABS, can correct for an oscillation of the sample, there is no way to compensate for information lost when the crystal is out of the beam as there is simply no diffraction signal recorded. An there is a second problem, even with the best aligned instrument and the best optically aligned sample there is no way to identify the best diffracting area within a sample or the least pathologic area within a pathologically twinned or disordered sample.

Today, we present a new package AGH II designed for the best possible alignment of the a given sample. The system consists of soft- and hardware to eliminate the shortcoming describes above.

The new AGH can be mounted on our KAPPA or the FIXED-CHI goniometer. However—in contrast to other automated goniometer heads in the market or in use at synchrotrons—the AGH complies with the IUCR requirements for goniometer heads and therefore, it can be mounted on many others goniometers. The magnet base of the AGH is ideal for the use with sample pins (typical height 22mm). Highly reliable, encoder controlled Piezo motors not only ensure high mechanical stability but also guarantee a long lifetime of the new tools.

Figure 1

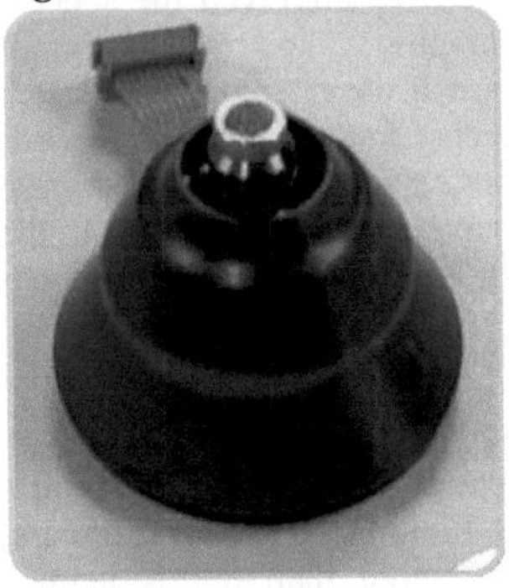

P85
Binary protein crystals for the construction of nanoparticle superlattices

T. Beck[1]
[1]RWTH Aachen University, Institute of Inorganic Chemistry, Aachen, Germany

In nature, biomolecules can function as a matrix to organize inorganic components into hybrid materials such as bone or nacre. These materials have inspired researchers to create novel composite materials based on biomolecules and inorganic building blocks such as nanoparticles.[1] To this end, for example, DNA is used as functional linker for the assembly of inorganic nanoparticles.[2] However, structuring of inorganic components into highly ordered materials still represents a major challenge, because the buildings blocks are not homogeneous in size and morphology. Here, we use two protein containers as atomically precise building blocks, engineered with opposite surface charge,[3] and construct a new type of biohybrid material.[4] Self-assembly of the charged protein containers yields binary structures with crystalline order, which were characterized with single crystal X-ray diffraction to high resolution. Moreover, prior to assembly of the charged protein containers, the cavity of the containers can be filled with metal oxide nanoparticles. The controlled assembly of these protein-nanoparticle composites yields highly ordered binary nanoparticle superlattices as free-standing biohybrid crystals, with up to a few hundred micrometers in size. Because the protein container is the primary building block with an atomically precise shell, the crystal lattice is solely defined by the protein shell and not the cargo particle. As a consequence, so far unrivaled long-range order and large domain sizes of nanoparticle superlattices could be achieved.[4] Moreover, because the structure and lattice parameters of the protein-nanoparticle crystals are independent of their nanoparticle cargo, binary protein materials based on charged protein containers may serve as a generally applicable matrix for the precise assembly of a variety of cargo.

References:
[1] M. A. Boles, M. Engel, D. V. Talapin, Chem. Rev. 2016, 116, 11220-11289.
[2] S. Y. Park, A. K. R. Lytton-Jean, B. Lee, S. Weigand, G. C. Schatz, C. Mirkin, Nature 2008, 451, 553-556.
[3] T. Beck et al., Angew. Chem. Int. Ed. 2015, 54, 937-940.
[4] M. Künzle, T. Eckert, T. Beck, J. Am. Chem. Soc. 2016, 138, 12731-12734.

Figure 1

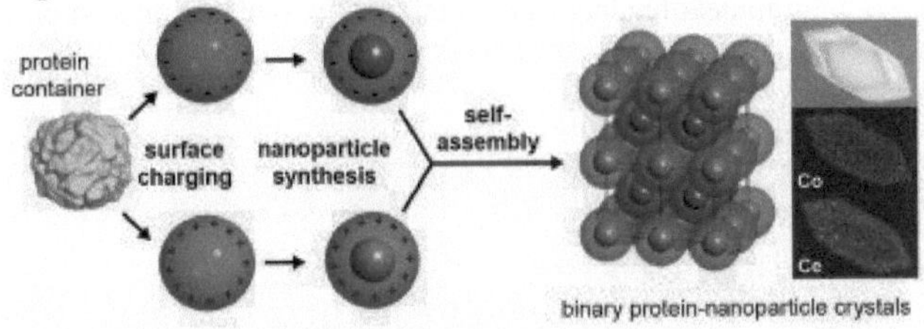

P86

PROTEUM3 – a new software pipeline for the collection and processing of crystallographic data for structural biology

M. Adam[1], T. Stuerzer[1], J. Kaercher[2]
[1]Bruker AXS GmbH, Karlsruhe, Germany
[2]Bruker AXS Inc., Madison WI, United States

In-house macromolecular crystallography has benefitted greatly from recent advances in X-ray sources, such as the METALJET and introduction of fast, single-photon sensitive CPAD detectors (PHOTON II). This requires most efficient, easy to use, comprehensive software with proven engines, for both the control of the hardware and the fast, high quality processing of data. The new PROTEUM3 software suite fulfils both needs. The software allows the user to control all Bruker instruments: our current generation of D8 QUEST and D8 VENTURE system but also offers full access to older CCD based systems. Numerous components are seamlessly integrated. Even better, from quality assessment to phasing, PROTEUM3 is easy to use and lets you launch the underlying modules with a single mouse click. The PROTEUM3 suite guides users through the entire experiment with minimum input and maximum graphical feedback.

The new PROTEUM3 offers a number of enhanced features with an increased level of comfort with several major improvements we will discuss in more detail within the presentation:

1. Multi-core processors for significantly faster data processing
2. The XPRESSO data processing pipeline has been completely rewritten to better fit the needs of structural biologists. Mount a crystal on the diffractometer, start XPRESSO and leave all decisions on data collection parameters, data integration scaling and phasing to XPRESSO. XPRESSO is based on the semi-automation pipeline, which suggests proper defaults whenever possible
3. The new Determine Phases plugin provides full GUI integration of the SHELX programs for *de novo* structure determination using SAD, SIR, SIRAS, MIRAS, and MAD. It provides easy parameter set up, graphical feedback, fast and reliable phasing, phase extension, and chain tracing
4. Integration of COOT for density map display
5. For pathologically twinned samples the newly written scaling interface. The plugin, featuring SADABS and TWINABS, handles single crystals and twins with graphical feedback and intelligent parameter suggestion.

Figure 1

P87

HKL2MAP: Improvement of SAD/MAD phasing results by means of data selection

F. Dall'Antonia[1], S. Storm[1], T. Schneider[1]
[1]EMBL Hamburg, Hamburg, Germany

Crystallographic phasing based on the presence of anomalous scatterers, as nowadays mostly employed in terms of the SAD and MAD methods, is highly dependent on the accurate measurement of Bijvoet differences. Errors of integrated diffraction intensities – which would propagate into the anomalous differences – are reduced statistically by collecting highly redundant reflection data, allowing for the averaging of multiple intensity observations of unique Bragg reflections. A possible downside of this practice is a prolonged exposure of the crystal in the X-ray beam, risking severe systematic error due to radiation damage. Moreover, when multiple datasets of a crystal (or from multiple isomorphous crystals) are to be merged for the sake of data multiplicity, there may be negative outliers with a globally reduced intensity-over-error, and their inclusion can deteriorate the combined data accuracy.

Here we present data collection and phasing scenarios where the deletion of data frames after intensity analysis and/or the removal of entire datasets from a multiple collection, prior to combined scaling, provide better unmerged datasets as starting points to the SHELXC/D/E phasing pipeline[1]. As a tool for our studies as well as for future routine application, we developed a software program that computes frame-based analyses and offers optional truncation of multiple XDS-ASCII or HKL2000-format data files. This program is also part of the new HKL2MAP[2] distribution 0.5.c, which features an extended GUI design with a separate panel for data preparation prior to SHELXC usage.

We introduce the new HKL2MAP GUI version, demonstrate its practical application in terms of SAD/MAD data selection and discuss the beneficial effect on phase/map quality for hard-to-solve structures.

References:
[1] Sheldrick, G. M. (2010). Acta Cryst. D66, 479-485.
[2] Pape, T. & Schneider, T. R. (2004). J. Appl. Cryst. 37, 843-844.

P88

Use of cadmium ions for High-throughput experimental phasing

S. Panneerselvam[1], A. Burkhardt[1], A. Meents[1]
[1]DESY, DESY, Hamburg, Germany

High-throughput crystallography experiments demand rapid data collection and robust experimental phasing procedures, which are suitable for various target proteins. Recent developments in sample preparation, beamline instrumentation have improved the success rate in macromolecular crystallography. The availability of microfocus beamlines at modern synchrotron sources and free electron lasers (FELs) providing intense microbeams between 1 and 20 µm in size , enable us to push the limits of the crystal size required for successful diffraction experiments [1]. But still the "phase problem" remains a big hurdle in macromolecular crystallography.

Cadmium ions are known to promote the crystal growth as well as improve the crystal quality of different proteins [2].Hence, cadmium salts are used as a component in multiple crystallization and additive screens. Currently there are more than 800 structures deposited in the Protein Data Bank (PDB) which contain cadmium ion as a ligand. Here we report the use of cadmium ions for high throughput experimental phasing with multiple test samples.

Protein samples such as Lysozyme, Thaumatin, Ferritin, Proteinase K, Ribonuclease A and Glucose isomerase were used. Heavy atom derivatization was performed by both co-crystallization and soaking procedures. X-ray diffraction data sets were collected at the beam line P11, PETRA III, DESY, Germany[3] . Experimental phasing methods such as single wavelength anomalous dispersion (SAD), single isomorphous replacement (SIR), single isomorphous replacement with anomalous signal (SIRAS) and multi-wavelength anomalous dispersion (MAD) were carried out with *SHELX-CDE* programs[4] through *HKL2map* package[5]. The results clearly suggest that cadmium ions provide dual benefits in improving the crystal quality and suitable for multiple experimental phasing methods.

References:
[1] Helliwell, J. R. & Mitchell, E. P. *IUCrJ* **2,** 283–91 (2015).
[2] Trakhanov, S., Kreimer, D. I., Parkin, S., Ames, G. F. & Rupp, B. *Protein Sci.* **7,** 600–604 (1998).
[3] Burkhardt, A. *et al. Eur. Phys. J. Plus* **131,** 56 (2016).
[4] Sheldrick, G. M. *Acta Cryst* **66,** 479–485 (2010).
[5] Pape, T. & Schneider, T. R. *J. Appl. Crystallogr.* **37,** 843–844 (2004).

P89

'Top-hat' beams for high-quality data collection on the EMBL beamline P14 at PETRA III

A. Polyakova[1], G. Bourenkov[1], I. Karpics[1], S. Fiedler[1], F. Liu[1], P. Vella[2], R. Schnell[2], G. Schneider[2], T. R. Schneider[1]
[1]EMBL, Hamburg, Hamburg, Germany
[2]Karolinska Institutet, Medical Biochemistry and Biophysics, Stockholm, Sweden

A recent addition to the beamline optics on the P14 beamline at PETRA III for macromolecular crystallography is a transfocator (Vaughan et al, 2011), containing several packs of compound refractive lenses (CRLs). With the CRLs in the beam path, beams with increased flux and variable focal properties across a wide energy range can be produced. On P14, we use the CRLs to produce homogeneous "top-hat" beams to match the size of crystals in the range of 100-500 µm with an up to 50-fold increase in flux with respect to the unfocussed beam, resulting in a lifetime of 2-5 min. for a cryogenically cooled crystal.

The adjustment of the beam size and flux using the transfocator is achieved by changing of the number of CRLs in the beam path. Systematic characterization of the beam properties as obtained by employing different combinations of CRLs have shown that the these can be used to produce large "top-hat" beams with homogeneous profiles for X-ray energies ranging between 7 – 15 keV on the P14 beamline.

We also show how high-quality native and anomalous data can be rapidly collected with the homogeneous "top-hat" beams generated using the CRLs at various energies, enabling structure solution by SAD or MAD phasing. For example, challenging crystals of proteins from major bacterial pathogens such as *M. tuberculosis* were studied in collaboration with the groups of Gunter Schneider and Robert Schnell (Karolinska Institute, Sweden). Several structures were successfully solved by native SAD phasing despite low anomalous signal and low solvent content of the crystals.

The CRLs are available to P14 users through the beamline control software MxCuBE (Gabadinho et al, 2010, Karpics et al., 2016), enabling automatic adjustment of the focal conditions for data collection with a homogeneous beam profile from crystals of a wide range of sizes.

References:
[1] Vaughan, G. B. M., et al. (2011). X-ray transfocators: focusing devices based on compound refractive lenses. *J. Sync. Rad.*, *18*(Pt 2), 125–133.
[2] Gabadinho, J., et al. (2010). MxCuBE: a synchrotron beamline control environment customized for macromolecular crystallography experiments. *J. Synch. Rad.*, *17*(5), 700–707.
[3] Karpics, I. et al. (2016). Graphical user interface and experiment control software at the MX beamlines at EMBL Hamburg . Proc. Of the 11th NOBUGS Meeting, https://doi.org/10.17199/NOBUGS2016.91

P90

P13 and P14, the EMBL Beamlines for Macromolecular Crystallography at PETRA III

T. R. Schneider[1], G. Bourenkov[1], G. Pompidor[1], M. Cianci[2], I. Bento[1], J. Kallio[1], I. Karpics[1], A. Polyakova[1], S. Fiedler[1]
[1]EMBL c/o DESY, Hamburg, Germany
[2]University Polytechnic Of Marche, Department of Food, Environment and Agricultural Studies , Ancona, Italy

EMBL is operating two beamlines for macromolecular crystallography on PETRA III (DESY, Hamburg). Both beamlines are fully tunable and provide a wide range of beam conditions. High flux X-ray beams with adjustable dimensions between 5 and 200 µm are available in the energy range between 4 and 18 keV. To demonstrate the capability of the beamlines, we will describe and discuss typical experiments such as:

- Structure solution via S-SAD phasing using 4 keV X-rays on P13.
- Structure solution using SAD phasing at 6.5 keV on multiple crystals with linear dimensions < 10 µm.
- Structure solution by molecular replacement from data collected using serial helical scans on micro-crystals presented to the beam at cryogenic or at room temperature in CrystalDirectTM plates.
- Rapid (< 3 min) data collection using a CRL-collimated X-ray beam with a top-hat profile.

As a CrystalDirect Harvester system has been installed at EMBL Hamburg in November 2016, we will present first results with crystals harvested with this system. Applications for beamtime can be made at smis.embl-hamburg.de.

P91

A Peltier-cooled microscope stage for protein crystal post-crystallization treatment

A. Dietl[1], C. Kieser[1], T. R. M. Barends[1]
[1]Max Planck Institute for Medical Research, Biomolecular Mechanisms, Heidelberg, Germany

Low internal order found in the crystals of many proteins and especially large macromolecular assemblies is a major limitation of diffraction resolution and might result in high mosaicity and diffuse scattering. On the one hand, this problem has been approached by various *pre-crystallization* techniques such as the design of new crystallization constructs, heterologous expression of homologous proteins, additive screening, seeding or limited proteolysis. On the other hand, *post-crystallization* treatments such as crystal dehydration, annealing or crosslinking can be used to improve already existing crystals. Here we describe the construction and application of a temperature-controlled microscope stage that was used in the post-crystallization treatment for crystals of hydrazine synthase[1] from the anaerobic ammonium-oxidizing (anammox) bacterium *Kuenenia stuttgartiensis*. The crystals of this multienzyme complex initially showed severe diffuse scattering and high mosaicity when conventional cryoprotectants were used. Successful cryoprotection accompanied by improved diffraction resolution and lower mosaicity could be achieved by soaking the crystals in highly concentrated betaine solutions. Since crystals otherwise dissolved rapidly it turned out to be beneficial to carry out this process at low ambient temperature. The application of a custom-built Peltier-cooled microscope stage allowed us to greatly optimize this procedure. Protein crystals could be cooled slowly using defined temperature gradients while being immersed in cryoprotectants to temperatures below 273 K. Moreover, this method enabled prolonged crystal soaking in ligand solutions and the preparation of a xenon derivative.

References

[1] A. Dietl et al.; Nature **(2015)**, 527, 394

New Crystal Structures and Structure Systematics

P92
Structure and crystal chemistry of calcium salts of substituted acetic acids

R. Kaden[1], H. Pöllmann[1]
[1]Martin Luther University Halle/Wittenberg, Mineralogy/Geochemistry, Halle, Germany

Acetic acid and its derivatives are used in different fields of foods, pharmacy, cosmetics, dyes, pesticides and microporous polymers and in building materials chemistry for the setting control of cements. The number and type of different substitutes (eq. 1) affect different acid strengths and interactions with cations.

$H_{3-a}X_aCCOOH$ with a = 1, 2, 3; X = F, Cl, Br, I, CH_3, CH_2, OH eq. 1

On example of the calcium compounds crystallised from aqueous solution the influence of substitutes X on bonding and coordination of Ca^{2+}, involvement of H_2O and additional interactions of X with Ca^{2+} and H_2O are systematically examined. Besides filling the gap of lacking structural data with own results (sc-XRD, PXRD) the thermal stability of the materials is analysed (TG/DSC, N_2-flow) in relation to crystal chemistry.

Number and type of substitutes lead to different hydrate stages, unit cell dimensions and interlayer distances of the crystallised layered structures. The configuration of the Ca^{2+}-coordination polyhedra dominantly depends on the number ***n*** of coordinative bonded H_2O molecules and is similar for ***n*** = 1 (c,h,j,k,o), 2 (f,i,m), 3 (d,e,g) and 4 (l,n), Fig. 1. Generally, ***n*** increases with the number of substitutes. The interlayer distances ***dil*** are mainly related to number of substitutes and atomic arrangement of the unit cell content. The 3 groups of substitutes affect:

I) non-polar: only carboxyl group interacts with Ca^{2+} and H_2O

II) polar (halogeno): additional Coulombic forces between the electronegative halogens and Ca^{2+} and H_2O. Sum of bond valences of all Hal···H_2O interactions increases with number and electronegativity of substitutes.

III) polar (α-hydroxyl): -OH form covalent bonds with Ca^{2+} having bond valences similar to that of the carboxyl group.

The thermal stability of the hydrated phases is dominated by the hydrate stage decreasing with ***n***. The anhydrous phases tend to be less stable with the number of substitutes. For the halogenoacetates the found interactions Hal···Ca^{2+} lead to Ca halides as decomposition products.

References:
[1] Watanabé, Matsui (1978) Acta Cryst. **B34**, 2731
[2] Klop et al. (1984) Acta Cryst. **C40**, 51
[3] Le Page et al. (1978) J. Polymer Sci. **16**, 2265
[4] Karipides, Peiffer (1988) Inorg. Chem. **27**, 3255
[5] Kaden, Poellmann (2015) Proc. 37th Int. Conf. Cem. Micr., 27
[6] Singh et al. (2014) Inorg. Chim. Acta **419**, 13

Figure 1

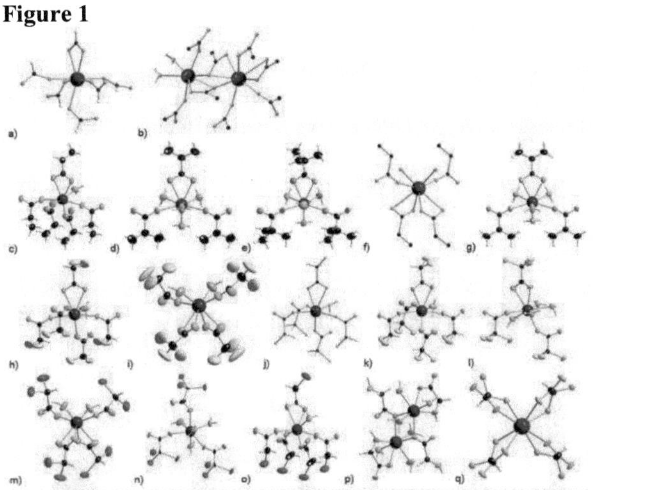

Fig. 1. Coordination spheres of the Ca^{2+} with the direct bonded ligands and coordinative water molecules of calcium a) formate [1], b) acetate hydrate [2], c) propionate hydrate, d) iso-butyrate hydrate, e) pivalate hydrate, f) acrylate hydrate [3], g) methacrylate hydrate, h) difluoroacetate hydrate, i) trifluoroacetate hydrate, j) monochloroacetate hydrate [4], k) dichloroacetate hydrate [5], l) trichloroacetate hydrate [6,5], m) dibromoacetate hydrate, n) tribromoacetate hydrate, o) monoiodoacetate hydrate, p) glycolate bromide hydrate, q) glyoxylate.

P93
Freehand experiment with rock candy crystals

H. Preuß[1]
[1]Ruhestand, Hameln, Germany

Rock candy crystals are everywhere available in discounters or supermarkets and can be used for an optimal inductive entree to crystallography with a freehand experiment. With nearly 1 cm size the crystals can be kept in the hand and examined by eyes. With little assistance of the teacher one can learn stepwise fundamental properties of crystals:

It is not possible to find two equal pieces; the crystals differ in size and details of their shape and also in colour (white or brown) because they were not made with tools but have grown individually in a hot solution of sugar during evaporation of water and slow cooling.

One has to find out the common properties: the shape is characterized by plain faces meeting together in strait edges joint in corners. The faces are differently shaped, two of them are parallel rectangles at opposite sides. The faces are arranged in bundles with edges preferably parallel to the sides of the rectangles. One bundle consists of eight faces, the other of six.

For more details one has to put the two pieces together with the rectangle faces. There are four possibilities, only one of them leads to a position in what all faces and edges of one crystal find a parallel corresponding partner at the other piece. This observation directly gives Steno"s law of constant angles just for a special substance. The corresponding parallel faces are not congruent, and the length relations of edges are not equal. Important is the orientation in space. The same situation one finds turning one of the pieces by 180° (360°/**2**) around the axis of the eight faces bundle. This means: the axis of the bundle is a **two**fold rotational axis). The study of the orientation of all corresponding parallel faces leads to inversion as further property of symmetry. The meaning can be demonstrated with hands in parallel position with opposite direction of the fingers. At last on can find a symmetry plane perpendicular to the twofold axis (symbol: 2/m) relating the faces at opposite ends of the bundle axis.

One should note the study with a sketch as orthogonal projection in three directions (Fig. 1).

Lit.: Heinz H. W. Preuss, Freihand-Gruppenversuch mit Kandiskristallen, DD 28 DPG Spring Meeting Bochum 2009

Figure 1

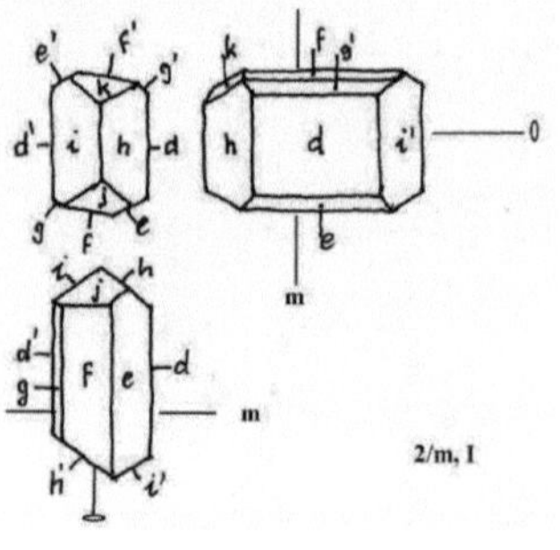

P94
Data Mining Force Field: Further Insights into Crystal Structure Prediction (CSP)

L. Kuleshova[1], D. W. M. Hofmann[1,2]
[1]FlexCryst, Uttenreuth, Germany
[2]CRS4, Bioscience, Pula, Italy

Even if crystal structure prediction of small organic molecules currently become more and more reliable [1], a lot of challenging problems still have to be solved in both "ab initio" and "force field" methods (for example, the impact of external conditions on prediction). The other problems are specific for force field approach, for example, calculations of the salts and charged particles. Here we show the further perspectives of Data Mining approach, concentrating on energy and structural aspects.

In the talk of Dr. Detlef Hofmann it was presented an efficient way to include the external conditions (temperature and pressure) into calculations and its impact on energy accuracy and structural reliability of prediction. The progress has been achieved by tacking into account heat capacity and thermal expansion in optimization of Force Field. Here we will report the further results of Force Field optimization via introducing of new atom types including charged one. As a pilot study we have choose two classes: carboxylic acids and amino-acids, which are widely presented in CCDC and often exist as internal salts. This very often caused problems in correct energy estimations and as consequence - failure in crystal structure prediction. For them in addition to the standard atom types, ammonium group is required, and carboxylic group has to be introduced. From the CCDC we selected all known crystal structures with these groups and used them as training set. The details of the procedure can be found in International Tables C [2].

In the result experimental and predicted energy correlates perfectly, Fig 1, and more than 58% of structures have been predicted in ranks 1-20, Fig 2. The results of screening for molecular salts of propranolol with dicarboxylic acids are presented.

References:
[1] Reilly, Anthony M., et al. Acta Cryst. B: 72.4 (2016): 439-459.
[2] Hofmann, D. W. M., and L. N. Kuleshova. International Tables for Crystallography (submitted)

Fig.1. E_pred vs E_exp

Fig.2. Overlay of proline structures

Figure 1

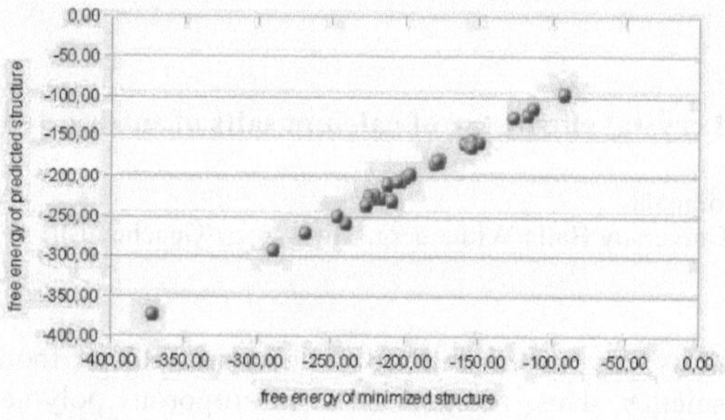

Figure 2

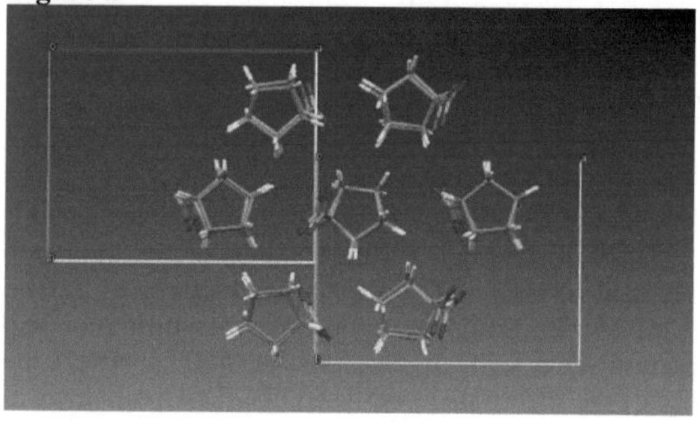

P95
A simple way for manually constructing decorated Penrose tilings

W. Steurer[1]
[1]ETH Zurich, Zurich, Switzerland

The potential is demonstrated of Kurt Bruckner's 'addition algorithm', which is based on the substitution rule for the generation of the Robinson triangle tiling, a variant of the Penrose tiling. The artist Kurt Bruckner developed his straightforward approach intuitively for the creation of quasiperiodic ornaments. This versatile method can be used for the construction of achiral, homochiral and racemic quasiperiodic ornaments, as well as for the generation of decorated two-level (two-color) Penrose tilings (Fig. 1). Thereby, the underlying tiling is always the same kind of Penrose tiling, which is invariant under the action of specific mirror and black/white mirror operations in contrast to unit tiles that are decorated in specific ways. Compared to the underlying classical substitution method the advantage of Kurt Bruckner's approach is its simplicity and versatility for the creation of decorated tilings. Using a vector graphics editor large, and arbitrarily complex quasiperiodic ornaments can be easily generated manually.

Figure 1

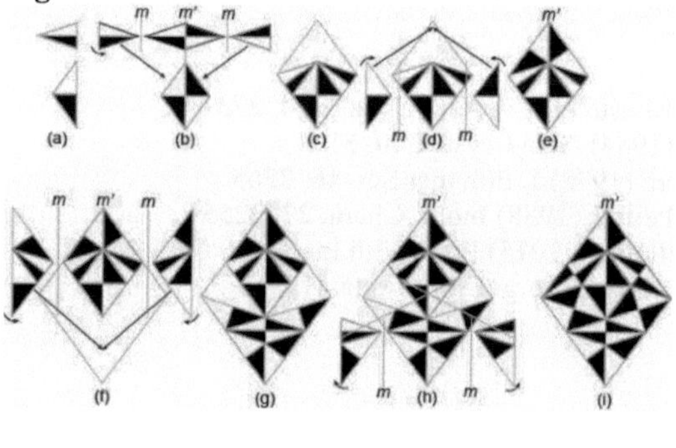

P96

Modelling Disorder with DSR in ShelXle and Olex2

D. Kratzert[1]
[1]Albert-Ludwigs-Universität Freiburg, Anorganische Chemie, Freiburg, Germany

One of the remaining challenges in single-crystal structure refinement is the proper description of disorder in crystal structures. DSR[1] performs semi-automatic modelling of disordered moieties using SHELXL[2]. It contains a database with 120 molecular fragments and their corresponding stereochemical restraints and a fitting procedure to place these fragments on the desired position in the unit cell. The program is also suitable for speeding up model building of well-ordered crystal structures. Writing a special DSR command into the SHELXL .res file of the target structure instructs DSR on where to place and how to orient a molecular fragment from the fragment database in the unit cell.

Recent features to DSR such as the Graphical user interfaces (GUIs) for DSR in ShelXle[3] (figure 1) and Olex2[4] are presented. With the new GUI, the user has full control over every aspect of the model building while keeping the procedure as simple as possible. Also additional checks were added to keep restraints in the databse consistent and to avoid erroneous refinement models or brainless automation.

Figure 1

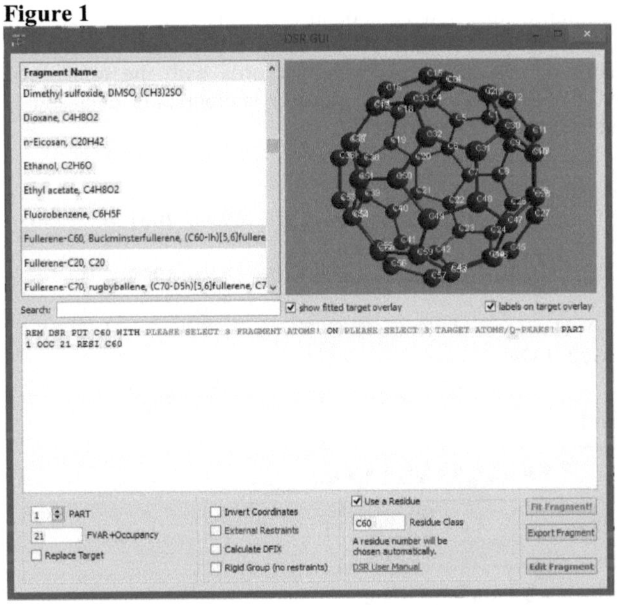

P97

SYNTHESIS AND CHARACTERIZATION OF A $[Li_{0+x}Mg_{2-2x}Al_{1+x}(OH)_6][Cl\cdot mH_2O]$ $(0\leq x\leq 1)$ SOLID SOLUTION

A. Niksch[1]
1Institut für Geowissenschaften und Geographie, Mineralogie/Geochemie, Halle, Germany

Layered double hydroxides consist of alternate positively charged mixed metal hydroxide layers and negative charged interlayer anions. The stoichiometry is normally formulated as $[M^{z+}_{1-x}M^{3+}_{x}(OH)_2]^{p+}[(A^{n-})_{p/n}\cdot mH_2O]$ with z = 2 or 1, M = mono-, bi- and trivalent metallic elements, A = organic or inorganic anion and m = amount of interlayer. In case of $[LiAl_2(OH)_6][x\cdot mH_2O]$ the only possible ratio of Li : Al is 1 : 2 due to the site occupancy of the Atoms **[1]**. The Mg : Al ratio in a $[Mg_yAl(OH)_6][x\cdot mH_2O]$ LDH varies between 2 : 1 and 5 : 1 because Mg^{2+} and Al^{3+} atoms occupy the same position. Due to a similar ionic radius between Li^+ and Mg^{2+} and the chemical similarities, there is a possibility to get a solid solution, containing Al^{3+}, Li^+ and Mg^{2+} in the main layer.

The syntheses were done by mixing solutions of LiCl, $MgCl_2\cdot 6H_2O$ and $AlCl_3\cdot 6H_2O$, adding NaOH until a pH of 9.5 was reached and heating it up to 100 °C, 120 °C,140 °C and 160 °C in an autoclave for 10h. The products were filtered, washed and dried (RH 35%). Starting with $[Mg_2Al(OH)_6][Cl\cdot 0.55H_2O]$ the amount of Li was raised and the amount of Mg was reduced in 10 mole% steps until 100 mole% Li and the pure Li- LDH was reached. XRD investigations showed two different mineral phases in the area of 10 - 80 mole% Li. The synthesis with 90 mole% Li showed a single mineral phase and a different lattice parameter ***a*** (5.10 Å) compared to the pure $[LiAl_2(OH)_6][Cl\cdot 0.51H_2O]$ LDH (5.08 Å). Syntheses between 90 and 100 mole% Li showed a straight increasing lattice parameter ***a*** from 5.08 Å to 5.10 Å like theoretically calculated (Tab. 1). Analysis by ICP-OES displayed, that all Mg was bound to the solid solution. Calculations suggest that the Mg^{2+} ions occupy both structural positions with 5 % on the Li^+ and 5 % on the Al^{3+} position. The amount of interlayer water was investigated by TGA and is ~0.5 mole. Synthesis with temperatures of 100 °C, 120 °C and 140 °C produced nearly similar solid solutions while using temperatures exceeding 140 °C resulted in a formation of AlO(OH). The solid solution with the highest Mg content and the chemical formula $[Li_{0,9}Mg_{0,2}Al_{1,9}(OH)_6]Cl\cdot 0,5H_2O$ was produced at 120 °C, 10 h synthesis time, pH 9.5 and a s/w ratio of 1/15.

References:
[1] Williams, G.R., et. al.: New insights into the intercalation chemistry of $Al(OH)_3$, Dalton Trans, 40, 2011

Figure 1

Tab. 1: theoretical and real measured/fitted lattice paramete ***a*** and ***c*** for the solid solutions with X = 0.9 – 0.98 and $[LiAl_2(OH)_6][Cl0.51H_2O]$ at X = 1 (120°C/10h/pH 9.5)

X	theoretical lattice parameter a [Å]	measured lattice parameter a [Å]	measured lattice parameter c [Å]	measured cell volume [Å]³	space group
0.9	5.0962	5.1004(4)	15.3512(1)	345.84(4)	$P6_3/m$
0.92	5.0922	5.0978(3)	15.3602(3)	345.69(4)	$P6_3/m$
0.94	5.0881	5.0975(1)	15.3563(7)	345.56(6)	$P6_3/m$
0.96	5.0840	5.0906(9)	15.3497(1)	344.48(3)	$P6_3/m$
0.98	5.0799	5.0886(8)	15.3550(3)	344.33(1)	$P6_3/m$
1	5.0759	5.0783(8)	15.3483(4)	342.78(9)	$P6_3/m$

P98

Maximum vacancy ordering in mullite

P. Klar[1], N. de la Pinta[1], G. A. Lopez[2], I. Etxebarria[2], T. Breczewski[2], G. Madariaga[1]
[1]Universidad del País Vasco UPV/EHU, Departamento de Física de la Materia Condensada, Bilbao, Spain
[2]Universidad del País Vasco UPV/EHU, Departamento de Física Aplicada II, Bilbao, Spain

Since the first observation of (diffuse) satellite reflections in the reciprocal space of mullite $Al_{4+2x}Si_{2-2x}O_{10-x}(vacancy)_x$, different models for the vacancy distribution were presented.

The semiquantitative model by Angel *et al.* (1991) modulated harmonically the oxygen occupancy assuming a "maximally ordered" structure. More recently a disordered superspace model (Birkenstock *et al.*, 2015) used again harmonically modulated site occupation parameters with a very small amplitude so that the vacancy distribution was almost uniform (disordered).

X-ray diffraction of a commercial 2/1 mullite sample with sharp satellite reflections yielded an ordered superspace model (Klar *et al.*, 2017) that represents a well ordered block structure (Fig. 1). However, this ordering scheme accounts for only 10 vol% and long-range ordered domains coexist within a mainly disordered sample. The refinement ($wR_{obs,main}$ = 0.034, $wR_{obs,satellites}$ = 0.075) used different scaling factors for main and satellite reflections, respectively. A similar block pattern was also observed in small domains with HRTEM (Ylä-Jääski & Nissen, 1983) in analogy to the X-ray study.

The different models for mullite thus can be unified under the observation that order and disorder are simultaneously present, though the disorder generally seems to dominate. Hence in the model of Birkenstock *et al.* occupancy maxima and minima coincide with *vacancy-free blocks* and *vacancy blocks* of the ordered superspace model, respectively, but the modulation amplitude is much smaller because a superposition of an ordered and disordered polymorph is modeled. Assuming maximum order gives therefore a similar result but with a higher modulation amplitude.

Financial support by FEDER (No. MAT2015-66441-P) and the Basque Government (project No. IT-779-13) is highly appreciated.

References:
[1] P. Klar, N. de la Pinta, G. López, I. Etxebarria, T. Breczewski, G. Madariaga (2017), *Acta Cryst. B, submitted*
[2] J. Ylä-Jääski & H.-U. Nissen (1983), *Phys. Chem. Miner.* 10, 47–54
[3] J. Birkenstock, V. Petricek, B. Pedersen, H. Schneider, R. Fischer (2015), *Acta Cryst. B*, *71*, 358–368.
[4] R. Angel & C. Prewitt (1987), *Acta Cryst. B*, 43, 116–126

Fig. 1: Block model of ordered mullite derived from the ordered superspace model. Vacancies (squares) group together in *vacancy blocks* (linked squares) that alternate with *vacancy-free blocks* along **a** and **c**.

Figure 1

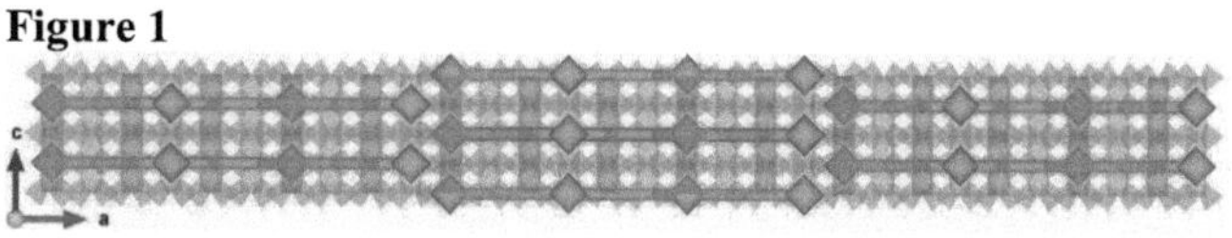

P99

Na_2HfSe_3: The First Ternary Sodium Hafnium Selenide

F. Lissner[1], B. Hack[1], T. Schleid[1]
[1]Institut für Anorganische Chemie, Universität Stuttgart, Stuttgart, Germany

Keywords: Hafnium, Sodium, Selenides, Crystal Structure

With the intention to prepare $NaHf_2N_2SeCl$ in analogy to $NaZr_2N_2SCl$ [1], orange-red, plate-like single crystals of Na_2HfSe_3 (monoclinic, *C2/m*; *a* = 689.47(6), *b* = 1195.18(9), *c* = 725.42(6) pm, *β* = 108.654(3)°; *Z* = 4) were obtained upon oxidation of hafnium metal with selenium and sodium azide (NaN_3) in the presence of fluxing NaCl (molar ratio: 7:6:2:3) in evacuated silica tubes at 850 °C within seven days. The crystal structure is isotypic with the analogous sodium zirconium sulfide Na_2ZrS_3 [2] (as well as Na_2ZrSe_3 [3]) and can be described as a stuffed $AlCl_3$-type arrangement with all cations (Na^+ and Hf^{4+}) in octahedral coordination of the Se^{2-} anions, which build up a cubic closest packed anionic partial lattice (Fig. 1). The metal-selenium distances range from 288 to 307 pm for all three crystallographically different Na^+ cations, and from 269 to 271 pm for Hf^{4+}.

On the Wyckoff position 2*a*, a disorder of $(Na1)^+$ and Hf^{4+} cations occurs within the $[HfSe_{6/2}]^{2-}$ layers (Fig. 1 and 2), so that the real composition of this new compound is calculated as $Na_{2-x}Hf_{1+x}Se_3$ with x = 0.098(2) correlating with an oxidation state only +3,732 for hafnium instead of +4, which accords with the orange-red colour. CSD-430952 contains more crystallographic data and is available at the FIZ Karlsruhe.

References:
[1] F. Lissner, Th. Schleid, *Z. Anorg. Allg. Chem.* **2001**, *627*, 2307–2309.
[2] F. Lissner, Th. Schleid, *Z. Anorg. Allg. Chem.* **1999**, *625*, 195–196.
[3] N. A. Harringer, A. Kolb, K. O. Klepp, *Z. Naturforsch.* **2001**, *56b*, 1322–1324.

Figure 1

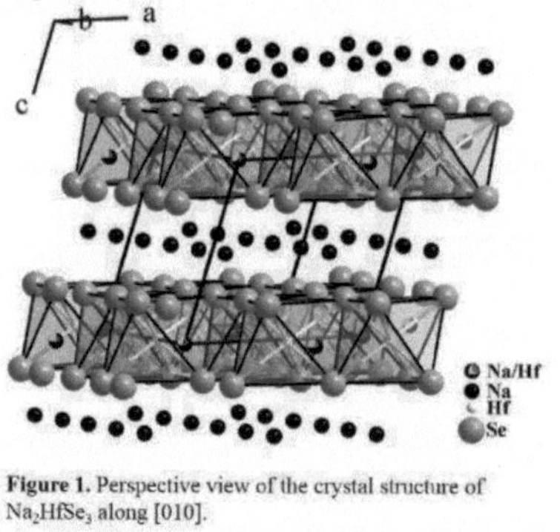

Figure 1. Perspective view of the crystal structure of Na_2HfSe_3 along [010].

Figure 2

Atom	Site	*x/a*	*y/b*	*z/c*	U_{eq} /pm²
Na1[a)]	2*a*	0	0	0	134(9)
Na2	2*d*	0	½	½	203(16)
Na3	4*h*	0	0.1601(4)	½	203(12)
Hf	4*g*	0	0.33183(4)	0	139(2)
Se1	4*i*	0.28161(18)	0	0.77873(16)	118(3)
Se2	8*j*	0.25315(13)	0.17829(7)	0.22245(12)	132(2)

a) occupation ratio: $(Na1)^+$ / Hf^{4+} = 0.804(4) / 0.196(4).

Figure 2. Fractional atomic coordinates and equivalent isotropic displacement parameters for Na_2HfSe_3.

P100

Single Crystals of the Trigonal High-Temperature Modification of $Cs_3Bi_2Cl_9$

B. Schulz[1], T. Schleid[1]
[1]Inorganic Chemistry, Stuttgart, Germany

Compounds with the empirical formula $Cs_3M_2Cl_9$ (M = As, Sb, Bi or Fe) are well known in literature [1–4] and the crystal structure of the high-temperature modification of $Cs_3Bi_2Cl_9$ has been first described by *Meyer* and *Schönemund* [4] on the basis of X-ray powder diffraction data with the space group *P*321 (no. 150). Now it was possible to obtain single crystals of this compound as by-product from reactions of strontium dichloride, bismuth trichloride, bismuth sesquioxide and cesium chloride as flux. The reactions were carried out in silica ampoules with a molar ratio of 4:1:1:8 for the binary chlorides and tempering for periods of 3 days at 850 °C.

The crystal structure of the trigonal high-temperature form of $Cs_3Bi_2Cl_9$ (CSD-432491) was determined by single-crystal X-ray diffraction at room temperature as crystallizing in the space group *P m*1 (no. 164) with the lattice parameters a = 767.12(5) pm and c = 945.83(6) pm for Z = 1. The crystal structure comprises two crystallographically distinguishable Cs^+ cations, which both are cuboctahedrally coordinated by twelve chloride anions (d(Cs–Cl) = 376–392 pm). Owing to their stereochemically active $6s^{2-x}p^x$ lone pairs, the Bi^{3+} cations reside in distorted $[BiCl_6]^{3-}$ octahedra (d(Bi–Cl) = 256 pm and 284 pm, 3 × each), which build up double layers with the formula 2D-$\{[Bi(Cl1)_{3/2}(Cl2)_{3/1}]_2\}^{3-}$ spreading out parallel to the (001) plane. Moreover, trigonal $Cs_3Bi_2Cl_9$ exhibits a cubic closest packed ABC structure for its 3D-$\{[CsCl_{12/4}]_3\}^{6-}$ part. With not much different Cs^+–Cl^- (d = 369–398 pm) or Bi^{3+}–Cl^- (d = 260–283 pm) distances and similar coordination spheres ($[CsCl_{12}]^{11-}$ and $[BiCl_{3+3}]^{3-}$), the orthorhombic room-temperature phase of $Cs_3Bi_2Cl_9$ crystalizes in the space group *Pnma* (no. 62) with the lattice parameters a = 1868.4(4) pm, b = 764.4(2) pm and c = 1322.7(3) pm [5], the stacking sequence ABACBC and a by 2% denser arrangement for Z = 4.

References:
[1] J. L. Hoard, L. Goldstein, *J. Chem. Phys.* **1935**, *3*, 117–122.
[2] K. Kihara, T. Sudo, *Z. Kristallogr.* **1971**, *134*, 155–158.
[3] M. T. Kovsarnechan, J. Roziere, *J. Inorg. Nucl. Chem.* **1978**, *40*, 2009–2011.
[4] G. Meyer, A. Schönemund, *Z. Anorg. Allg. Chem.* **1980**, *468*, 185–192.
[5] K. Kihara, T. Sudo, *Acta Crystallogr.* **1974**, *B30*, 1088–1093.
[6] R. Fischer, E. Tillmanns, *Acta Crystallogr.* **1988**, *C44*, 775–776.

Figure 1

Fractional atomic coordinates for trigonal $Cs_3Bi_2Cl_9$.

Atom	*Wyckoff* site	*x/a*	*y/b*	*z/c*	U_{eq}/pm²
Cs1	1*a*	0	0	0	657(5)
Cs2	2*d*	1/3	2/3	0.33648(14)	570(3)
Bi	2*d*	1/3	2/3	0.81252(5)	311(2)
Cl1	3*e*	1/2	0	0	751(12)
Cl2	6*i*	0.17212(7)	−x/a	0.6639(3)	604(7)

$U_{eq} = {}^1/_3\,[U_{33} + {}^4/_3\,(U_{11} + U_{22} - U_{12})]$ [6].

Figure 2

Three- or two-dimensional frameworks of trigonal $Cs_3Bi_2Cl_9$ (*top*: cuboctahedrally coordinated Cs^+ cations, *bottom*: octahedrally coordinated Bi^{3+} cations).

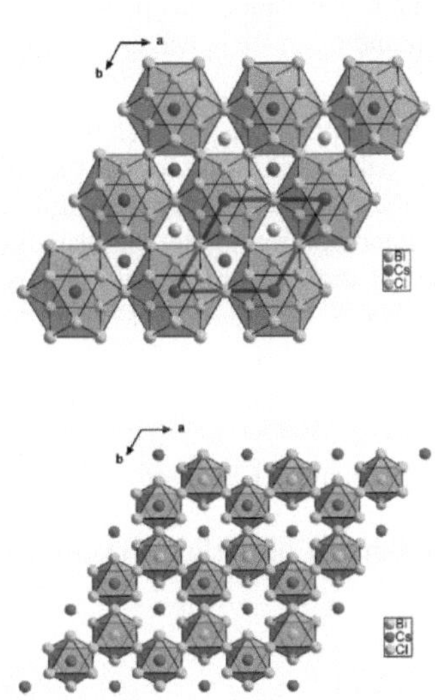

P101

Crystal structure and properties of double molybdate $Pr_2Zr_3(MoO_4)_9$

I. Svoboda[1], Y. Tushinova[2], A. Sarapulova[3], B. Bazarov[2], O. Chimitova[2], J. Bazarova[2], H. Ehrenberg[3]
[1]Technische Universität Darmstadt, Darmstadt, Germany
[2]Baikal Institute of Nature Management, Siberian Branch, Russian Academy of Sciences, Ulan-Ude, Russian Federation
[3]Karlsruhe Institute of Technology, Institute for Applied Materials – Energy Storage Systems (IAM-ESS), Eggenstein-Leopoldshafen, Germany

Double praseodymium-zirconium molybdate of $Pr_2Zr_3(MoO_4)_9$ composition are grown by flux method under spontaneous nucleation conditions from the mixture $Rb_2Mo_2O_7$ – $RbPrZr_{0.5}(MoO_4)_3$. Crystal structure was determined using single crystal diffractometer (Mo*K*α radiation, STOE, STADI 4). Parameters of the trigonal unit cell are: a = b = 9.8304(3)Å, c = 58.661(2) Å, V = 4909.33(30) Å3, Z = 6, space group *R-3c*. The crystal structure is composed of three types of polyhedra: MoO_4 tetrahedra, ZrO_6 octahedra, and nine-vertex PrO_9 (Fig. 1). All three types of polyhedra are bonded among themselves by common oxygen vertices of bridging MoO_4 tetrahedra forming an openwork three dimensional structure. The obtained composition belongs to analogs of the natural mineral kosnarite.

Fig. 1 Three-dimensional carcass of $Pr_2Zr_3(MoO_4)_9$ structure, projections on (010) and (001)

In comparison to $Pr_2Hf_3(MoO_4)_9$ structure analogues obtained earlier [1], the average interatomic distances are increasing parallel with increasing ionic radius of tetravalent element. Nd substitution decreases all distances in the structure $Nd_2Zr_3(MoO_4)_9$ [2], except Mo tetrahedron bonds which become much longer. From the data presented in the Table we can summarize that in case Hf→Zr substitution unit cell volume increasing leads to high values of **a** and **b** parameters. Contrarily, decreasing unit cell volume in case Pr→Nd substitution is due to **c** parameter decreasing.

Table 1. Crystallographic data of obtained single crystals

Thus, we can observe influence of trivalent element substitution and tetravalent element substitution on the crystal structure metric. The presented row of compound can possess interesting physical properties due to presence of Rare earth elements. Wide variations of substitutions in these compounds determine the possibility of their use as matrices to fix nuclear and other toxic waste.

During investigation of the electro-physical characteristics of $Pr_2Zr_3(MoO_4)_9$ (on the ceramic pellets) it was found that the temperature dependence of the dielectric characteristics did not form pronounced peaks in heating and cooling mode, the resistivity did not depend on the frequency of the measuring field and was reduced at high temperatures 550-900 K from $\rho = 10^8\ \Omega \times$ cm to $\rho = 10^5\ \Omega \times$ cm.

References
[1] B. G. Bazarov et al., Journal of Structural Chemistry. Vol. 50, No. 3, pp. 566-569, 2009
[2] R. F. Klevtsova et al., Russ. J. Struct. Chem. (Engl. Transl.), 2000, 41, 280

Figure 1

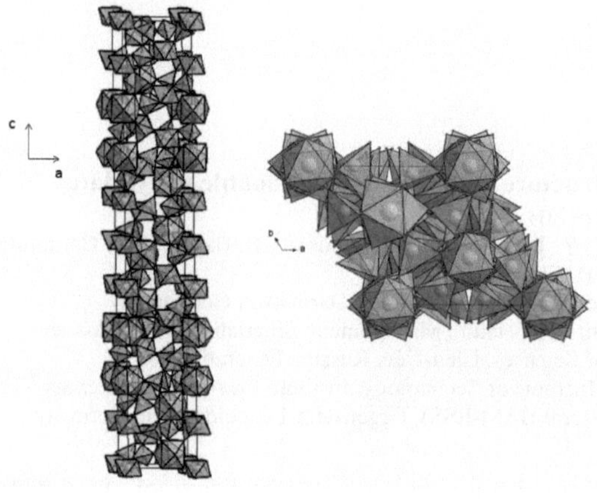

Figure 2

Compounds	Unit cell parameters	Ionic radius, Å	Average interatomic distances
$Pr_2Hf_3(MoO_4)_9$ [1]	a=b= 9.8001(1) c= 58.7095(8) V = 4883.15(10) Å^3	Hf 0.71 Pr 1.179	Mo-O 1.7555, 1.763 Hf-O 2.069, 2.070 Pr-O 2.492
$Nd_2Zr_3(MoO_4)_9$ [2]	a=b= 9.804(1) c= 58.467(12) V= 4866.85(141) Å^3	Zr 0.72 Nd 1.163	Mo-O 1.759, 1.768 Zr-O 2.074, 2.075 Nd-O 2.479
$Pr_2Zr_3(MoO_4)_9$	a=b=9.8304(3) c=58.661(2) V = 4909.33(30) Å^3	Zr 0.72 Pr 1.179	Mo-O 1.757, 1.764 Zr-O 2.076, 2.078 Pr-O 2.497

P102

SYNTHESIS AND CHARACTERIZATION OF NEW MATERIAL, TYPE SILLENITE $BI_{12}TIO_{20}$

H. AIT OULAHYANE[1], A. CHAGRAOUI[2], L. LOUBBIDI[1], L. BOURJA1[1], S. VILLAIN[2], A. MOUSSAOUI[1], O. AIT SIDI AHMED[1], A. TAIRI[1]
[1]Hassane II University, Chemistry, Casablanca, Morocco
[2]toulon var university, Laboratory IM2NP, toulon, France

The sillenite structure has a number of interesting properties, such as piezoelectric, electro-optical, elasto-optical and photoconductive properties [1, 2]. Due to these characteristics, the sillenite phase is useful for many advanced applications such as image processing applications, coherent light amplification and many optical techniques [3, 4].

Our work focuses on the synthesis and characterization of new materials based on bismuth oxide. Indeed, the investigation of the ternary Bi_2O_3-MgO-TiO_2 system allowed us to stabilize isotype compositions in the sillenite phase $Bi_{12}TiO_{20}$, of formulas $Bi_{0.9}Ti_{0.1}O_{1.55}$, $Bi_{0.8}Mg_{0.1}Ti_{0.1}O_{1.65}$ and $Bi_{0,9}Mg_{0.05}Ti_{0.05}O_{1.5}$. These compositions were prepared solidly at 800 °C. and characterized by X-ray powder diffraction.

The structural study was carried out by Rietveld software. The synthesized compositions crystallize in the space group cubic crystal system I23 with a mesh parameter a = 10.17 Å. The bi^{3+} bismuth atoms occupy the site 24f, as for the magnesium Mg^{2+} and titanium Ti^{4+} atoms occupy the same site 2a. On the other hand, the oxygen atoms O^{2-} (1) are located in the site 24f while the oxygen atoms O^{2-} (2) and O^{2-} (3) occupy the same site 8c.

References:
[1] I.F. Vasconcelos, M.A. Pimenta, A.S.B. Sombra, Optical properties of Bi12SiO20 (BSO) and Bi12TiO20 (BTO) obtained by mechanical alloying, J. Mater. Sci. 36 (2001) 587.
[2] I.V. Kityk, A. Majchrowski, J. Ebothe, B. Sahraoui, Nonlinear optical effects in Bi12TiO20 nanocrystallites embedded within a photopolymer matrix, Opt. Commun. 236 (2004) 123.
[3] J.F.Carvalho, R.W.Franco, C.Magon, L.A.O.Nunes, F. Pellegrini, A.Hernandes, Vanadium Characterization in BTO: V Sillenite Crystals, Mater.Res.2 (1999).
[4] J.F.Carvalho, R.W.Franco, C.Magon, L.A.Nunes, A.Hernandes, Optical and magnetic characterization of pure and vanadium doped Bi12TiO20 sillenite crystals, Opt. Mater. 13 (1999) 333.

P103

Pillarplexes – Organometallic Host-Compounds

P. Altmann[1], A. Pöthig[1]
[1]Technische Universität München, Catalysis Research Center, Garching bei München, Germany

Recently, we presented a novel approach for the design of supramolecular host systems employing organometallic rather than purely organic compounds.[1] These so-called pillarplexes consist of two macrocyclic calix[4]imidazolylidene[2]pyrazolate ligands and eight metal centers (Ag, Au) coordinated between them. This innovation is accompanied by numerous advantages compared to the current benchmark systems, such as intrinsic photoluminescence of the host, a higher shape selectivity and affinity as well as an easily tunable solubility via simple anion exchange reactions.

Figure 1: Left: Solid-state molecular structure of the cation of [Au8(LMe)2](PF6)4 shown in capped sticks with one molecule 1,8-diaminooctane cocrystallized within the cavity. Right: Top view of the host–guest complex in the space filling model (N-bound H atoms were not refined).

Single-crystal X-ray diffraction was the key analytic method, since the elucidation of the molecular structure initially revealed the potential application as a supramolecular host. Owing to the exact determination of the molecular and void dimensions, we additionally were able to explain a multitude of properties of this novel compound class:

1) The structure exhibits a pillar-shaped constitution with a cavity that is of perfect size to host long, linear molecules, but too small for six-membered aromatics.

2) The co-crystallized guest molecule (1,8-diaminooctane) in the void proves the ability of the complex to act as host-compound.

3) The structure reveals several very short intramolecular Au–Au contacts of around 3 Å within the complex responsible for the photoluminescent behavior of the pillarplex and therefore applicable for a quantification of host-guest interactions via luminescence quenching.

4) The fact that the uptake of the linear guest was determined to be solely enthalpy driven whereas the entropy of the process is negative could be explained by the required all-anti conformation of the diaminooctane molecule as was found in the molecular structure.

Based on this structural data, we were able to test the pillarplexes successfully in supramolecular applications. Based on this stepstone we envision a multitude of further promising studies towards mechanically interlocked compounds and molecular machines.

References:
[1] P. J. Altmann, A. Pöthig, *J. Am. Chem. Soc.* **2016**, *138*, 13171–13174.

Figure 1

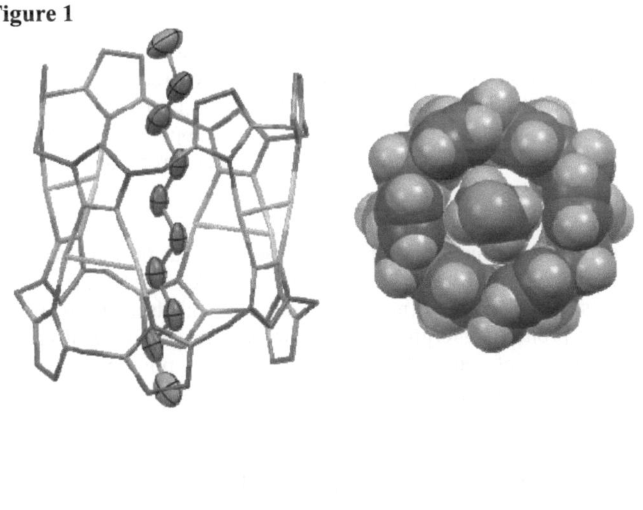

P104

Refuting the existence of bicarbonates of higher valent metal ions and crystal structures of new carbonates

C. Rincke[1], S. Bette[1,2], H. Schmidt[1], R. E. Dinnebier[2], W. Voigt[1]
[1]TU Bergakademie Freiberg, Institute of Inorganic Chemistry, Freiberg, Germany
[2]Max-Planck-Institute for Solid State Research, Stuttgart, Germany

Introduction

Within the scope of the CO_2-research the formation of bicarbonates is of particular interest. While bicarbonates of the alkali metal ions (Na – Cs) are known[1], there are only a few hints for the existence of solid bicarbonates of higher valent metal ions.

Objectives

$Ni(HCO_3)_2$ was described by Isaacs[2], but the thermal analysis of this phase gave rise to doubts, that it was really $Ni(HCO_3)_2$. Furthermore, the crystal structures of two bicarbonates of Ho^{3+} and Gd^{3+} are published.[3,4]

According to Ponizovskii[5] $Mg(HCO_3)_2$ can be synthesized at 0°C and 18 bar. In subsequent literature, there are no hints of the existence of a solid $Mg(HCO_3)_2$ but of some double salts of Mg^{2+} containing HCO_3^- anions, like $AMgH(CO_3)_2{\cdot}4H_2O$ (A = K, Rb).[6] The cesium analogue, $CsMgH(CO_3)_2{\cdot}4H_2O$, is not known, yet.

Methods

We synthesized the phase, described by Isaacs as $Ni(HCO_3)_2$, and determined the structure from XRPD patterns by means of global optimization methods. Aside from that, we analysed the synthesis routes of published structures of $Ho(HCO_3)_3{\cdot}6H_2O$ and $Gd(HCO_3)_3{\cdot}5H_2O$.

Approaches were made to isolate $Mg(HCO_3)_2$ at 0°C and CO_2-pressures up to 56 bar and to synthesize $CsMgH(CO_3)_2{\cdot}4H_2O$. The structure of the latter phase was determined from X-ray SCD data.

Results

The phase described as $Ni(HCO_3)_2$ in literature was revealed as $Ni_{12}(CO_3)_8(OH)_8{\cdot}(5\text{-}7)H_2O$ with a cubic lattice, containing zeolite-like water.[7] The syntheses of the rare earth bicarbonates[8] and $Mg(HCO_3)_2$ failed. Instead of the expected double salt $CsMgH(CO_3)_2{\cdot}4H_2O$ structure analysis of the product leads to $Cs_2Mg_4(CO_3)_5{\cdot}10H_2O$.

Conclusion

We refuted the existence of bicarbonates of higher valent metal ions and carried out the determination of structures of new carbonates.

References:
[1] V. Cirpus, *Diss.*, Köln **1997.**
[2] T. Isaacs, *Mineral. Mag.* **1963**, *33*, 663–678.
[3] N. G. Furmanova, L. V. Soboleva, L. M. Belyaev, *Kristallograf.* **1981**, *26*, 312–315
[4] W. J. Rohrbaugh, R. A. Jacobson, *Inorg. Chem.* **1974**, *13*, 2535–2539.
[5] A. M. Ponizovskii, N. M. Vladimirova, S. N. Stavrov, *Vopr. Kompleksn. Pererabotki Rassolov Morsk. Tipa i Polucheniya Rapn. Stroit. Materialov (Simferopol: Krymizdat) Sb.* **1963**, 133–136.
[6] G. Gloß, *Diss.*, Berlin **1937**.
[7] C. Rincke, S. Bette, R. E. Dinnebier, W. Voigt, *Eur. J. Inorg. Chem.* **2015**, *36*, 5913–5920.
[8] C. Rincke, H. Schmidt, W. Voigt, *Z. Anorg. Allg. Chem.*, submitted.

P105
New ruthenates and iridates with garnet-like crystal structures.

H. Poddig[1], J. Hunger[1], S. Kamusella[2], T. Doert[1]
[1]Technical University, Inorganic Chemistry II, Dresden, Germany
[2]Technical University, Solid State Physics, Dresden, Germany

Introduction

Compounds of the second- and third-row late transition metals are currently under investigation due to their strong spin-orbit coupling and the resulting physical properties [1]. With the approach of hydrothermal synthesis new iridates and ruthenates with interesting properties could be synthesized [2, 3].

Objectives

Attempts to synthesize new ruthenates and iridates in high oxidation states require a strongly oxidizing environment. We therefore used alkaline H_2O_2 solutions as reaction media for the oxidation of *M*(III) starting materials under hydrothermal conditions.

Materials & methods

$RuCl_3$, $Ir(OAc)_3$, $Fe(NO_3)_3$, $Ca(NO_3)_2$, H_2O_2 and NaOH were used as starting materials for hydrothermal syntheses. The products were characterized by single-crystal- and powder-X-ray diffraction, ^{57}Fe-Mössbauer spectroscopy, FT-IR spectroscopy, and quantitative EDS.

Results

Starting from $Ca(NO_3)_2$, $Fe(NO_3)_3$ and $Ir(OAc)_3$ or $RuCl_3$, black powders were obtained as products. The crystallites were large enough to determine the crystal structures of the compounds with the general formula $Ca_3Fe_{3-x-y+z}M_{2-z}O_{12-u-v-w}(OH)_u(H_2O)_v$ (*M* = Ru, Ir) by single crystal diffraction (Fig. 1). The Ru compound adopts the garnet structure in space group $Ia\bar{3}d$ (No. 230) with a lattice parameter of a = 12.489(1) Å and Z = 8. The octahedral site (*Wyckoff* sites 16*a*) is mixed occupied by Ru (52%) and Fe (48%), whereas the tetrahedral site (24*d*) is partially occupied by Fe (56%). The Ir compounds crystallizes in space group $I\bar{4}3d$ (No. 220), a *t*2 subgroup of $Ia\bar{3}d$, with a = 12.540(1) Å; the octahedral site is occupied by Ir (69%) and Fe (31%). The symmetry reduction splits the 24*d* site of the garnet in two non-equivalent tetrahedral sites, 12*a* and 12*b*, which are both partially occupied by Fe (72% and 18%, respectively), in the Ir compound.

Conclusion

New ruthenates and iridates with garnet-like crystal structures were synthesized by hydrothermal synthesis in strongly oxidizing environments.

References:

[1] W. Witczak-Krempa et al., *Annu. Rev. Condens. Matter Phys.* **2014**, *5*, 57–82.
[2] K. Sardar et al., *Chemical Science* **2011**, *2*, 1573.
[3] C. I. Hiley et al., *Angew. Chem. Int. Ed.* **2014**, *53*, 4423–4427.

Figure 1: Octahedral site of Ir and Fe surrounded by the two different tetrahedral sites of Fe.

Figure 1

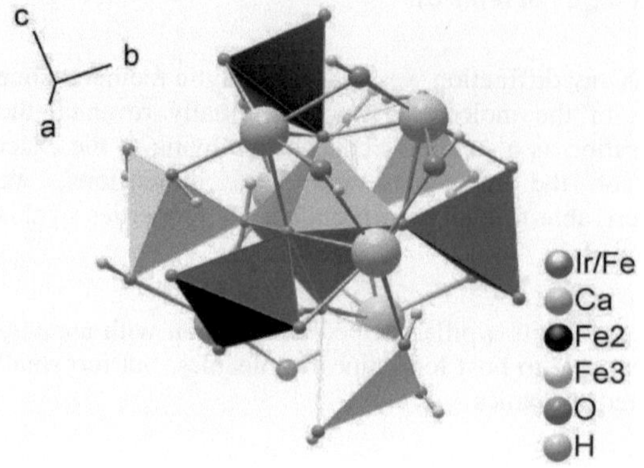

P106
Bi$_x$TeI Family: Modular Design of Topological Insulators

A. Zeugner[1], M. Kaiser[1], T. Doert[1], M. Ruck[1], A. Isaeva[1]
[1]TU Dresden, Inorganic Chemistry II, Dresden, Germany

Introduction

Topological Insulators (TIs) remain an extensively investigated field in solid state chemistry. Regarding their potential use in spintronics or quantum computing, especially 2D-TIs are of great interest. Therefore, wide studies on layered structures with extended spin orbit coupling are performed.

Objectives

Layered structures with topological properties are found in the Bi-Te-I system. The compound BiTeI consists of $^2_\infty[BiTe_{3/3}I_{3/3}]$ triple layers and shows giant Rashba spin-splitting.[1] Bi_2TeI can be described as stacking of the two triple layers with an additional $[Bi_2]$ layer. Bi_2TeI is both a 2D-TI and a so-called topological crystalline insulator.[2,3] Recently, we discovered the consequential compound Bi_3TeI with an alternating stacking of the triple layers with $[Bi_2]$ layers.[4]

Materials & Methods

We describe the crystallographic characterization of Bi_3TeI by X-ray diffraction on single crystals and powdered samples. The elucidation of inversion twinning was taken into account in the course of structure refinement and was supported by electron diffraction and high-resolution transmission electron microscopy.

Results

Structure determination on Bi_3TeI revealed a further structure comprising $^2_\infty[BiTe_{3/3}I_{3/3}]$ triple layers. Bi_3TeI crystallizes in the polar space group *R*3*m* (no. 160). Thus, the real structure of Bi_3TeI is very prone to intergrowth of polar domains. In the real crystal correlation of atomic positions occurs pointing towards polysynthetic twinning.

Conclusion

Obviously, different stacking motifs of the triple layers with $[Bi_2]$ layers lead to a family of structures $[Bi_2]_m[BiTe_{3/3}I_{3/3}]_n$ (Fig.) similar to $[Bi_2]_m[Bi_2Te_3]_n$. Therefore, the modular design of stacking sequences may result in further members of this family and allow for the optimization of properties.

References:

[1] K. Ishizaka, M. S. Bahramy, *et al.*, *Nat. Mater.* **2011**, *10*, 521–526.

[2] P. Tang, B. Yan, W. Cao, S.-C. Wu, C. Felser, W. Duan, *Phys. Rev. B* **2014**, *89*, 041409.
[3] I. P. Rusinov, T. V. Menshchikova, A. Isaeva, S. V. Eremeev, Y. M. Koroteev, M. G. Vergniory, P. M. Echenique, E. V. Chulkov, *Sci. Rep.* **2016**, *6*, 20734.
[4] A. Zeugner, M. Kaiser, P. Schmidt, T.V. Menshchikova, I.P. Rusinov, A.V. Markelov, W. Van den Broek, E.V. Chulkov, T. Doert, M. Ruck, A. Isaeva, *Chem. Mater.* **2017**, *accepted.*

Fig.: Bi_xTeI family.

Figure 1

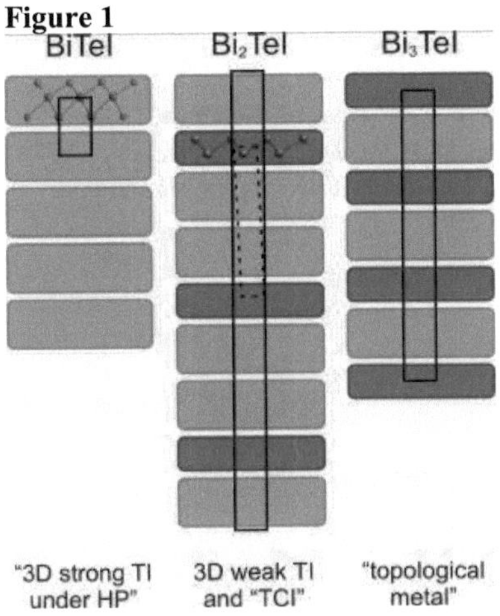

P107

Synthesis and crystal structures of the new ternary borides $Fe_3Al_2B_2$ and $Ru_9Al_3B_8$

S. Hirt[1], F. Hilfinger[1], H. Hillebrecht[1]
[1]University Freiburg, Institut für Anorganische Chemie, Freiburg, Germany

The new compounds Fe3Al2B2 and Ru9Al3B8 were obtained from mixtures of the elements.

Fe3Al2B2 (*P2/m*, *Z* = 2, *a* = 5.7236(1), *b* = 2.8567(1), *c* = 8.7229(4)Å), ß = 98.570(3), 893 refl, 44 param. R1 = 0.025, wR2 = 0.065) is characterized by tetrameric units of BFe6 prisms in trans configuration (B–B:1.772/1.803Å; B–B–B: 114.6°). This motif is similar to Mo2IrB2 [1]. These units are separated by Al atoms. Fe3Al2B2 represents the second ternary phase. Fe2AlB2 [2] contains linear zig-zag chains of B atoms. The structure of Fe3Al2B2 is obtained by an additional *bcc*-like layer FeAl perpendicular to the zig-zag chains in Fe2AlB2.

Fig.1: Crystal structure of Fe3Al2B2 (left) and Ru9Al3B8 (right).

Ru9Al3B8 (*P* 2*m*, *Z* = 4, *a* = 9. 0779(3), *c* = 2.9132(1)Å), 1030 refl, 27 param. R1 = 0.017, wR2 = 0.041) contains zig-zag chains of BRu6 prisms and isolated BRu6 prisms. Three of chains are connected to columns running in direction [001]. The chains are connected by the trigonal prisms around the isolated B atoms. The B–B distance within the chain is 1.749 Å and the B–B–B angle 112.8°. The combination of zig-zag chains and isolated B atoms in Ru9Al3B8 is similar to Ni10.6Ga0.4B8 [3].

The crystal structures of Fe3Al2B2 and Ru9Al3B8 are further examples for the well-known tendency in borides, than boron prefers the bonding to the transition metals and avoids the interaction with main group metals. Therefore, the trigonal-prismatic surrounding is made by the transition metal, (Fe, Ru) while the capping positions are occupied by Al or further B atoms. Fe3Al2B2 represents the second ternary phase in the system Fe/Al/B and might be of interest as magnetocaloric material like Fe2AlB2 [4].

References:

[1] P. Rogl, F. Beneshovsky, H. Nowotny, *Monatsh. Chem.* **1972** *103*, 965–989.
[2] M. Ade, H. Hillebrecht, *Inorg. Chem.* **2015**, *54*, 6122–6135.
[3] M. Ade, D. Kotzott, H. Hillebrecht, *J. Solid State Chem.* **2010**, *183*, 1790–1797.
[4] S. Hirt, F. Yuan, Y. Mozharivskyj, H. Hillebrecht, *Inorg. Chem.* **2016**, *55*, 9677–9684.

Figure 1

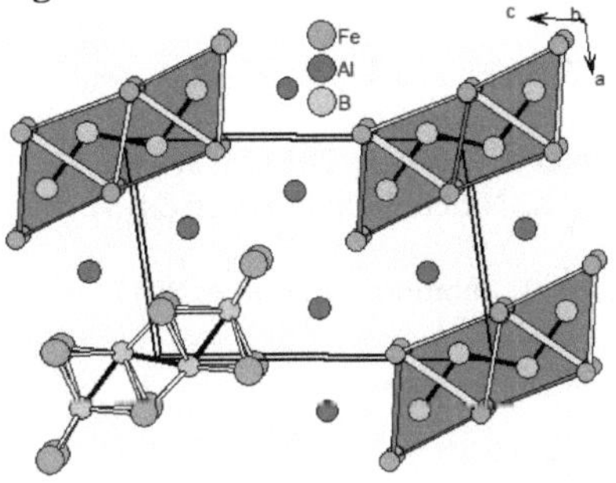

Figure 2

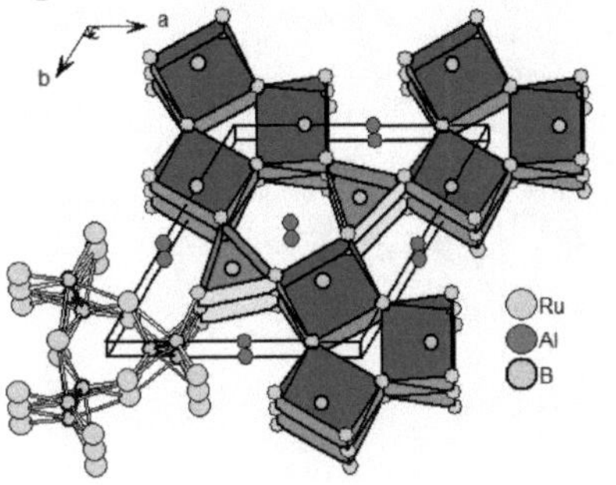

P108

Synthesis, crystal structures and properties of the new 2-D hybride perovskite $C(NH_2)_3HC(NH_2)_2PbI_4$

M. Daub[1], H. Hillebrecht[1]
[1]University Freiburg, Institut für Anorganische Chemie, Freiburg, Germany

Since a few years so-called hybride perovskites, i.e. ternary lead halides with organic cations had become a rapidly growing fields of research [1-3]. The reason is that solar cells based on dyes were developed within e few myears to an efficiency of more than 20%. Nevertheless, the instability versus humidity and its long term stability are still a challenge.

Recently we have shown, that combinations of specific organic cations can enable the formation of 2D perovskites with enhanced stability ($GuMA_2Pb_2I_7$ [4]). Now we show, that the combination of guanidinium $C(NH_2)_3$ and formamidinium $HC(NH_2)_2$ in $GuFoPbI_4$ results in a new type of 2D-perovskites.

Single crystals of $GuFoPbI_4$ were obtained from solutions of the binary halides in HI or DMF. Large crystals (HI) are dark-brown, single phase powders (from DMF) orange. The crystal structure of $GuFoPbI_4$ (*I2/m*, *Z* = 2, *a* = 14.3147(2), *b* = 12.7591(2), *c* = 25.7377(4)Å), β = 101.459(3), 4256 refl, 197 param. R_1 = 0.025,

$wR_2 = 0.065$) represents a 2-D perovskite with a new motif derived from Ruddlesden-Popper phases $(AX)_m(A''BX_3)_n$ and $m = n = 1$. The layers are formed by corner-sharing PbI_6 octahedra. According to the cut-out of the 3-D perovskite the structure of $GuFoPbI_4$ can be assigned as a (210) representative. So one octahedron is linked in trans-position and two in cis-orientation.

GuFoPbI4 was characterized by DTA-TG measurements, vibrational spectroscopy and optical spectroscopy. The optical band gap was determined as 2.34 eV in agreement to the colour. Similar to other Pb-containing 2-D perovskites there is a significant excitonic peak (2.31 eV).

Fig.1: Crystal structure of $GuFoPbI_4$, projection in direction [010]

References:
[1] A. Kojima, K. Teshima, Y. Shirai and T. Miyasaka, *J. Am. Chem. Soc.* **2009**, *131*, 6050.
[2] S. D. Stranks, H. J. Snaith et al., *Science*, **2013**, *342*, 341.
[3] P. Gao, M. Grätzel and M. K. Nazeeruddin, *Energy Environ. Sci.* **2014**, *7*, 2448.
[4] M. Daub, H. Hillebrecht et al,. submitted *Advances* **2017**

Figure 1

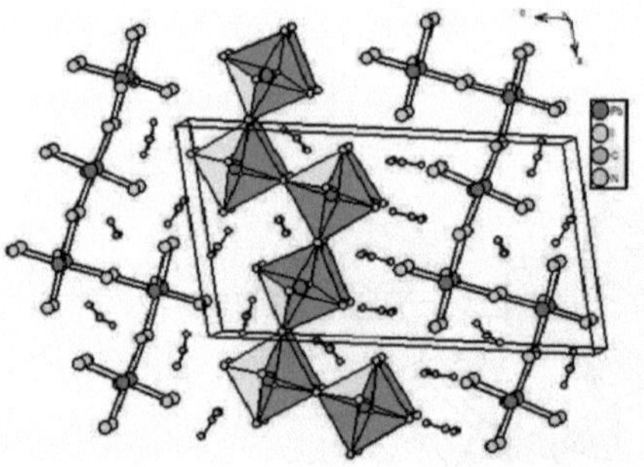

P109

Ordering of ferrocene guest molecule in the pentaphosphaferrocene-based supramolecule

E. Peresypkina[1], A. Virovets[1], M. Scheer[1]
[1]University of Regensburg, Institute of Inorganic Chemistry, Regensburg, Germany

An inorganic analogue of ferrocene, pentaphosphaferrocene, $[Cp^RFe(\eta^5\text{-}P_5)]$ ($Cp^R = \eta^5\text{-}C_5R_5$, R = Me (Cp*), CH_2Ph (Cp^{Bn}), PhC_4H_9 (Cp^{BIG})), is able to coordinate Cu^+ cations resulting in either coordination polymers or giant supramolecules [1 and references therein]. One of the self-assembled spherical supramolecules, $[(Cp^{Bn}Fe(\eta^5\text{-}P_5)_{12}Cu_{20\text{-}n}X_{20\text{-}n}]$ (X = Cl, Br, $1.5 > n > 1.2$), with icosahedral copper-phosphorous core, was proved to be able to qualitatively include ferrocene into central cavity (Fig.) [2]. We found that ferrocene plays a role of the template in the formation of the supramolecule [2]. Possible reason is the presence of π-π-stacking interactions between aromatic Cp ligands of the Cp_2Fe guest and planar P_5-rings of the inorganic host. At T=123K, the guest Cp_2Fe molecule is in staggered conformation and is uniformly disordered over six positions imposed either by (i) the presence of 12 P_5-rings equally available for host-guest interactions or by (ii) high T_h (*m*-3) site symmetry of the *Pm*-3*n* space group. The nature of the disorder, static or dynamic, cannot be therefore revealed in such highly symmetrical crystalline phase.

To break through this uncertainty, we prepared an analog, $(Cp_2Fe)@[(Cp^*Fe(\eta^5\text{-}P_5)_{12}Cu_{20\text{-}n}Br_{20\text{-}n}]$, that crystallizes in lower symmetrical *R*-3 space group, and performed temperature-dependent single crystal X-ray diffraction study at $85 < T < 243$ K. We have found that at higher temperatures the ferrocene molecule retains staggered conformation and is also disordered over six positions, but with different relative weights. At lower temperatures, the guest molecule demonstrates partial ordering down to T = 85K, where it becomes ordered, and remains ordered down to T = 5K without any other structural changes. Therefore, the disorder seems to be of the dynamic nature.

The work is supported by ERC grant AdG339072-SELFPHOS. The diffraction experiment at 5K was performed by Dr. Anja Burkhardt, the P11 beamline at light source PETRA III at DESY.

References:
[1] E. Peresypkina, C. Heindl, A. Virovets, M. Scheer (2016) *Structure and Bonding*, online, DOI 10.1007/430_2016_2.
[2] F. Dielmann C. Heindl, F. Hastreiter, E.V. Peresypkina et al (2015) *Chem.-A Eur. J.*, **21**, 6208.

Figure 1

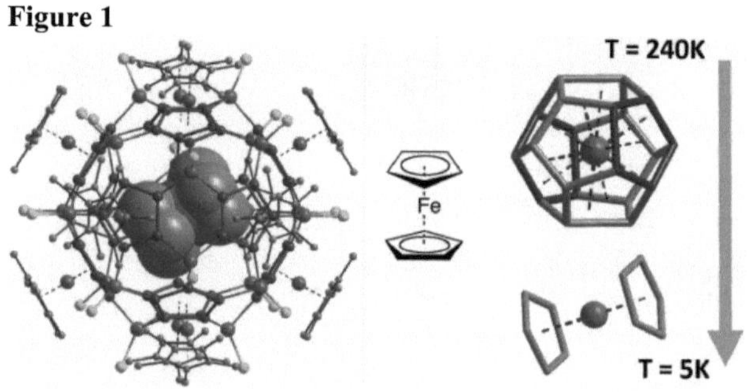

P110

Crystal structure and hydrate water content of synthetic hellyerite, $NiCO_3 \cdot 5.5\ H_2O$

S. Bette[1], C. Rincke[2], R. E. Dinnebier[1], W. Voigt[2]
[1]Max-Planck-Institute for Solid State Research, Stuttgart, Germany
[2]TU Bergakademie Freiberg, Institute for Inorganic Chemistry, Freiberg, Germany

Question
Nickel carbonate salts are relevant both for geochemical modelling of potential risks of Ni-contamination in aquatic systems[1] and as precursor materials for catalysts[2]. Beside gaspeite ($NiCO_3$) and various nickel hydroxide carbonates[3], hellyerite ($NiCO_3 \cdot 5.5/6.0\ H_2O$) occurs naturally[4] and can be obtained artificially[5] as well. Since its discovery the content of hydrate water, is discussed controversially[1, 4, 5] and the crystal structure is also unknown.

In order to clarify the controversies about the composition of hellyerite the mineral was synthesized, characterized via chemical und thermal analyses and the crystal structure was determined.

Methods
Pure, artificial hellyerite was synthesized in aqueous solution at 0°C and subsequently characterized by spectroscopic methods and SEM imaging. The phase composition was derived from chemical and thermal analyses as well as from the crystal structure that was solved from XRPD data employing simulated annealing as method for global optimization and Rietveld refinement.

Results
The measured XRPD pattern of the synthetic sample was in complete agreement with reference data of natural hellyerite. The

crystal structure solution, however, was not unambiguous. Two almost identical structure models with the space groups *C2/m* and *P2/m* having identical lattice parameters were obtained. The *c*-centered cell is related to a phase composition of $NiCO_3 \cdot 6\ H_2O$ and the primitive cell to $NiCO_3 \cdot 5.5\ H_2O$. By inspection of low intensity reflection and applying the systematic extinction rules the $NiCO_3 \cdot 5.5\ H_2O$ structure model was evaluated as valid. The findings are in accordance with complementary chemical and thermal analyses.[6]

Conclusions

The crystal structure of nickel(II) carbonate hydrate, hellyerite, was solved 64 years after the first synthesis[5] and 57 years after its discovery in nature[4]. The phase composition was corrected from $NiCO_3 \cdot 6\ H_2O$ to $NiCO_3 \cdot 5.5\ H_2O$.

References:
[1] Wallner et al. *Acta* **2002**, *382*, 289.
[2] Sun et al. *Alloys Compd.* **2013**, *581*, 217.
[3] Frost et al. *Raman Spectrosc.* **2008**, *39*, 1250.
[4] Williams et al. *Mineral.* **1959**, *44*, 533.
[5] Rossetti-Françeois *Rend.* **1952**, *234*, 840.
[6] Bette et al. *Z. Allg. Anorg. Chem.* **2016**, *642*, 652.

Fig 1. Illustration of the main building blocks of the crystal structure of syn. hellyerite.

Figure 1

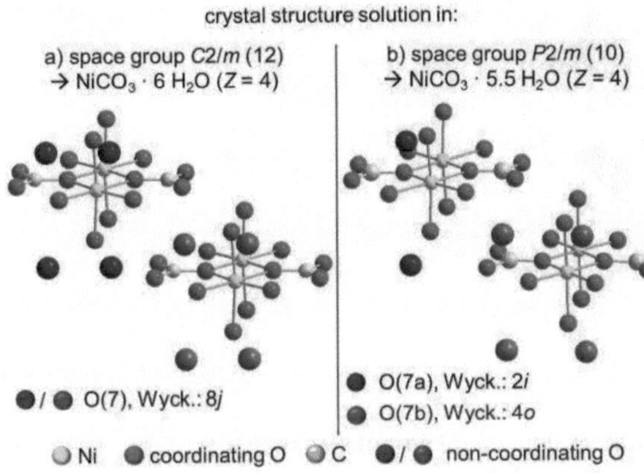

P111

$SrSc_2[SeO_3]_4$ and $BaSc_2[SeO_3]_4$: Two Alkaline-Earth Metal Oxoselenates(IV) with the Smallest Rare-Earth Element

S. Greiner[1], T. Schleid[1]
[1]University of Stuttgart, Institute for Inorganic Chemistry, Stuttgart, Germany

Introduction:

Among the alkaline-earth metal scandium oxoselenates(IV) $AE$$Sc_2[SeO_3]_4$ it was possible to synthesize two new compounds, one with strontium and one with barium. Beside these many $ASc[SeO_3]_2$ representatives with alkali metals (*A* = Na, K, Rb and Cs) [1] were already known in literature.

Results:

$SrSc_2[SeO_3]_4$ (CSD-432357) crystallizes in the monoclinic space group $P2_1/c$ with the lattice parameters of a = 903.41(5), b = 1567.85(9), c = 826.98(5) pm and β = 92.106(3)°. The Sr^{2+} cations reside in eightfold coordination of oxygen atoms, which belong to six different $[SeO_3]^{2-}$ anions. A ninth contact at larger distance turning one more of them into a bidentate ligand should be not forgotten to mention. For both crystallographically distinct Sc^{3+} cations octahedral coordination spheres of oxygen atoms from six terminal $[SeO_3]^{2-}$ anions are found. Two $[SrO_9]^{16-}$ units are connected via a common O3···O3 edge to form $[Sr_2O_{14/1}O_{2/2}]^{26-}$ bipolyhedra. These doubles are connected via common edges and vertices with the $[ScO_6]^{9-}$ octahedra erecting the three-dimensional network 3D-$\{[SrSc_2O_{12}]^{16-}\}$ of $SrSc_2[SeO_3]_4$. In contrast, $BaSc_2[SeO_3]_4$ (CSD-432358) crystallizes in the triclinic space group *P*-1 with a = 536.97(3), b = 710.02(4), c = 791.68(5) pm, α = 85.654(2), β = 89.906(2) and γ = 84.881(2)°. The Ba^{2+} cations are coordinated by oxygen atoms with six short and four long distances forming $[BaO_{10}]^{18-}$ polyhedra. For the two distinguishable Sc^{3+} cations again octahedra with six oxygen atoms from six terminal $[SeO_3]^{2-}$ anions occur. The $[BaO_{10}]^{18-}$ polyhedra are linked via edges, which include the four long Ba^{2+}–O^{2-} distances, to generate 1D-$\{[BaO_{6/1}O_{4/2}]^{16-}\}$ strands running parallel to [100]. These 1D-$\{[BaO_{6/1}O_{4/2}]^{16-}\}$ strands get further connected via common O···O edges to $[ScO_6]^{9-}$ octahedra forming the three-dimensional network 3D-$\{[BaSc_2O_{12}]^{16-}\}$ of $BaSc_2[SeO_3]_4$. The anionic frameworks of both alkaline-earth metal(II) scandium(III) oxoselenates(IV) are filled up with Se^{4+} lone-pair cations for charge compensation in such a way that isolated ψ^1-tetrahedral $[SeO_3]^{2-}$ units form.

References:
[1] S. Y. Song, K. M. Ok, *Inorg. Chem.* **2015**, *54*, 5032–5038.

Figure 1. Three-dimensional frameworks of $SrSc_2[SeO_3]_4$ (*top*) and $BaSc_2[SeO_3]_4$ (*bottom*).

Figure 1

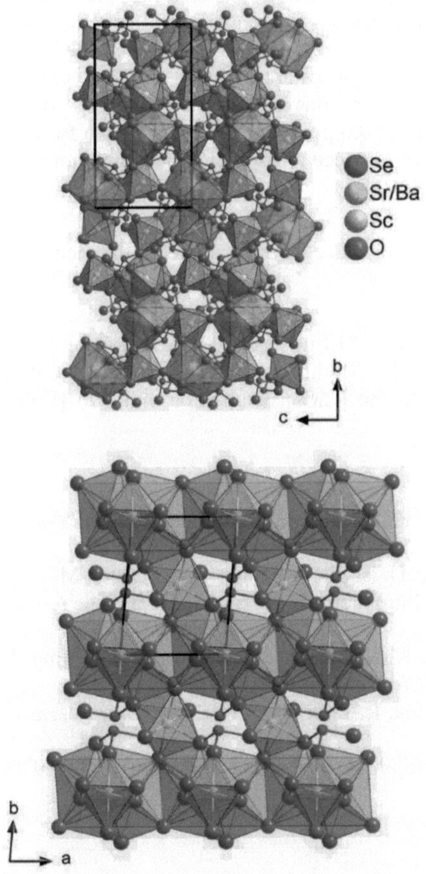

P112

$Li_7Eu_3Nb_2O_{12}$ with a Garnet-Related Structure Containing Mixed-Valent Europium

D. Rudolph[1], C. Funk[1], T. Schleid[1]
[1]University of Stuttgart, Institute for Inorganic Chemistry, Stuttgart, Germany

Red-brown single crystals of $Li_7Eu_3Nb_2O_{12}$ occurred in syntheses using Eu_2O_3 and LiH in Nb ampoules with lithium halide flux at 750 °C. Its crystal structure is a garnet-related arrangement, which has been already observed for the analogous compounds $Li_7La_3Zr_2O_{12}$ [1, 2], $Li_7La_3Hf_2O_{12}$ [3] and $Li_7La_3Sn_2O_{12}$ [4] with Li^+-cation conducting properties.

The crystal structure of $Li_7Eu_3Nb_2O_{12}$ (CSD-432460) was determined by single-crystal X-ray diffraction and is isotypic with the three above-mentioned compounds. It crystallizes in the tetragonal space group $I4_1/acd$ (Figure 1). Assuming that niobium is pentavalent, europium must show a mixed valence state. At least one of the two crystallographically distinct sites for Eu has to be occupied by both Eu^{2+} and Eu^{3+}. Applying the concept of bond-valence sums [5], the value for Eu1 on *8b* is 1.92, but 2.31 for Eu2 on *16e* indicating that Eu^{3+} should be preferentially located at this position. The bond-valence sum of niobium equals 4.72, suggesting an oxidation state of five as assumed.

The crystal structure (Figure 2) is a garnet-related structure type, where extra voids in the garnet-type structure are filled up with lithium. The crystal structure consists of trigonal dodecahedra $[(Eu1)O_8]$ and $[(Eu2)O_8]$ as well as $[NbO_6]$ octahedra. The $(Li1)^+$ and $(Li3)^+$ cations occupy tetrahedral positions in the oxygen framework, whereas $(Li2)^+$ resides in octahedral interstices, while in the ideal cubic garnet just one tetrahedral position is filled with small cations. The Eu1–O distances are 256.5(4) and 264.6(3) pm (4× each) and the Eu2–O separations range from 245 to 268 pm for *C.N.* = 8. The averages of 260.6 pm for Eu1 and 254.4 pm for Eu2 support the assignment of the second site being occupied not only with Eu^{2+}, but also with Eu^{3+}.

Figure 1

chemical formula	$Li_7Eu_3Nb_2O_{12}$
space group	$I4_1/acd$ (no. 142)
crystal system	tetragonal
a / pm	1298.87(9)
c / pm	1235.08(9)
c/a	0.951
Z	8
ρ_{calc} / g·cm^{-3}	5.625
molar volume, V_m / cm^{-3}·mol^{-1}	156.85(1)
diffractometer	κ-CCD (Bruker-Nonius)
$F(000)$	3104
range in ±h, ±k, ±l	17, 17, 16
μ / mm^{-1}	19.99
total number of reflections	12803
number of unique reflections	604
R_{int}, R_σ	0.048, 0.015
structure solution and refinement	SHELX-97
R_1, wR_2	0.022, 0.036
goodness of Fit	1.294
residual electron densities / e·Å^{-3}	1.31, –0.97
CSD number	432460

Figure 2

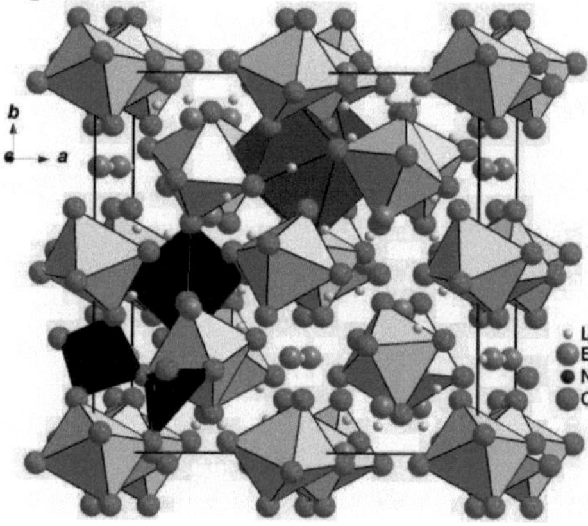

Figure 1. Crystallographic data and structure refinement for $Li_7Eu_3Nb_2O_{12}$.

Figure 2. Crystal structure of $Li_7Eu_3Nb_2O_{12}$ as viewed along [001] with all $[NbO_6]$ octahedra (light grey), two trigonal $[EuO_8]$ dodecahedra (dark grey), one $[LiO_6]$ octahedron and two $[LiO_4]$ tetrahedra (black).

References:

[1] J. Awaka, N. Kijima, H. Hayakawa, J. Akimoto, *J. Solid State Chem.* **182** (2009) 2046–2052.
[2] A. Logéat, T. Köhler, U. Eisele, B. Stiaszny, A. Harzer, M. Tovar, A. Senyshyn, H. Ehrenberg, B. Kozinsky, *Solid State Ionics* **206** (2012) 33–38.
[3] J. Awaka, N. Kijima, K. Kataoka, H. Hayakawa, K. Ohshima, J. Akimoto, *J. Solid State Chem.* **183** (2010) 180–185.
[4] J. Percival, E. Kendrick, R. I. Smith, P. R. Slater, *Dalton Trans.* **2009** (2009) 5177–5181.
[5] N. Brese, M. O'Keeffe, *Acta Crystallogr.* **B47** (1991) 192–197.

P113

Precious and coinage metal complexes with adamantanecarboxylates

P. Schlender[1], M. Kaiser[1], J. Hunger[1]
[1]Technische Universität Dresden, Inorganic Chemistry, Dresden, Germany

Introduction

Precious and coinage metal carboxylates show a large structural diversity, including paddlewheels[1], chains[2], clusters[3] and coordination polymers[4]. These compounds exhibit metal coordination by carboxylate ligands and lewis basic solvent molecules, which may be replaced or removed to enable the complexes for catalytic activity. Here we present three new compounds with adamantanecarboxylates:

$[Ni\{Ad(COO)(COOH)\}_4(H_2O)_2]$ **(1)**,
$[Cu_2(AdCOO)_4(C_4H_8O_2)_2]\cdot C_4H_8O_2$ **(2)**
and $[Ag_2\{Ad(COO)_2\}(H_2O)]\cdot 2H_2O$ **(3)**.

Materials & methods

(1) and **(3)** were prepared in aqueous solutions from 1,3-dicarboxyadamantane and $NiCl_2\cdot 6H_2O$ or $AgOC_2H_5$. **(2)** was prepared from a solution of copper acetate and 1-carboxyadamantane in 1,4-dioxane. The solutions were allowed to evaporate in air. Single crystals grew within a few days to two weeks.

Results

(1) crystallizes as green plates in the orthorhombic space group *Pccn* (a = 1063.3(2) pm, b = 2359.4(4) pm, c = 900.9(2) pm). **(1)** consists of strands along ***c***, where each nickel cation is planar coordinated by four units of hydrogenadamantane-1,3-dicarboxylate. Due to hydrogen bonds between the apical aqua ligands along ***a***, these strands build up a weak 2D coordination polymer.

(2) grows as millimetre sized, green-blue crystals in the monoclinic space group *P21/n* (a = 2259.97(1) pm, b = 1370.94(1) pm, c = 3441.92(2) pm, β = 96.2510(4)°) . The structure exhibits the paddlewheel motif many copper carboxylates own. **(2)** shows a strong UV absorbance at 265 nm.

(3) crystallizes as colorless plates in the triclinic space group *P-1* (a = 721.64(5) pm, b = 1024.30(6) pm, c = 1043.03(7) pm, α = 97.728(2)°, β = 100.050(2)°, γ = 108.035(2)°). Infinite double chains of silver cations are interconnected by 1,3-adamantanedicarboxylate anions, building up a 2D-polymeric network. Despite short Ag-Ag distances from 271 to 323 pm, ELI-D calculations do not support any significant Ag-Ag interactions.

References:
[1] Griebel, F. Leistner, H. Krautscheid, R. Kirmse, *Z. anorg. allg. Chem.* **2006**, *632*, 866–870.
[2] Z. Zheng, Z. Zheng, M.-L. Tong, X.-M. Chen, *Polyhedron* **2013**, *52*, 1159–1168.
[3] C. Jin, Y.-Y. Wang, W.-H. Zhang, A. S. Lermontov, E. K. Lermontova, Q.-Z. Shi, *Dalton Trans.* **2009**, 10181–10191.
[4] J. Xia, Z. Su, L. Yang, C. Zhang, S. Wang, S. Shi, L. Wang, Y. Fan, *Inorg. Chim. Acta* **2014**, *418*, 106–111.

P114

Exploration on Structural Chemistry and Ion-exchange Property of Uranium Borates Mixed with Oxo-anion $[PO4]^{3-}$

Y. Hao[1], E. V. Alekseev[1,2]
[1]IEK-6/Forschungszentrum Jülich GmbH, IEK-6, Jülich, Germany
[2]RWTH Aachen, Institut für Kristallographie, Aachen, Germany

Introduction

Borates possess a variety of chemical structural types, based on two building units, BO3 triangles and BO4 tetrahedra. The number of remarkable structures range from zero dimensional (0D) clusters over one dimensional (1D) chains to two dimensional (2D) sheets and three dimensional (3D) frameworks. In combination with the PO4 anion group, the new family of borophosphates or borate-phosphates with complex structural arrangements arises. In addition, borophosphate glasses have found high interest due to their potential technological applications for nuclear waste vitrification.

Objectives

U^{VI} adopts three O-coordination types, UO6-8. The introduction of U^{VI} into the B-P-O system will lead to innumerable novel structures. However, from the current databank, only four uranyl borophosphates or borate-phosphates were reported. Our research interest is to focus on the *X*-U-P-B-O system with the different methods, to further understand the uranium chemistry.

Materials & methods

UO2(NO3)2·6H2O, KOH, Cs(OH) ·xH2O, H3PO3, H3BO3, H3PO4, NH4H2PO4, Sr(NO3)2;

Hydrothermal and High temperature flux methods

Results

Three alkali-metal uranyl borophosphates were obtained from mild hydrothermal conditions, K5(UO2)2[B2P3O12(OH)]2(OH)(H2O)2 (1), K2(UO2)12[B(H2PO4)4](PO4)8(OH)(H2O)6 (2) and Cs3(UO2)3[B(PO4)4](H2O)0.5 (3). **1** features a 3D open framework, composed of two corner sharing 1D anionic BP chains [B2P3O13] and PU chains (UO2)(PO4)3. **2** is based on a 3D open framework, {(UO2)12[B(PO4)4](PO4)8}, which contains UP tunnels linked by (UO2)4B(PO4)4 clusters. **3** is an extraordinary porous material, which is constructed by [B(PO4)4] clusters and UO7 pentagonal bipyramids, in which nanoscale cages U12P24B8 are present. **3** possess an exceptional ion exchange property with A^{I}, A^{II}, A^{III} to A^{IV} cations at 20-90 °C.

The first salt-inclusion uranyl borate-phosphate, [Sr3(PO4)2][Sr5(UO2)(PO4)2(B5O9)2] (**4**), was obtained through the high temperature flux method. **4** features a 3D uranyl borate-phosphate framework, which is based on corrugated borate layers, linked with PO4 tetrahedra and UO6 polyhedra.

Conclusion

More uranyl borophosphates or borate-phosphates can be prepared with appropriate methods, which have promising structure dependent applications within the field of actinide chemistry.

P115

ROY – A new approach to co-crystals and the effect of deuteration

J. Lehnen[1]
[1]Ruhr-Universität-Bochum, Inorganinc Chemistry I, Bochum, Germany

ROY (5-methyl-2-[(2-nitrophenyl)amino]-3-thiophenecarbonitrile) is known for its numerous colorful (Red Orange and Yellow) crystalline modifications with 10 explored polymorphs until today [1],from which 7 have been structurally characterized.[1a,b] In all known polymorphs an intramolecular hydrogen bond exists between the amino and nitro group, which makes a strong intermolecular hydrogen bond impossible. The molecules are mainly bound by van der Waals forces.[2]

Figure 1: Structure of ROY with intramolecular hydrogen bond.

Most attempts to synthesize multicomponent crystals like co-crystals are made by building intermolecular hydrogen bonds. In 2016 ROY was screened against 342 potential co-former, which led to a list of energetically favorable combinations for possible hydrogen bonds between the co-former and ROY, however, no co-crystal could be synthesized. The fact that ROY has more than just this active site to build intermolecular interactions may have been neglected. In our research the aim was to find co-crystals, synthesized *via* solid state grinding, based on intermolecular interactions build up by supramolecular synthons involving halogen bonds (XB).

Furthermore it is interesting to know if ROY can be functionalized in the smallest possible way, *via* deuteration, and what effect that has on the crystallization and co-crystallization of the different polymorphs. The effect of the exchange of hydrogen with deuterium was explored greatly in the last years.[3-5] In the case of ROY the H/D exchange is highly interesting due to the intramolecular interaction between the nitro and amino group. Will the intramolecular hydrogen bond still exist after deuteration or will the nitro group and the deuterated amino group be available for intermolecular interactions? In our experiments we deuterated ROY *via* solvent deuteration and explain the effect of H/D exchange on the crystalline structure.

References:
[1] (a) G. A. Stephenson, T. B. Borchardt, S. R. Byrn, J. Bowyer, C. A. Bunnell, S. V. Snorek, L. Yu, *J. Pharm. Sci.* **1995**, *84*, 1385. (b) L. Yu, G. A. Stephenson, C. A. Mitchell, C. A. Bunnell, S. V. Snorek, J. J. Bowyer, T. B. Borchardt, J. G. Stowell, S. R. Byrn, *J. Am. Chem. Soc.* **2000**, 122, 585.
[2] L. Yu, *Crystals*. Acc. Chem. Res. 2010, *43*, 1257.
[3] K. Merz, A. Kupka, *Cryst. Growth Des.* **2015**, *15*, 1553.
[4] A. Kupka, V. Vasyleva, D. W. M. Hofmann, K. Merz, *Cryst. Growth Des.* **2012**, *12*, 5966.
[5] G. V. Shishkin, S. V. Shishkina, A. V. Maleev, R. Zubatyuk, V. Vasyleva, K. Merz, *ChemPhysChem* **2013**, *14*, 847.

Figure 1

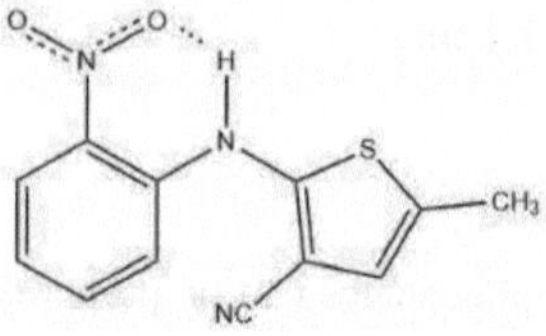

P116
Structural Comparison of the Matlockite-Type Pair EuFBr and EuHBr

D. Rudolph[1], M. E. Bohem[1,2], T. Schleid[1]
[1]University of Stuttgart, Institute for Inorganic Chemistry, Stuttgart, Germany
[2]Fraunhofer Institute for Manufacturing and Automation IPA, Stuttgart, Germany

EuFBr [1,2] and EuHBr [3] were first described by Beck, but the structure determination was solely done by X-ray powder diffraction. Now the structure determination by single-crystal X-ray diffraction on EuFBr and EuHBr could be carried out, thus allowing for the determination of precise atomic coordinates and comparison of bond lengths within these two compounds.

Colorless platelets of EuFBr occurred in reactions using Al, EuF_3 and $EuBr_2$ in molar ratios of 3 : 3 : 1 in Nb ampoules at 877 °C. Red-brown platelets of EuHBr could be obtained by reacting Eu, NaH and NaBr in molar ratios of 1 : 0.75 : 3 in Nb ampoules at 900 °C.

Like all fluoride and hydride halides *Ln*F*X* [1,2] and *Ln*H*X* [3] (*X* = Cl, Br, I) of the divalent lanthanoids (*Ln*), EuFBr (CSD-432469) and EuHBr (CSD-428572) crystallize in the matlockite-type structure [4] of PbFCl in the tetragonal space group *P4/nmm*. This crystal structure can be described as a layered structure (Figure 1). The Eu^{2+} cations are coordinated by four F^-/ H^- and five Br^- anions forming a square antiprism, where one of the bromide anions acts as a cap. The Eu–Br distance for this capping Br^- anion is always larger than the other four Eu–Br contacts.

The unit-cell parameters for EuHBr are smaller than for EuFBr leading to a cell volume reduced by approximately 2 % (Figure 2). This shrinking of the unit-cell dimensions also arises in slightly smaller bond lengths Eu–H (250.3(1) pm) as compared to Eu–F (252.2(1) pm). The four short Eu–Br distances are 318.0(1) and 320.9(1) pm for EuHBr and EuFBr, respectively, while the Eu–Br separations to the capping Br^- anion are the same for both compounds (334.7(2) pm).

These results clearly show the analogy of the fluoride and hydride anions in analogous compounds. The small deviations, especially the contraction of the unit cell and thus smaller bond lengths for the hydride compound, can be explained by a higher polarizability of the hydride as compared to the fluoride anion, leading to a higher covalency of the Eu–H bond or a smaller ionic radius of the hydride anion [5].

Figure 1. View at the crystal structure of EuFBr and EuHBr along [100].

Figure 2. Crystallographic data for EuFBr and EuHBr. The fractional atomic coordinates are ($^1/_4$,$^1/_4$,*z*) for Eu and Br at 2*c* and ($^3/_4$, $^1/_4$,0) for F/H at 2*a*.

References:
[1] H. P. Beck, *J. Solid State Chem.* **17** (1976) 275–282.
[2] H. P. Beck, *J. Solid State Chem.* **23** (1978) 213–217.
[3] H. P. Beck, A. Limmer, *Z. Naturforsch.* **37b** (1982) 574–578.
[4] W. Nieuwenkamp, J. M. Bijvoet, *Z. Kristallogr.* **1932**, *81*, 469–473.
[5] R. D. Shannon, *Acta. Crystallogr.* **A32** (1976) 761–767.

Figure 1

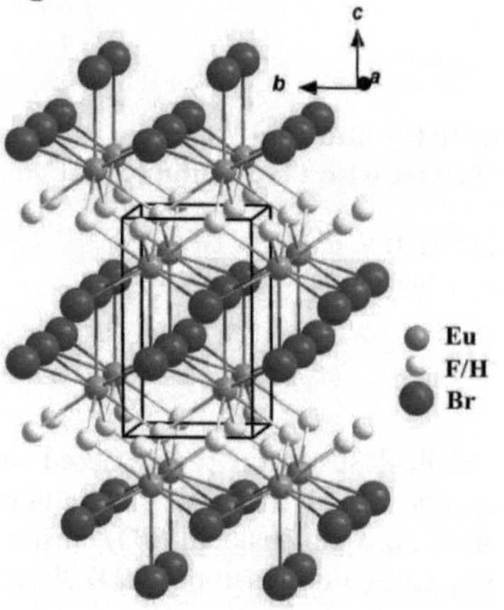

Figure 2

chemical formula	EuFBr	EuHBr
space group	*P4/nmm*	
crystal system	tetragonal	
a / pm	421.56(3)	419.64(3)
c / pm	730.45(6)	721.97(6)
c/a	1.733	1.720
Z	2	
$V_{unit\ cell}$	129.81(2)	127.14(2)
total no. of reflections	1698	1502
no. of unique reflections	117	116
R_{int} / R_σ	0.085 / 0.029	0.094 / 0.038
R_1 / wR_2	0.020 / 0.040	0.025 / 0.059
GooF	1.029	1.158
CSD number	432469	428572
z(Eu)	0.18951(8)	0.18892(9)
z(Br)	0.6477(2)	0.6526(2)
d(Eu–F/H) / pm (4×)	252.2(1)	250.3(1)
d(Eu–Br) / pm (4×)	320.9(1)	318.0(1)
d(Eu–Br) / pm (1×)	334.7(2)	334.7(2)

P117
The Crystal Structure of Tetra-*n*-propylammonium Permanganate

M. Conrad[1], J. Bauchert[1], T. Schleid[1]
[1]University of Stuttgart, Institute for Inorganic Chemistry, Stuttgart, Germany

Abstract:
The baryte-type structure of ammonium permanganate $(NH_4)[MnO_4]$ was first described in 1966 by *Müller* and *Krebs* [1] based on X-ray powder diffraction data and confirmed about 20 years later in 1985 by Chang and Jansen [2] via single crystal X-ray diffraction. However, there were only few reports on crystal structures of its alkyl derivatives up to this day [3].

Objectives:
Various alkyl derivatives of ammonium permanganate have been investigated as potentially suitable precursors for the synthesis of octahedral molecular sieves based on porous MnO_2.

Materials and methods:
To an aqueous solution of barium permanganate $Ba[MnO_4]_2$, an equimolar amount of an aqueous solution of tetra-*n*-propylammonium sulfate $[N(n\text{-}C_3H_7)_4]_2[SO_4]$ was added. Afterwards, the solution was concentrated, separated from the barium-sulfate precipitate and dried overnight in a desiccator over silica gel yielding deep purple crystals of tetra-*n*-propylammonium permanganate, $[N(n\text{-}C_3H_7)_4][MnO_4]$.

Results:
The crystal structure of tetra-*n*-propylammonium permanganate was determined by single-crystal X-ray diffraction at 100 K. It crystallizes in the orthorhombic space group $Pna2_1$ with the lattice parameters a = 1299.84(8), b = 1216.23(8) and c = 980.52(6) pm for Z = 4. The crystal structure contains a single crystallographically unique Mn^{7+} cation position surrounded by four crystallographically distinguishable oxide anions forming virtually ideal and isolated $[MnO_4]^-$ tetrahedra (d(Mn–O) = 160 – 163 pm). Additionally, the structure comprises a crystallographically unique nitrogen atom, which is surrounded by four *n*-propyl groups forming nearly ideal and isolated $[N(n\text{-}C_3H_7)_4]^+$ tetrahedra with typical N–C (149 – 156 pm) and C–C distances (151 – 155 pm) as well.

References:
[1] A. Müller, B. Krebs, *Naturwissenschaften* **1965**, *52*, 494–494.
[2] F. M. Chang, M. Jansen, *Acta Crystallogr.* **1985**, *C41*, 1693–1694.
[3] D. Whang, S. K. Chung, K. Kim, *Acta Crystallogr.* **1991**, *C47*, 2672–2674.

Figure 1: The tetrahedral components $[N(n\text{-}C_3H_7)_4]^+$ (*bottom*) and $[MnO_4]^-$ (*top*) in the crystal structure of tetra-*n*-propylammonium permanganate.

Figure 2: View at the extended unit cell of tetra-*n*-propylammonium permanganate along [001].

Figure 1

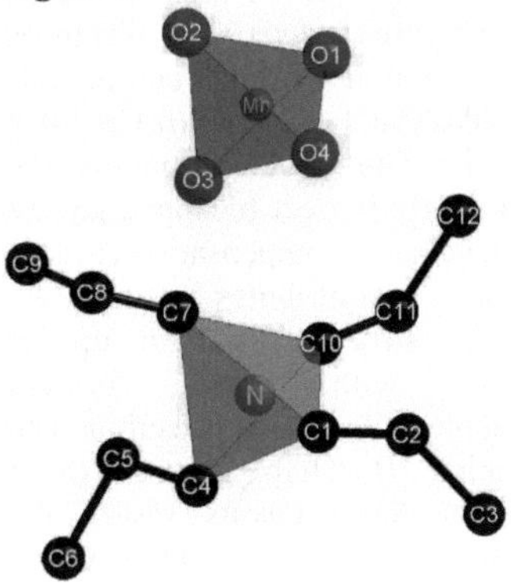

Figure 2

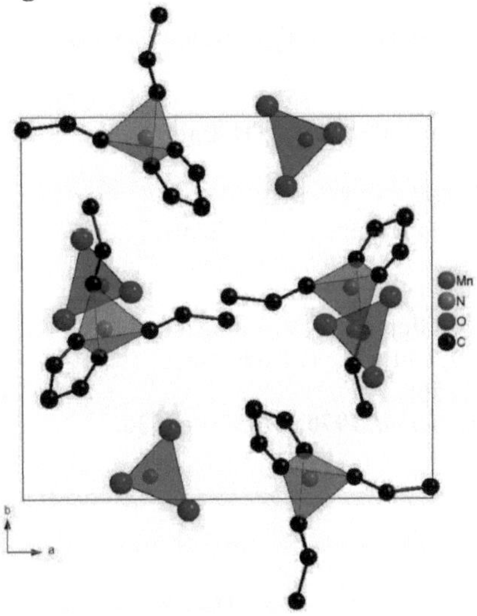

P118
$Cu(H_2O)_4[B_{10}Cl_{10}] \cdot 5\ H_2O$: Crystal Structure of a Copper(II) Decachloro-*closo*-Decaborate Hydrate with an Astonishing Coordination Feature

F. Kleeberg[1], T. Schleid[1]
[1]University of Stuttgart, Institute of Inorganic Chemistry, Stuttgart, Germany

Introduction

Shortly after the successful structure determination of the $[B_{10}H_{10}]^{2-}$ anion in $Cu_2[B_{10}H_{10}]$ by *Lipscomb et al.* [1], reports about partial and complete halogenation products of this cluster were published [2]. Up to now, however, halogenated derivatives of $[B_{10}H_{10}]^{2-}$ clusters seem to be a very special field of boron chemistry, thus articles containing $[B_{10}Cl_{10}]^{2-}$ anions are quite rare in literature [3, 4].

Structure Description

The structure determination of the green title compound based on single-crystal X-ray diffraction data resulted in a triclinic structure with P-1 as space group and the lattice parameters a = 885.24(5), b = 954.19(6), c = 1545.08(9) pm, α = 74.426(2), β = 82.663(2), γ = 85.794(2)° for Z = 2. The crystal structure contains two crystallographically unique Cu^{2+} cations surrounded by four water molecules of hydration (d(Cu–O) = 191–196 pm) and two chlorine atoms of two adjacent $[B_{10}Cl_{10}]^{2-}$ dianions each ($d(Cu\text{–}Cl_{Cluster})$ =

288–297 pm), forming a significantly Jahn-Teller distorted $[Cu(H_2O)_4(Cl_{Cluster})_2]^{2+}$ octahedron (Figure 1, *left*). The distances are in a good agreement with those found in other compounds containing Cu^{2+}, Cl^- and H_2O, just like $CuCl_2 \cdot 2\ H_2O$ (d(Cu–O) = 195 pm, d(Cu–Cl) = 293 pm) [5]. The structure entails the $[B_{10}Cl_{10}]^{2-}$ anion in the shape of a nearly perfect bicapped square antiprism, signing responsible for the charge compensation (Figure 1, *right*). Each B_{10} cluster contributes to two $[Cu(H_2O)_4(Cl_{Cluster})_2]^{2+}$ units, so the structure is built up by electroneutral zigzag chains with the formula $^1_\infty\{[Cu(H_2O)_4][B_{10}Cl_{10}]_{2/2}\}$ running along the [011] direction. The chain-forming Cl atoms at the linking $[B_{10}Cl_{10}]^{2-}$ anions exhibit 1,6-orientation (Figure 2). Between these chains reside the remaining zeolitic crystal water molecules (O5w–O9w) interconnected via classical hydrogen bond interactions with oxygen-to-oxygen distances d(Ow···O/Ow) = 26 7–292 pm.

References:
[1] W. N. Lipscomb, A. R. Pitochelli, M. F. Hawthorne, *J. Am. Chem. Soc.* **1959**, *81*, 5833–5834.
[2] W. H. Knoth, H. C. Miller, J. C. Sauer, J. H. Balthis, Y. T. Chia, E. L. Muetterties, *Inorg. Chem.* **1964**, *3*, 159–167.
[3] V. V. Avdeeva, A. V. Vologzhanina, L. V. Goeva, E. A. Malinina, N. T. Kuznetsov, *Inorg. Chim. Acta* **2015**, *428*, 154–162.
[4] E. A. Kravchenko, A. A. Gippius, A. V. Vologzhanina, V. V. Avdeeva, E. A. Malinina, O. Ulitin, N. T. Kuznetsov, *Polyhedron* **2016**, *117*, 561–568.
[5] A. Engberg, *Acta Chem. Scand.* **1970**, *24*, 3510–3526.

Figure 1

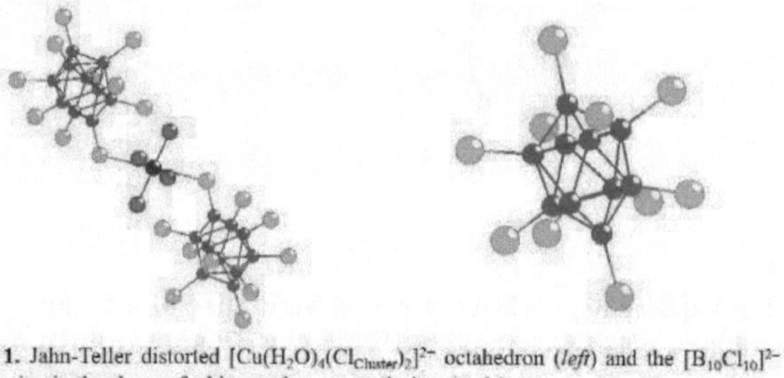

Figure 1. Jahn-Teller distorted $[Cu(H_2O)_4(Cl_{Cluster})_2]^{2-}$ octahedron (*left*) and the $[B_{10}Cl_{10}]^{2-}$ cluster anion in the shape of a bicapped square antiprism (*right*).

Figure 2

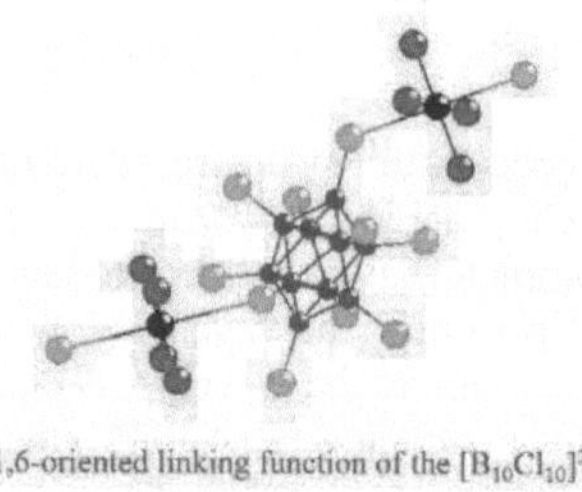

Figure 2. The 1,6-oriented linking function of the $[B_{10}Cl_{10}]^{2-}$ cluster anion.

P119

The Crystal Structure of the Anhydrous Cesium Thiosulfate $Cs_2[SO_3S]$

C. Funk[1], T. Schleid[1]
[1]University of Stuttgart, Institute for Inorganic Chemistry, Stuttgart, Germany

Thiosulfates are known since decades for their common applications in the fixation of photographs [1], where mostly the water-containing sodium thiosulfate was used. *Csordás et al.* first published in 1960 the single-crystal structure of water-free $Na_2[SO_3S]$ [2]. This structure could be subsequently confirmed by neutron-scattering experiments at 120 K [3]. Much later *Röhr et al.* published the structure of anhydrous potassium thiosulfate ($K_2[SO_3S]$) in 2013 [4]. So far, no crystal structures of a water-free alkali-metal thiosulfates are found with lithium, rubidium or cesium [4]. Now it was possible to obtain anhydrous $Cs_2[SO_3S]$ as a side-product. It crystallizes monoclinically in the space group $P2_1/c$ with a = 646.94(5), b = 1096.57(8), c = 951.02(7) pm and β = 96.025(3)° for Z = 4 (Figure 1a) (CSD-432518). Three oxygen atoms (O1, O2 and O3) and one sulfur atom (S2) coordinate each central S1 particle tetrahedrally to form isolated $[SO_3S]^{2-}$ anions. The S–S distance is found to be 199.4(3) pm, which is in accordance to the S–S distances in $Na_2[SO_3S]$ and $K_2[SO_3S]$ [2, 4]. The distances between the central S1 atom and the oxygen ligands range from 145.8(5) to 147.7(5) pm, which is likewisely matching with the literature data [2, 4]. There are two independent sites for the cesium cations (Table 1). The $(Cs1)^+$ cation is coordinated by six $[SO_3S]^{2-}$ tetrahedra, while the $(Cs2)^+$ cation exhibits a surrounding of only five of them (Figures 1b and 1c). The Cs–S distances are ranging from 354.4(2) to 374.7(2) pm and the Cs–O contacts cover broad intervals from 303.0(6) to 363.7(6) pm in order to provide an eightfold coordination (5 O + 3 S) for $(Cs1)^+$, but a tenfold one (7 O + 3 S) for $(Cs2)^+$.

References:
[1] A. Seyewetz, G. Chicandard, *Bull. Soc. Chim. Fr.* **1895**, *13*, 11–19.
[2] E. Sándor, L. Csordás, *Acta Crystallogr.* **1961**, *14*, 237–243.
[3] S. T. Teng, H. Fuess, J. W. Bats, *Acta Crystallogr.* **1984**, *C40*, 1785–1787.
[4] A. J. Lehner, L. V. Schindler, C. Röhr, *Z. Naturforsch.* **2013**, *68b*, 323–337.
[5] R. X. Fischer, E. Tillmanns, *Acta Crystallogr.* **1988**, *C44*, 775–776.

Figure 1. a) Crystal structure of $Cs_2[SO_3S]$ as viewed along [010], b) coordination sphere of $(Cs1)^+$ with six $[SO_3S]^{2-}$ tetrahedra, c) coordination sphere of $(Cs2)^+$ with five $[SO_3S]^{2-}$ tetrahedra.

Figure 1

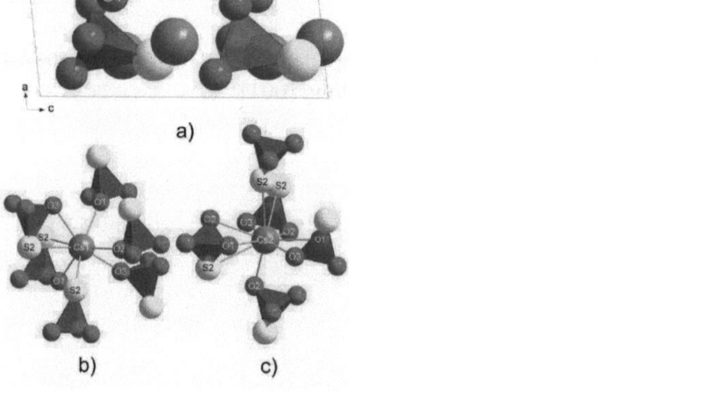

Figure 2

Table 1. Atom positions and equivalent isotropic displacement parameters (U_{eq} / pm²) for $Cs_2[SO_3S]$.

Atom	Site	*x/a*	*y/b*	*z/c*	U_{eq}*
Cs1	*4e*	0.71359(7)	0.21768(4)	0.01144(5)	332(2)
Cs2	*4e*	0.21296(7)	0.09024(4)	0.30322(5)	364(2)
S1	*4e*	0.2793(3)	0.07428(16)	0.71677(17)	269(4)
O1	*4e*	0.4801(8)	0.1336(5)	0.7119(6)	459(15)
O2	*4e*	0.1180(9)	0.1302(6)	0.6184(6)	558(17)
O3	*4e*	0.7031(8)	0.0574(5)	0.3124(6)	424(14)
S2	*4e*	0.1934(3)	0.09476(18)	0.91063(19)	397(5)

*$U_{eq} = {}^1/_3\,[U_{22} + 1/\sin^2\beta\,(U_{11} + U_{33} + 2\,U_{13}\cos\beta)]$ [5].

P120

Crystal structure determination of $[NiBr_2(4\text{-cypy})_2]_n$ and $[NiBr_2(4\text{-cypy})_1]_n$ by X-ray powder diffraction

M. Heine[1], L. Fink[1], M. U. Schmidt[1]
[1]Goethe-Universität Frankfurt, Institut für Anorganische und Analytische Chemie, Frankfurt am Main, Germany

Reaction of nickel bromide with 4-cyanopyridine (4-cypy) leads to the formation of two new coordination polymers with the composition [NiBr2(4-cypy)2]n and [NiBr2(4-cypy)1]n. Their crystal structures were determined by X-ray powder diffraction. In these compounds the NiII ions are octahedrally coordinated by two 4-cypy ligands and two bridging bromide ions to form single chains in [NiBr2(4-cypy)2]n and two dimensional networks in [NiBr2(4-cypy)1]n. In [NiBr2(4-cypy)2]n the 4-cypy molecules act as monodentate ligands, they are bonded *via* the N atom of the pyridine ring (Npy) to the Ni atom, whereas in [NiBr2(4-cypy)1]n 4-cypy acts as a bidentate ligand: Both N atoms coordinate to the NiII atom, resulting in a nearly linear Ni – py – C≡N – Ni bridge. Furthermore Warren reflection profiles led to a modeling of disordered 4-cypy groups. The best fit was obtained by implementing a head-to-tail-orientation disorder of the 4-cypy groups (see Fig. 1). This behavior hasnt been observed yet in similar coordination compounds of the stoichiometry [MX2(4-cypy)1]n (with M = Mn, Co, Ni, Cu and X = Cl, Br).

Figure 1: Preliminary crystal structure of [NiBr2(4-cypy)1]n, projected along a, showing the disordered 4-cypy groups (light and dark blue: represent the two possible orientations of 4-cypy; green: Ni; orange: Br).

Keywords:

X-ray powder diffraction, crystal structures, coordination polymers, condensed networks, substituted pyridine ligands, transition-metal halides, disordered structures

References:
[1] Zhao et al., **2017**, submitted

Figure 1

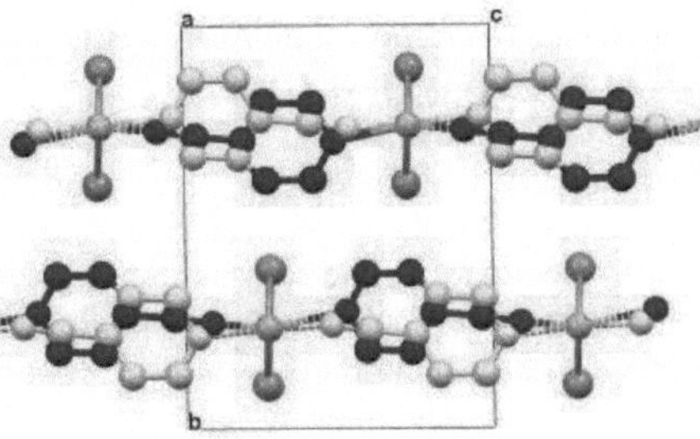

P121

A series of coordination polymers employing 2,2"-bipyridine-3,3",6,6"-tetracarboxylate as a ligand.

A. Kochel[1]
[1]University of Wrocław, Poland, Department of Chemistry, Wocław, Poland

In the design of new magnetic/luminescent/semicoductors materials the appropriate choice of the organic ligand is of crucial importance. It was shown that inclusion of such features as multiple carboxylic groups or pyridyl moieties extends the ligand coordination capabilities and promotes formation of 1D-3D motifs in the resulting d-electron metal complexes [1] One of the designed ligands incorporating all these useful properties is 2,2"-bipyridine-3,3",6,6"-tetracarboxylic acid (*bptcH4*). The flexibility of the applied organic ligand may lead to even more interesting metal-organic frameworks in the future. Moreover, a targeted modification of the ligand is currently applied to induce formation of new molecular topologies in the resulting materials.

In this communication we extend the field of *bptcH4* complexes chemistry, reporting on the synthesis, crystal structures and properties of new manganese(II)/copper(II) coordination polymers. Coordination polymers forms polymeric double chains and are characterized by X-ray diffraction studies, IR spectra, elemental analysis, TGA and SQUID magnetic properties measurements .

Fig. 1. One of the motives coordination polymers for Mn(II) compounds.

References:
[1] (a) S. Qiu, Guangshan Zhu, Coord. Chem. Rev. 253 (2009) 2891; (b) X. Huang, J. Dou, D. Li, D. Wang, J. Coord. Chem. 60 (2007) 319; (c) X. Cao, X. Zheng, M. Chen, X. Xu, T. Sun, E. Wang, J. Coord. Chem. 65 (2012) 754; (d) J. E. Beves, E. C. Constable, C. E. Housecroft, M. Neuburger, S. Schaffner, CrystEngComm 10 (2008) 344; (e) E. C. Constable, G. Zhang, C. E. Housecroft, J. A. Zampese, CrystEngComm 13 (2011) 6864.

Figure 1

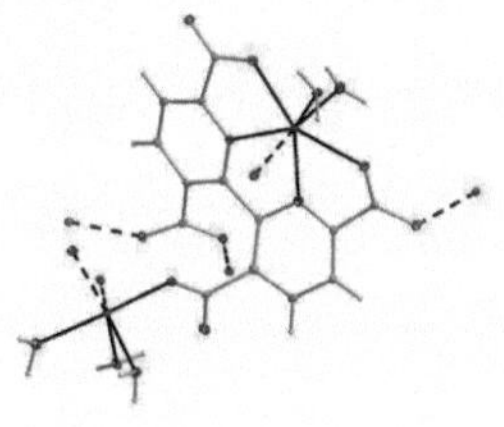

P122

Amine Complexes of Mercury(II) Halogenides

C. Döring[1]
[1]Technische Universität Braunschweig, Institut für Anorganische und Analytische Chemie, Braunschweig, Germany

Amine complexes of mercury(II) halogenides were synthesized by dissolving and crystallizing HgX_2 (X = chlorine, bromine or iodine) in a primary or secondary amine or azaaromatic. Different molecular and ionic forms were obtained, in which both the halogen and the diamine can act as bridging ligands in some cases.

In contrast to halogengold-amine complexes[1,2], where a particular coordination geometry is associated with the oxidation state of the metal, the mercury complexes show a variety of coordination geometries starting with tetrahedral (Fig. 1) through trigonal-bipyramidal to octahedral and trigonal-prismatic (Fig. 2).

All complexes were obtained as colourless crystals. They display a variety of forms (molecular, ionic, polymeric) and coordination geometries.

The structures of the products were analysed by X-ray structure determination and their packing was investigated with reference to hydrogen bonds. In structures with N-H groups, classical hydrogen bonds between amine H atoms and the free halogenides or those bound to mercury are generally formed. In some cases the diamines or the halogens bridge two mercury centres. This leads to chains which form a central element of the packing. These are further linked by hydrogen bonds to form ribbons, layers or three-dimensional networks.

Fig. 1. Chain of tetrahedrally coordinated mercury(II) centres bridged by iodine atoms.

Bond lengths and angles [Å, °]: Hg1-I1 2.657(6), Hg1-I2 2.801(5), Hg1-I2A 2.976(6), Hg1-I2A-Hg1A 91.35(2).

Fig. 2. Three ethylenediamine ligands in trigonal-prismatic coordination at one mercury(II) centre. Bromide ions are involved in hydrogen bonds.

Bond lengths and angles [Å, °]: N14-H03···Br2 2.709(18)/169(3), N24-H08···Br1 2.68(2)/155(4), N31-H09···Br1 2.73(2)/146(3), N34-H012···Br2 2.78(2)/153 (3).

References:
[1] C. Döring, P. G. Jones, *Acta Crystallogr., Sect. C: Cryst, Struct. Commun.* **2013**, *C69*, 709-711.
[2] C. Döring, P. G. Jones, *Z. Anorg. Allg. Chem.* **2016**, *642*, 930-936.

Figure 1

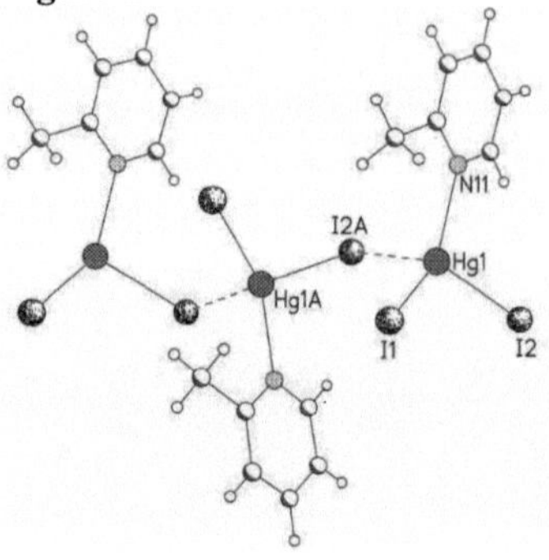

Figure 2

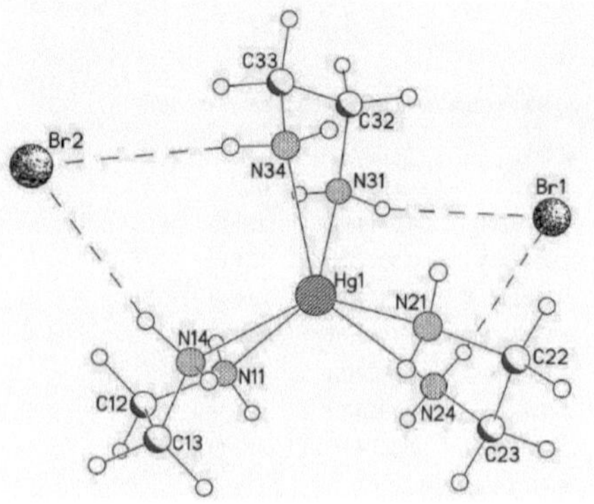

P123

Intermolecular interactions study within two 4-(aminomethyl)benzoic acid proton transfer compounds by means of Hirshfeld surface analysis

A. DIREM[1], A. Altomare[2], A. Moliterni[2], N. Benali-cherif[1]
[1]LASPI2A Laboratory, Faculté des sciences et technologie, University of Khenchela, Khenchela, Algeria
[2]Istituto di Cristallografia, Via G. Amendola, 122/o, 7016, Bari, Italy

The crystal engineering is considered as the understanding of intermolecular interactions in the context of crystal packing and the use of such understanding in the design of new solids with desired physical and chemical properties [1]. It relies on non-covalent forces (hydrogen bonding, π–π stacking, C–H...π and van der Walls) for the construction of crystalline materials from molecules and/or ions [2]. These interactions are very useful in devising gas-storage devices, sensors, optical switches and solar cells [3]. Being one of the several types of non-covalent interactions in many organic and inorganic species, the hydrogen bonds resulting in aggregation and control self-assembly are well-studied [4]. Whereas, the non-classical interactions are less-investigated. As a contribution to the study of this kind of interactions, we report herein in detail the Hirshfeld surface analysis [5] of two 4-(aminomethyl) benzoic acid proton transfer compounds, namely 4-(ammoniomethyl) carboxybenzene nitrate and 4-(ammoniomethyl) carboxybenzene perchlorate [6]. This study revealed that the two structures are dominated by H...O/O...H and H...H contacts with the highest contribution of approximately 50% and 20%, respectively. Other close contacts are also present, including weak C...H/H...C contacts (with about 10 %).

References: :
[1] Desiraju, G. M. (1989). *Crystal engineering, the design of organic solids.* Elsevier, Amsterdam.
[2] a) Meléndez, R. E. & Hamilton, A. D. (1998). *Top Curr. Chem.* **198**, 97-129. b) Aakeröy, C. B. & Beatty, A. M. (2001). *Aust. J. Chem.* **54**(7), 409-421.
[3] Lehn, J. M. (1990). *Angew. Chem. Int. Ed. Engl.* **29**, 1304–1319.
[4] a) Desiraju, G.R. (2002). *Acc. Chem. Res.* **35**, 565-573. b) Liu, J. Q., Wang, Y. Y., Ma, L. F., Zhang, W. H., Zeng, X. R., Zhong, F., Shi, Q. Z. & Peng, S. M. (2008). *Inorg. Chim. Acta.* **361**, 173-182. c) Biswas, C., Drew, M. G. B., Escudero, D., Frontera, A. & Ghosh, A. (2009). *Eur. J. Inorg. Chem.* 2238-2246.
[5] a) Spackman, M. A. & Jayatilaka, D. (2009). *Cryst. Eng. Comm.*, **11**, 19–32. b) Spackman, M. A. & McKinnon, J. J. (2002). *Cryst. Eng. Comm.* **4**, 378–392.
[6] Direm, A., Altomare, A., Moliterni, A. & Benali-Cherif, N. (2015). *Acta Cryst.* B**71**, 427–436.

P124

High temperature Cs_2S_3 and the incommensurate modulated phase $K_{37}Te_{28}$: New binary alkali chalcogenides

P. Stüble[1], A. Berroth[1], F. Wortelkamp[1], C. Röhr[1]
[1]Universität Freiburg, Institut für Anorg. und Analyt. Chemie, Freiburg, Germany

A new potassium telluride $K_{37}Te_{28}$ and a new h.t. polymorph of Cs_2S_3 were found as by-products in a comprehensive study on alkali chalcogenido metalates. Both were firstly yielded from ternary samples with the scope of extending the M_3Q_7 / M_4Q_8 cluster family [1], later on, they were reproduced from binary samples.

The new monoclinic form of Cs_2S_3 ['m-Cs_2S_3', new structure type, $P2_1/c$, a = 999.97(4), b = 1029.30(5), c = 2642.07(12) pm, β = 90.083(2)°, Z = 16, T = 100 K, $R1$ = 0.034] *(Fig. 1)* contains four crystallographically different $[S_3]^{2-}$ units with slightly varying bond lengths ($d_{S\text{-}S}$ = 205.7 - 208.3 pm) and angles (α = 109.3 – 111.6°), comparable to the long known orthorhombic polymorph (α = 106.0°, $d_{S\text{-}S}$ = 212.0 pm) [2].

Both modifications share a similar arrangement of the $[S_3]^{2-}$ units which can be derived from a b.c.c. packing. Suitable for a h.t. form, the molar volume of m-Cs_2S_3 (V_M=174.6·10^6 pm^3) slightly exceeds the one of o-Cs_2S_3 (V_M = 171.1·10^6 pm^3, $\Delta(V_M) \approx 2\%$). Raman studies, which are in good agreement with own Raman spectra, already suggested a yet unknown h.t. form [3]. Moreover, ongoing phase transition of m-Cs_2S_3 to o-Cs_2S_3 has been observed via powder-XRD.

The tetragonal compound $K_{37}Te_{28}$ [new structure type, $I4_1/amd$, a = 1938.17(6), c = 2644.34(11) pm, Z = 4, T = 295 K, $R1$ = 0.037] contains different anionic units: isolated Te^{2-}, neighbouring dumbbells $[Te_2]^{2-}$ similar to K_5Te_3 [4] as well as linear chains $^1_\infty[Te^{-1.125}]$ *(Fig. 2)*. The bond length within the chain ($d_{Te\text{-}Te}$ = 330.6 pm) is significantly larger than the ones observed in reference compounds, e.g. TlTe ($d_{Te\text{-}Te}$ = 308.6 pm) [5] or CuTe ($d_{Te\text{-}Te}$ = 305-315 pm) [6].

At 100 K, satellite reflections were observed and the structure was refined with an incommensurate modulation [$I4_1/amd(0,0,\gamma)\sigma0\sigma0$, q = 0.5143, wR(obs) = 0.038 (main) / 0.072 (all)], which mainly affects the linear chains and their squared anti-prismatic K-coordinations ($d_{Te\text{-}Te}$ = 299.7 - 357.1 pm).

References:
[1] P. Stüble, A. Berroth, C. Röhr. *Z. Naturforsch.* **71**, 485 (2016).
[2] P. Böttcher. *Z. Anorg. Allg. Chem.* **461**, 13 (1980).
[3] W. Bues, H. Ziemann. *Z. Anorg. Allg. Chem.* **456**, 54 (1979).
[4] I. Schewe, P. Böttcher. *Z. Naturforsch* **45b**, 417-422 (1990)
[5] K. Stöwe. *J. Solid State Chem.* **149**, 123 (2000).
[6] K. Stolze, A. Isaeva, F. Nitsche, U. Burkhardt, H. Lichte, D. Wolf, T. Doert. *Angew. Chem.* **125**, 895 (2013).

Figure 1
H.T. form of Cs_2S_3

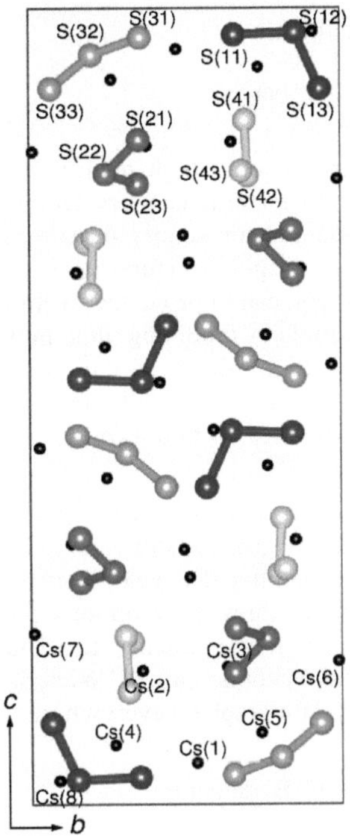

Figure 2
$K_{37}Te_{28}$ (295 K) Te ellipsoids: 99%, distances: pm

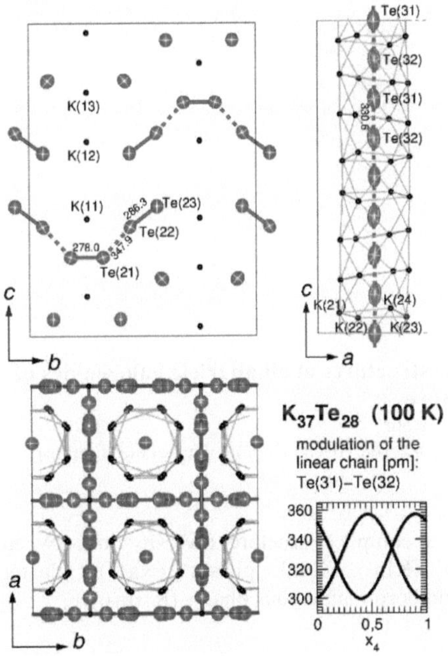

P125

Improving your single crystal data collection

A. Griffin[1]
[1]Rigaku Oxford Diffraction, Kent, United Kingdom

Performing data collections which are not only high quality but also efficient is a key requirement of diffractometer users around the world, either to enable high throughput, or simply maximise data quality in a given time. By using a rapid structural analysis before the start of your data collection, you can improve the overall strategy and data quality of your sample. Combining this **new** approach to data collection with the high speed of the latest instruments, e.g XtaLAB Synergy (Fig 1), gives crystallographers the tools to achieve the best possible structures for publication.

Ensuring the best possible use of the diffractometer hardware and the data it provides requires high performance software. The CrysAlisPro software package is under continual development in order to provide enhanced automation features, new options for sample screening, data collection and improvements to data processing algorithms. One of these algorithms, called "What Is This" offers an exciting solution for crystallographers everywhere.

The XtaLAB Synergy-S with HyPix-6000HE detector

Figure 1

P126

Crystal and electronic structures of alkali triele halogenides of the K_8In_{11}-type structure

M. Falk[1], A. El Addad[1], C. Röhr[1]
[1]Institut für Anorganische und Analytische Chemie, Universität Freiburg, Freiburg, Germany

Alkali trielides show a complex structure diversity in between simple intermetallics (KIn_4), Zintl phases (NaTl), cluster compounds ($RbGa_3$) and boron-analogous phases ($K_{17}In_{41}$).

The K_8In_{11}-type structure (R-3c) includes both polyanion clusters and distorted alkali cubes. Three M atom positions form isolated, strongly compressed pentacapped trigonal prismatic M_{11} clusters. Until 1997, only compounds with empty alkali cubes were known [1-4]. Since then some halogen containing phases have been reported [5].

In systematic studies, different compounds $A_8M_{11}X_{0.1\text{-}1}$ (A = K, Rb, Cs; M = Ga, In; X = Cl, Br, I) were synthesized from the elements and AX to investigate the effects of the halogenide content on the structure and to explore the "coloring" of the polyanionic cluster, i. e. the influence of different size and electronegativity of the trieles Ga and In. FP-LAPW DFT calculations were additionally performed.

The amount of incorporated X depends on the size of A only: with increasing ionic radius of A, the content of X is gradually raised (K: ~10%, Rb: 52-100%, Cs: 88-100%) and the deviation of the M_{11} cluster symmetry from D_{3h} to D_3 symmetry decreases due to twisting. The distortion of the A_8 cube along the c axis decreases from Cl to I compounds and is generally smaller for Ga-containing clusters.

The calculated Bader charges indicate no significant differences for the three M positions, but a clearly smaller Bader volume for M(2). In accordance, Ga occupies this position preferably, followed by M(1). M(3) shows significantly larger distances to all neighbouring atoms and thus prefers an In occupation.

The formula $A_8M_{11}X$ implies an M_{11}^{7-} cluster and the DOS plot shows a band gap. The arrow indicates that the additional electron in A_8M_{11} (M_{11}^{8-} cluster) populates the conduction band with p_M-p_M antibonding character, having an impact on the cluster geometry [6].

References:

[1] W. Blase, G. Cordier, *Z. Kristallogr.* **194**, 150-151 (1991).
[2] S. C. Sevov, J. D. Corbett, *Inorg. Chem.* **30**, 4875-4877 (1991).
[3] G. Cordier, V. Müller, *Z. Kristallogr.* **203**, 154-155 (1993).
[4] W. Blase, G. Cordier, V. Müller, U. Häußermann, R. Nesper, M. Somer, *Z. Naturforsch.* **48 b**, 754-760 (1993).
[5] R. W. Henning, J. D. Corbett, *Inorg. Chem.* **36**, 6045-6049 (1997).
[6] M. Cobián, P. Alemany, A. García, E. Canadell, *Inorg. Chem* **48**, 9792-9799 (2009) and references therein.

Figure 1

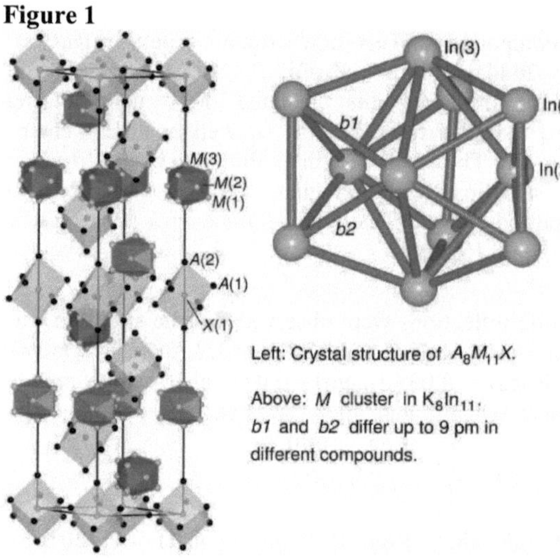

Left: Crystal structure of $A_8M_{11}X$.

Above: M cluster in K_8In_{11}. $b1$ and $b2$ differ up to 9 pm in different compounds.

Figure 2

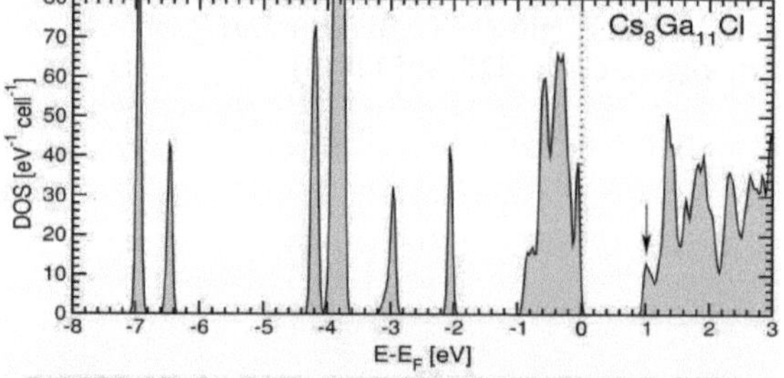

Total DOS of $Cs_8Ga_{11}Cl$. The arrow indicates the position of E_F in Cs_8Ga_{11}.

P127

The structure of α- and β-moganite

H. Grimmer[1], B. Delley[1]
[1]Paul Scherrer Institut, Research with Neutrons and Muons, Villigen PSI, Switzerland

Moganite, which is monoclinic at ambient temperature, undergoes a displacive transition to an orthorhombic phase at ≈570 K. Whereas the monoclinic phase may be considered as a-quartz that is Brazil twinned along {1 0 -1 1} at the unit-cell scale (cell-twinning), the orthorhombic phase cannot be interpreted as a Brazil cell-twin of β-quartz, in contrast to statements made in the literature. The shape of the oxygen tetrahedra in monoclinic moganite has been determined more reliably by density functional theory (DFT) calculations than by experiment: the differences between the various experimental results for the shape of the oxygen tetrahedra at ambient temperature are typically ten times larger than the differences between the DFT results. The DFT calculations suggest that the oxygen tetrahedra in moganite are very close in shape to the oxygen tetrahedra in α-quartz. Among the three DFT calculations considered, the most convincing results for the bond angles in moganite are obtained for the $DMol^3$ code with functional PBE.

Figure 1: The structure of monoclinic moganite obtained by DFT using the $DMol^3$ code with functional PBE.

Figure 1

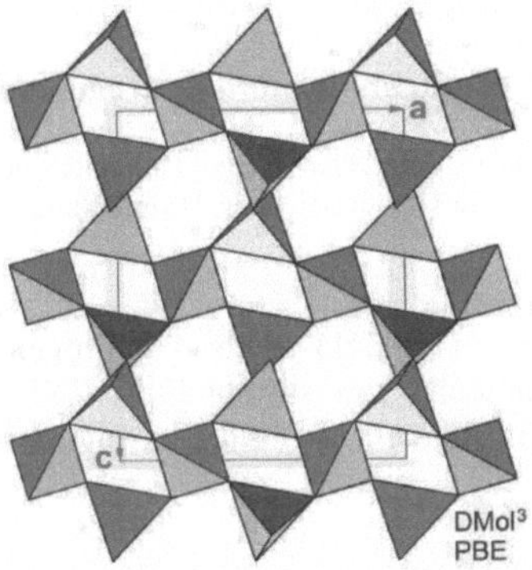

P128

Synthesis and crystal structure of the new sodium tellurido manganate $Na_2Mn_3Te_4$

M. Langenmaier[1], C. Röhr[1]
[1]Universität Freiburg, Anorg. und Analyt. Chemie, Freiburg, Germany

In the system *A*-Fe-Te mixed-valent Fe(II/III) cluster-containing alkali (*A*) tellurido ferrates are already known and were obtained recently. [1]

Analogous manganates (with one exception) are divalent only. Known compounds are the ortho salts $(Na/K)_6[MnTe_4]$ (Na_6ZnO_4-type [2]), the chain metalates A_2MnTe_2 (*A*=K,Rb,Cs; K_2ZnO_2-type [3]), layered tellurido manganates(II) $NaMnTe_2$ [4] and the mixed valent manganate $Na_3Mn_4Te_6$ [5] (defect variants of the $CaAl_2Si_2$-type). Similarily, $AMnTe_2$ ($TlFeS_2$-type [5]) and $A_2Mn_3Te_4$ ($Cs_2Zn_3S_4$-type [6]) are defect variants of the tetragonal $ThCr_2Si_2$-type layers.

The new compound, $Na_2Mn_3Te_4$, was yielded from stoichiometric mixtures of the elements at T_{max} = 900 K. The monoclinic crystal structure has been determined by means of X-ray single crystal data (*C*2/*m*, *mC*18, *a* = 17019.9(3), *b* = 438.74(1), *c* = 691.23(1) pm, β = 90.317(1) °; *R*1 = 0.0282) and contains one Na, two Mn and two Te sites. Mn(1) (2*a*, 2/*m*) is octahedrally coordinated by Te^{2-} ($[MnTe(1)_{4/4}Te(2)_{2/3}]$, gray polyhedra, *fig. 1*). These octahedra are egde-linked to form chains along the monoclinic *b* axis. Tetrahedra ($[MnTe(1)_{2/4}Te(2)_{2/3}]$, blue polyhedra) formed by Mn(2) (4*i*, *m*) and Te are connected via a common edge. The resulting dimers are corner-linked to build a chain along *b*. These polyhedra form puckered layers along [100] in which Na^+ ions (4*i*) are embedded and coordinated in a octahedrally manner (yellow polyhedra). Te forms an almost planar $3^3.4^2$ net (*y*=1/4, 3/4; light arrows in *fig. 1a*) and a puckered net indicated by dark arrows.

Refinements in the SG *C*2/*m* show a disordered Mn(1) (Δ ≈ 25 pm from -1) as seen by the 99% ellipsoids and the difference electron density map (*fig. 1c*) leading from an octahedron to a square bipyramid.

$Na_2Mn_3Te_4$ is compared with known tellurido manganates as well as with the related, but not isotypic, sulfide $Na_2Cu_2ZrS_4$ [7].

References:
[1] P. Stüble, A. Berroth, C. Röhr, *Z. Naturforsch.* **71**, 485 (2016).
[2] W. Bronger, H. Balk-Hardtdegen, *Z. Anorg. Allg. Chem.* **574**, 89 (1989).
[3] W. Bronger, H. Balk-Hardtdegen, D. Schmitz, *Z. Anorg. Allg. Chem.* **574**, 99 (1989).
[4] J. Kim, Ch. Wang, T. Hughbanks, *Inorg. Chem.* **37**, 1428 (1998).
[5] Ibid. **38**, 235 (1999).
[6] W. Bronger, H. Hardtdegen, M. Kanert, P. Müller, D. Schmitz, *ZAAC* **622**, 313 (1996).
[7] M. F. Mansuetto, J. A. Ibers, *J. Solid State Chem.* **117**, 30 (1995).

Figure 1

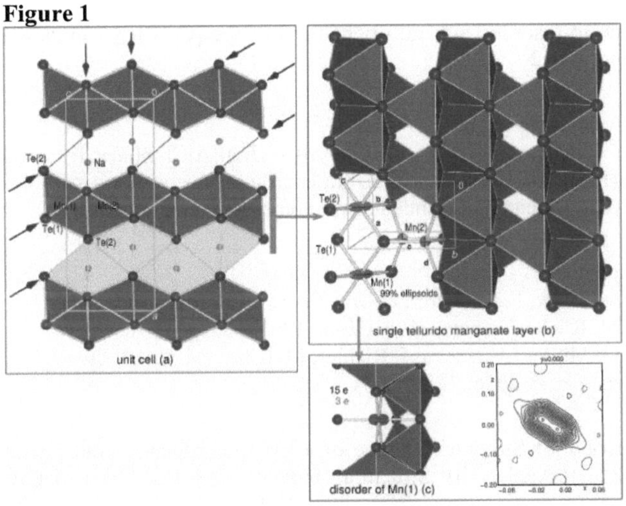

P129

Synthesis of Copper Coordinated Bismuth Cluster Compounds

M. Knies[1], M. Kaiser[1], U. Müller[1], T. Doert[1]
[1]TU Dresden, Chemistry and Food Chemistry, Dresden, Germany

1. Introduction
Bismuth is known for its ability to form homoatomic polycationic clusters with bismuth atoms in low oxidation states.[1] Square-antiprismatic Bi_8^{2+} polycations were found as ligands towards ruthenium ions through a η^4 coordination of one of its square faces,[2] but Bi_8^{2+} clusters can also act in varying hapticity coordinating Au^+ in $[Au(Bi_8)_2](AlCl_4)_5$, Fig. 1a.[3]

2. Objectives
In this work, we focused on substituting the gold ion in $[Au(Bi_8)_2](AlCl_4)_5$ by other group 11 transition metal ions. The crystal structures of two copper compounds are presented here.

3. Materials & Methods
The title compounds were synthesized from CuCl, Bi and $BiCl_3$ in the ionic liquid $[BMIm^+]Cl^-$ + 4 $AlCl_3$ (BMIm = 1-*N*-butyl-3-methylimidazolium) under inert gas at 180 °C. After sixty hours the mixtures are slowly cooled to room temperature resulting in the crystallization of air sensitive shiny black planks. The crystal structures were determined by single crystal X-ray diffraction.

4. Results
Two new compounds were obtained via this synthesis route: $[Cu(Bi_8)](AlCl_4)_2(Al_2Cl_7)$ and $[Cu(Bi_8)](AlCl_4)_3$, both containing $[Cu(Bi_8)]^{3+}$ units in which Cu^+ ions are η^4 coordinated by Bi_8^{2+} clusters. In contrast to the Au compounds, the Cu coordination is completed by one tetrahedral $AlCl_4^-$ anion in a η^1 coordination, Fig. 1b. The resulting $[Cu(Bi_8)(AlCl_4)]^{2+}$ units are separated by non-coordinated $AlCl_4^-$ and, in case of $[Cu(Bi_8)](AlCl_4)_2(Al_2Cl_7)$, also $Al_2Cl_7^-$ anions. The molecular structure of $[Cu(Bi_8)](AlCl_4)_3$ is prone to disorder, several orientations of the $[Cu(Bi_8)]^{3+}$ units are found, Fig. 2.

5. Conclusion
The successful synthesis of $[Cu(Bi_8)](AlCl_4)_2(Al_2Cl_7)$ and $[Cu(Bi_8)](AlCl_4)_3$ shows that the coordination chemistry of Bi_8^{2+} polycations can be extended toward other transition metals.

References:
[1] M. Ruck, F. Locherer, *Coord. Chem. Rev.* **2015**, *285*, 1–10.
[2] M. Ruck, *Z. Anorg. Allg. Chem.* **1997**, *623*, 1591–1598.
[3] U. Müller, A. Isaeva, J. Richter, M. Knies, M. Ruck, *Eur. J. Inorg. Chem.* **2016,** 3580–3584.

Figure 1 a) Structural motive of a Bi_8^{2+} coordinated gold ion in $[Au(Bi_8)](AlCl_4)_5$ **b)** Structural motive of a Bi_8^{2+} and $AlCl_4^-$ coordinated copper ion in $(CuBi_8)(AlCl_4)_2(Al_2Cl_7)$ and $(CuBi_8)(AlCl_4)_3$

Figure 2 Resolution of the statistical disorder of $[Cu(Bi_8)(AlCl_4)]^{2+}$ units in $(CuBi_8)(AlCl_4)_3$

Figure 1

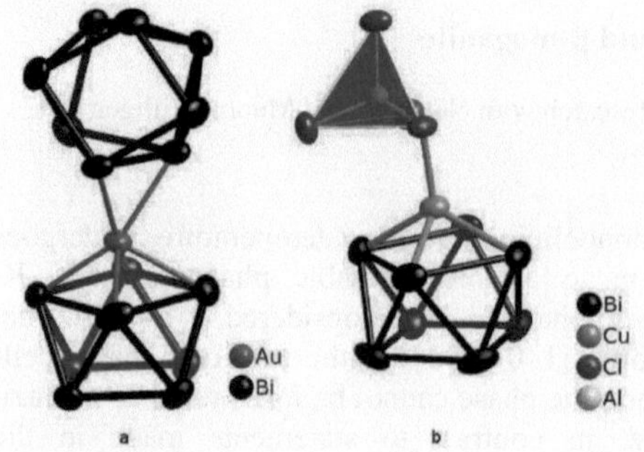

Figure 2

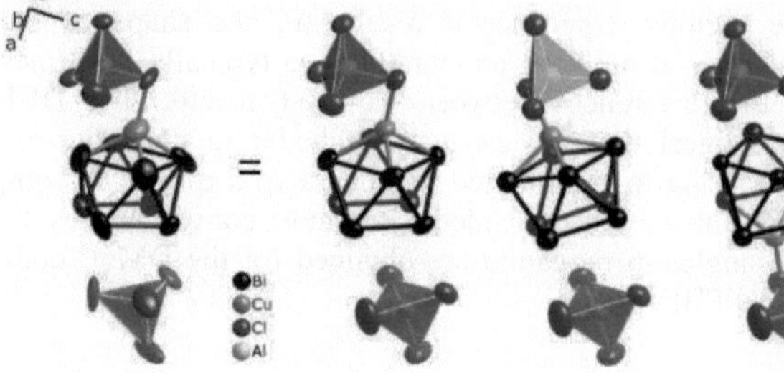

P130

Triethanolammonium ion conformation and hydrogen bonding

C. Schwalbe[1,2]
[1]Cambridge Crystallographic Data Centre, Cambridge, United Kingdom
[2]Aston University, School of Life & Health Sciences, Birmingham, United Kingdom

The triethanolammonium ion (TEA, **1**) offers 4 hydrogen donor groups, but its ethanol moieties can surround the $^+$NH donor (synclinal H-N-C-C torsion angles), hindering its intermolecular interaction.

Párkányi et al. surveyed neutral and cationic TEA moieties in crystal structures then available [1] finding only this conformation, called *endo* or, more picturesquely, lampshade [2]. According to their *ab initio* calculations [1] this form of neutral TEA is more stable than the fully *exo* conformation, and making just one leg *exo* cost ≈ 1 kcal mol^{-1}. A new search of the Cambridge Structural Database (version 5.37 with 3 updates) requiring R <= 10% and excluding entries with errors has found 47 independent structures of TEA salts comprising 68 independent ion pairs. In 58 of these all three H-N-C-C torsion angles are synclinal, occasionally synperiplanar. This *endo* conformation renders the TEA liable to disorder, which is found in 12 of the structures and particularly affects the inner methylene groups. With *endo* TEA only the very small fluoride ion (GIMYUY, [2]) and the blade-like cymantrenecarboxylate ion, (PEHDAJ, [3]) can engage in $^+$N-H...X^- interaction. Water is an acceptor twice and there is one apparent dimer [4] (with H...H clashes). Otherwise the NH remains unused. The other 10 ion pairs have one single antiperiplanar H-N-C-C torsion angle in the TEA, creating a narrow opening for a hydrogen bond acceptor to gain access to the $^+$N-H group. Generally this hydrogen bonding is to one OH group of another adjacent cation: forming centrosymmetric paired N-H...OH dimers 5 times, N-H...OH chains 3 times, $R_2^1(7)$ ring incorporating water once and leaving the NH unused in one special case, perhaps because one OH group in that TEA cation is acting as a ligand in a metal complex. The hydrogen bonding preferences of the $^+$N-H group are not well satisfied by intramolecular hydrogen bonding to OH groups since the $^{+}$N-H...OH angles never exceed 121°. However, the OH groups are extensively involved in intermolecular hydrogen bonding.

Reference:
[1] L. Párkányi et al. (1996). *J. Mol. Structure* **377**, 27-33.
[2] M.G. Voronkhov et al. (2013). *Zh. Strukt. Khim.* **54**, 192-195.
[3] A.B. Ilyukhin et al. (2013). *J. Mol. Structure* **1033**, 187-199.
[4] D-L. Long et al. (2004). *J. Amer. Chem. Soc.* **126**, 13880-13881.

Figure 1

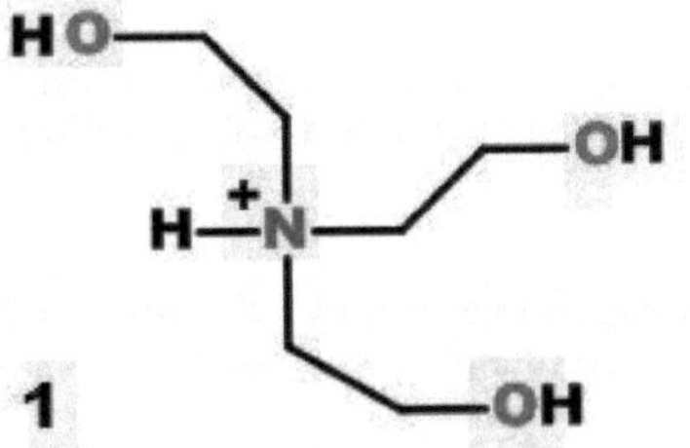

1

P131

Twinning by Merohedry of 1,3,5-Tribenzyl-4,6-tetrahydrothiazine-2-thione and -selenone

W. Frank[1]
[1]Heinrich-Heine-Universität Düsseldorf, Institut für Anorganische Chemie und Strukturchemie II, Düsseldorf, Germany

The isotypic title thiourea (**1**) and selenourea (**2**) derivatives are prepared by sulfur and selenium trapping, respectively, of the related diamino-N-heterocyclic carbene that is not stable at ambient temperature.[1] Crystals of the sulfur compound were grown from ether/hexane, crystals of the selenium compound from methylene chloride. Taking into account the symmetry of the diffraction patterns, the systematic absences, as well as the number of formula units in the unit cell, the members of the enantiomorphous pair $P3_1$ and $P3_2$ were identified as possible space group types for the crystals. For the crystals under investigation, by inspection of the anomalous dispersion (conventional Flack parameter)[2] $P3_1$ proved to be the right choice for the thione [−0.04(12)] and $P32$ [−0.002(16)] for the selenone. Crystal data and symmetry and anomalous dispersion informations are given in the table shown in Figure 1.

Crystals of **1** and **2** notoriously suffer from twinning by merohedry and, assuming additivity of the intensities of the reflections of the twin components, the twin law 0 1 0 1 0 0 0 0 −1 (expressed as the matrix that transforms the hkl indices of one component into the other) was taken into account. The fractional contributions of the twin components for the crystals under investigation were 0.376(3) and 0.624(3) for the thione and 0.310(4) and 0.690(4) for the selenone. There are no indications for additional racemic twinning in both cases. Details concerning intensity data and twin refinement are given in the table shown in Figure 2. The structures of the molecules in the crystals of **1** (left) and **2** (right) are shown in the lower part of Figure 2.

References:
[1] A. Makhloufi, W. Frank, C. Ganter, *Organometallics*, **2012**, *31*, 2001.
[2] H. D. Flack, *Acta Crystallogr.*, **1983**, *A39*, 876.

Figure 1

Compound	Thione **1**	Selenone **2**
Formula	$C_{24}H_{25}N_3S$	$C_{24}H_{25}N_3Se$
Unit cell parameter	a = 9.2617(4) Å, c = 20.5439 (9) Å	a = 9. 3268(6) Å, c = 20.6644(18) Å
V (Å^3) / Z	1526.14(11) / 3	1556.7(2) / 3
T (K)	291	291
Point Group Symmetry of single molecule	1	1
Reflection Conditions	000l: l = 3n	000l: l = 3n
Space Group of Crystal under Investigation	$P3_1$	$P3_2$
Flack-Parameter[2]	−0.04(12)	−0.002(16)

Figure 2

Compound	Thione **1**	Selenone **2**
$N_{ref.}/N_{uniq.}/N_{gt.}/N_{param.}$	21864/3951/3214/254	12575/3639/3310/254
R_{int}/R_{sigma}	0.0410/0.0309	0.0425/0.0313
R_1/wR_2 (gt.)	0.0553/0.1378	0.0457/0.1008
R_1/wR_2 (tot.)	0.0649/0.1437	0.0544/0.1045
S	1.076	1.077
Twin Components	0.376(3)/0.624(3)	0.310(4)/0.690(4)
Solution & Refinement	SHELX	SHELX

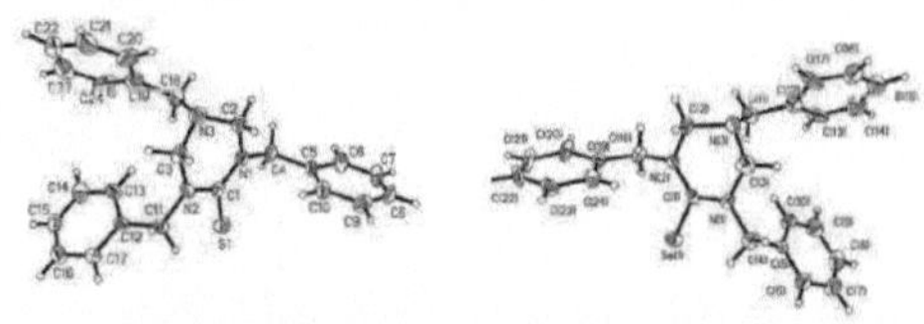

P132

Ternary trielides A_3M_5: A combined synthetic, crystallograficand bond theoretical study

C. Meyer[1], C. Röhr[1]
[1]Albert-Ludwigs-Universität Freiburg, Institut für anorg. und analyt. Chemie, Freiburg, Germany

Alkaline-earth trielides $A^{II}{}_xM^{III}{}_y$ are interesting intermetallic compounds to investigate. They reveal a rich structural diversity reaching from classic electron precise Zintl phases like $SrIn_2$[1] to `hypoelectronic` compounds like CaGa[2].

Belonging to a derived class of electron deficient structures, Ca_3Ga_5 and Sr_3In_5 (Ca_3Ga_5 structure type, SG $Cmcm$, fig. 1) are known as intergrowth structures consisting of KHg_2 and CrB building blocks[3,4]. Both, KHg_2 as well as CrB are common structure types of intermetallic phases, which have been studied systematically for mixed trielides [1,2].

By substitution of the triel elements against each other, ternary compounds $Ca_3Ga_{5-x}Al/In_x$ and $Sr_3In_{5-x}Ga/Al_x$ were synthesized and characterized by means of single crystal X-ray diffraction. The phase widths of the known binary phases seem to be limited by the radius ratio rA^{2+}/rM (rA^{2+}/rM = 0.866-0.950), which is not true for a small series of $Sr_3Al_{5-x}Ga_x$ (x = 1.4-2.6, rA^{2+}/rM = 1.010-1.013), which could be obtained although the binary phases do not exist[5]. The differences in electronegativity and size determine site preferences for each triel combination (fig 2). Combining the experimental results concerning the `coloring` of the polyanionic network with the calculated Bader charges and volumes leads to several important conclusions: In case of triel combinations with a large difference in electronegativity (e.g. Al and Ga) the most negatively charged and largest $M1$ position is taken by the more electronegative element, whereas in cases of combinations with large size difference (Al and In) the larger element occupies the $M1$

site. As a consequence, the neighboring position *M2* is taken by the other triel element to avoid homonuclear *M1*-*M2* bonding. This fact is confirmed by calculated total energies for pairs of compounds with `inverse coloring` (homo/ heteronuclear *M1*-*M2* bonds, ΔE = 0.08-0.27 eV).

References:
[1]: W. Harms, M. Wendorff, C. Röhr, *Z. Naturforsch.* **62b**, 177-194 (2007).
[2]: W. Harms, M. Wendorff, C. Röhr, *J. Alloys Compd.* **469**, 89-101 (2009).
[3]: D.-K. Seo, J. D. Corbett, *J. Am. Chem. Soc.* **123**, 4512-4518 (2001).
[4]: G. Cordier, H. Schäfer, M. Stelter, *Z. anorg. allg. Chem.* **539**, 33-38 (1986).
[5]: C. Meyer, C.Röhr, *Z. Kristallogr. Suppl.* **36**, 80 (2016).

Figure 1

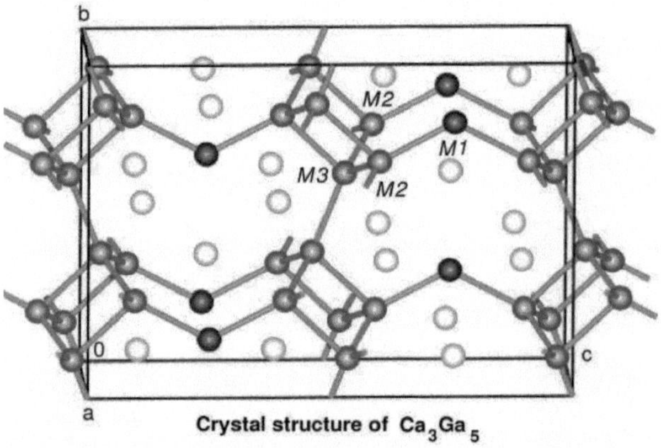

Figure 2

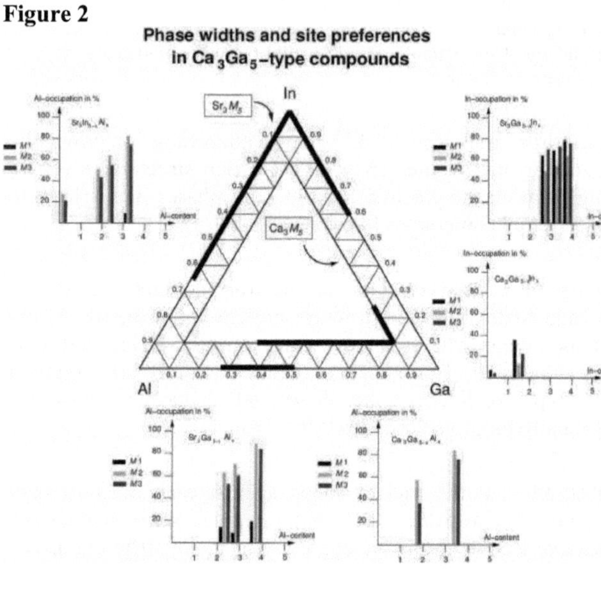

P133

Hydrogen bonds and π-π interactions are structure-directing tools of the 4-aminopyridium cation for the construction of a complex framework

G. J. Reiss[1], M. van Megen[1], W. Frank[1]
[1]Heinrich-Heine-Universität Düsseldorf, Institut für Anorganische Chemie und Strukturchemie, Düsseldorf, Germany

Aminopyridinium cations are known to be excellent structure-directing agents [1,2]. Especially the 4-aminopyridinium is an outstanding building block to form salt-type materials, as the hydrogen bond donor groups at both sides ensure a bridging connection of counter anions. We could, for example, show that the framework of the crystal structure of 4-aminopyridinium hydrogen (9-*phosphonononyl*)phosphonate has separated regions dominated by hydrogen bonds, π-π interactions and van-der-Waals forces, respectively.

The reaction of 4-aminopyridine (*apy*) with 1,2-ethanediphosphonic acid and recrystallization from methanol/water gives the addition compound
$(apyH)_8[HO_3P(CH_2)_2PO_3H]_2[O_3P(CH_2)_2PO_3]\ 10H_2O$ (space group *P*-1, a = 10.5907(3) Å, b = 10.6239(3) Å, c = 18.2795(4) Å, α = 88.201(2)°, β = 75.799(2)°, γ = 64.247(3)°, V = 1789.15(9) Å^3, Z = 1, wRref(F^2) = 0.0867, Rgt(F) = 0.0345).

One of the four crystallographically independent *apy*H cations, the anions and the water molecules form a hydrogen bonded framework with a complex pattern, resembling a distorted hexagonal packing of rods. The remaining three crystallographically independent *apy*H cations are stacked along [100]. Each of these three *apy*H cations has an individual orientation in the stack. This kind of stacking with 3-4 repetition units is a typical feature of *apy*H salts [4], as well as for related compounds [5]. The distances between the planes of the stacked *apy*H cations in the range of 3.43-3.54 Å indicate weak to medium strong π-π interactions.

References:
[1] M.-C. Hsien, H.-M. Kao, K.-H. Lii, *Chem. Mater.* **2001,** *13,* 2584-2588.
[2] C. R. Kennedy, S. Lin, E. N. Jacobsen, *Angew. Chem. Int. Ed.* **2016,** *55,* 12596-12624.
[3] M. van Megen, G. J. Reiss, W. Frank, *Acta Crystallogr.* **2016,** *E72,* 1456-1459.
[4] G. J. Reiss, P. B. Leske, *Z. Kristallogr. – NCS* **2014,** *229* 452-454.
[5] T. Zhang, Y.-Y. Jia, X.-L. Sun , *Z. Kristallogr. – NCS* **2016,** *231,* 295-297.

Figure 1

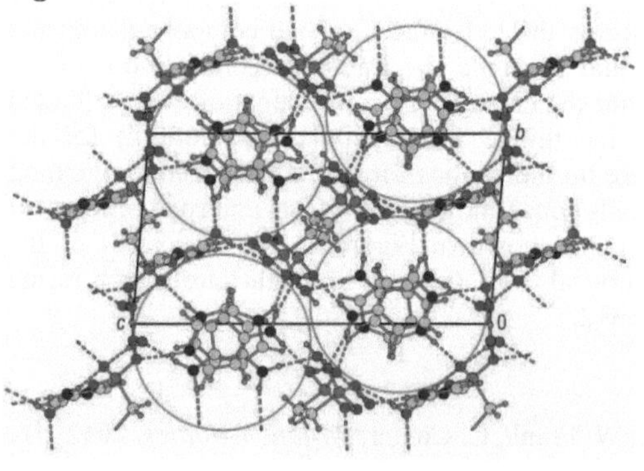

P134

The structure of Br@NaY, $Na_{45}H_{15}[Si_{139}Al_{53}O_{384}Br_7]$ * 247 H_2O, a NaY zeolite impregnated with NaBr

B. Marler[1], M. Müller[1], A. Wozniak[1], H. Gies[1]
[1]Ruhr University Bochum, Geology, Mineralogy and Geophysics, Bochum, Germany

Introduction: Zeolite/water systems offer a high potential as heat storage materials. Storage properties of a zeolite can be fine-tuned to meet specific requirements by choosing an appropriate Si/Al ratio, the type of cation, or by impregnating the zeolite with additional ions. Seidel et al. [1] published a study on the impregnation of NaY zeolite (Si:Al = 2.4) with NaBr and proposed a structure model based on ^{23}Na MAS NMR spectroscopic data: The silicate framework of NaY which contains small β-cages and large super-cages incorporates Br^- in the β-cages as $[Na_4Br]^{3+}$ clusters. Boddenberg et al. showed that the presence of these clusters has an influence on the water adsorption properties of NaY [2]. Here we present the structure analysis of an analogous material.

Materials & methods: A commercial zeolite NaY (Si:Al = 2.6) was used for the incorporation of NaBr into the structure of NaY (according to the procedure of Seidel et al.) yielding the zeolite Br@NaY: NaY was mixed with NaBr and heated at 700°C under high vacuum for 24 h. After cooling, the material was intensively washed whereby the salt was completely removed from the super-cages of the structure but remained in the small β-cages. Powder diffraction data were recorded from a Siemens D5000 diffractometer. The Rietveld structure refinement was performed using the Fullprof 2K program.

Results: The structure analysis (Chi^2 = 1.76, R(F) = 0.039) confirmed the existence of $[Na_4Br]^{3+}$ clusters. Br^- is tetrahedrally coordinated by Na^+ ions (Na1) with a bond distance of 2.79(1) Å (Fig. 1a). The other Na^+ ions (Na2), located in the super-cage, are bonded to three framework oxygen atoms of the 6-ring which interconnects β- and super-cage (Fig. 1b). Na2 is additionally coordinated by a water molecule (W2). The position in front of this 6-ring is only partially occupied by Na2 and W2 (ca. 40%) but predominantly by water molecule W1 (ca. 60%). In total, 5 different water sites exist in the super-cage, protons could not be located.

The $[Na_4Br]^{3+}$ clusters which are present in 86% of the β-cages reduce only slightly the total uptake of water (24wt% instead of 25wt%) but decrease the temperature of water desorption which is beneficial for storage of (low temp.) solar heat (Fig. 2).

References:
[1] A. Seidel et al., Solid State Nuc. Mag. Res. **9**, 129, 1997. [2] Boddenberg et al., PCCP **4**, 4172, 2002.

Figure 1

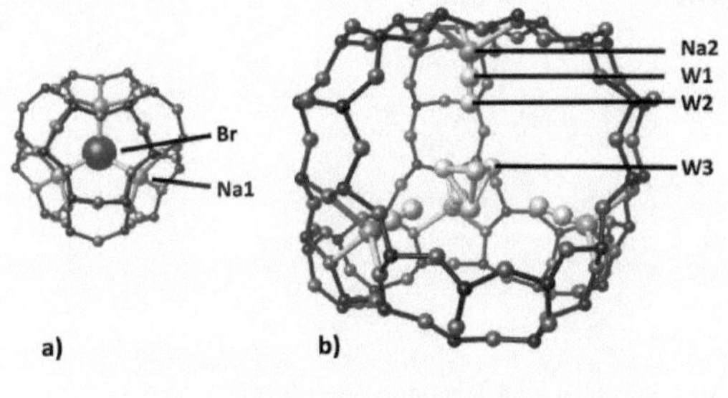

Fig. 1: The structure of Br@NaY: a) the $[Na_4Br]^{3+}$ cluster in the β-cage; b) arrangement of Na^+ ions (Na2) and water molecules (W1, W2, W3) in the super-cage. Water molecules W4 and W5 are omitted for clarity.

Figure 2

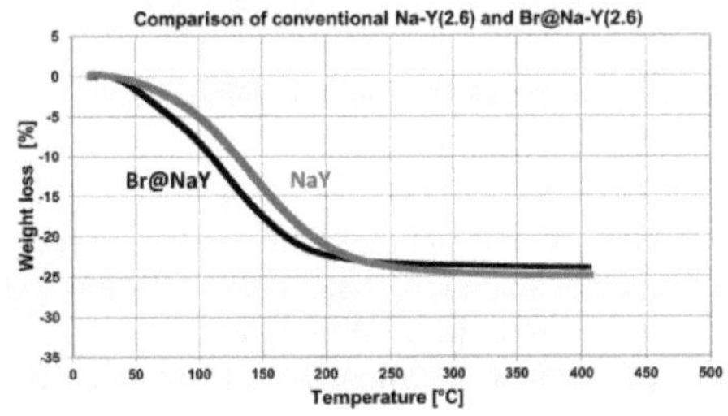

Fig. 2: Thermogravimetric curves showing the water desorption from conventional NaY and Br@NaY.

P135

Temperature-dependent structure and thermal stability of $|Na_8(MO_4)(H_2O)_2|[AlSiO_4]_6$ cancrinites

H. Petersen[1], L. Robben[1], T. M. Gesing[1]
[1]Universität Bremen, Institut für anorganische Chemie und KristallographieMAPEX Center for Materials and Processes, Bremen, Germany

Cancrinites (CAN) crystallizing in space group $P6_3$ consist of six-membered ring sheets formed by alternating AlO_4 and SiO_4 tetrahedra, stacked in an *ABAB*... sequence. This leads to the typical columns of small ε-cage as well as large 12-membered channels running parallel to the *c* axis. In these channels template cations and the anions are enclathrated, whereas additional template cations and the water molecules are found in the ε-cages. The thermal expansion of the CAN framework causes an increase in the sodium framework-oxygen distances [1,2] and by this a tilting of the TO_4 tetrahedra. The CAN thermal stability has been studied earlier [1–4]. Barrer et al. [4] describe theses thermal stabilities to be proportional to those of the respective template salt.

To clarify the template-framework interactions and the effect of the template on the CANs" stability a comparative study with CANs enclathered with MnO_4^{2-}, CrO_4^{2-}, SeO_4^{2-}, MoO_4^{2-} and WO_4^{2-} was carried out.

All compounds were prepared hydrothermally at 393 K (Mn) 473 K (Cr), 423 K (Se) and 503 K (Mo, W) using a NaAlO2, Na2SiO3, NaOH and NaMO4.

From the temperature-dependent powder X-ray diffraction data the thermal expansion behavior as well as the common water release of the structure could be observed. Between 915 K (W) and 980 K (Cr) a reconstructive phase transition to the nosean type structures (NOS, *ABCABC*.. stacking) could be observed for M = Cr, Mo, W, Se.

The CAN-NOS transition temperature could be linearly correlated with the ion size of the templates.

References:
[1] I. Hassan et al. Am. Mineral. 91 (2006) 1117
[2] G.D. Gatta et al. Am. Mineral. 101 (2016) 253
[3] I. Hassan Can. Mineral. 34 (1996) 893
[4] R.M. Barrer et al. J. Chem. Soc. (1970) 1516

P136

Crystal Engineering of layered hybrid compounds: a layered inorganic-organic hybrid material derived from the "short-chain amphiphile" sodium methanesulfonate and chloroauric acid

F. Thoelen[1], W. Frank[1]
[1]Heinrich-Heine Universität Düsseldorf, Institut für Material- und Strukturforschung, Düsseldorf, Germany

According to the properties and applications, methanesulfonates are widely used as precursor materials[1] and also will shown to be adequate compounds to generate layered inorganic-organic hybrid materials in terms of crystal engineering. Thin platelets of $(H_2O)_6Na_{11}(CH_3SO_3)_8[AuCl_4]_3 \cdot 2\ H_2O$ (**1**) were prepared by the slow evaporation of a solution of H_2O/MeOH containing sodium methanesulfonate and tetrachloridogold(III) acid at room temperature. Compound **1** crystallizes in the triclinic space group *P*-1 with the lattice parameters $a = 8.9868(6)$ Å, $b = 11.0876(7)$ Å, $c = 16.1312(9)$ Å, $\alpha = 76.455(5)$ °, $\beta = 81.254(5)$ °, $\gamma = 72.590(5)$ ° ; $Z = 1$, ($R_1 = 0.0468$, $wR_2 = 0.0496$, 7981 reflections and 450 parameters). Figure 1 shows the unit cell of the crystal structure of **1**, which is characterized by a rigorous hierarchic layered structure. Thereby it is possible to differentiate two different types of layers. A tenside-like bilayered array contains a $\{(Na_{10}[CH_3SO_3]_8)^{2+}\}_n$ cation with a hydrophilic and a hydrophobic layer region. The latter is formed by the methyl groups of the methanesulfonate anions and is disrupted by water molecules, i.e. in comparison to the parent structure of sodium methanesulfonate[2] the organic layer is broken up at definite places, so that hydrogen bonding is enabled to an inorganic polymeric strand layer ($\{(Na[AuCl_4]_3)^{2-}\}_n$), containing alternatively arranged planar tetrachloridoaurate ions and sodium cations (Figure 2). The different layers are connected to form a complex hydrogen bonding network. There are two types of hydrogen bonds, which seem to have significant influence on the formation of the different building blocks of this structure. The strongest hydrogen bonds of O–H···O type (D···A distances from 2.838(8) Å to 2.909(8) Å) are present between the aqua ligands of the sodium ions and those water molecules, that extend into the organic layer. Medium and weak O–H···Cl hydrogen bonds (D··· A distances from 3.324(6) Å to 3.497(6) Å) are present between chlorido ligands of the tetrachloridoaurate(III) ions and additional water molecules, which are located between the aforementioned layers. In conclusion, the new compound is classified as a novel layered inorganic-organic hybrid material with tensidic bilayered regions that are connected to layers of pure inorganic components by hydrogen bonding.

Reference:
[1] U. Betke, K. Neuschulz, M. S. Wickleder, *Chem. Eur. J.* **2011**, *17*, 12784-12801.
[2] C. H. Wei, B. E. Hingerty, *Acta Crystallogr.* **1981**, B*37*, 1992–1997.

Figure 1

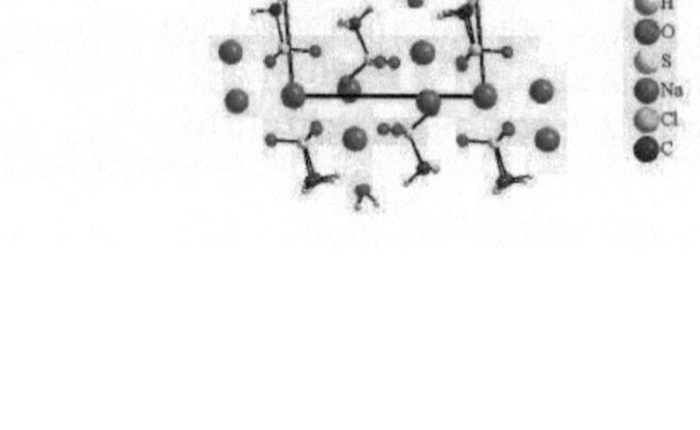

Figure 2

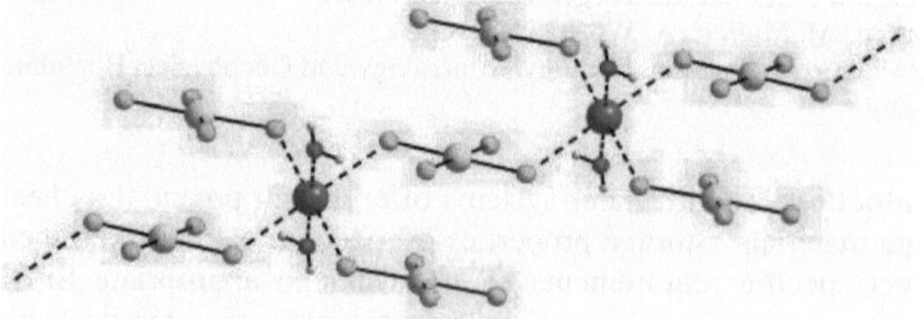

P137

Synthesis and Structural Characterization of Li3K3Y7(BO3)9

S. Bräuchle[1], H. Huppertz[1]
[1]University of Innsbruck, Innsbruck, Austria

Over the last couple of years, the synthesis and characterization of new borates are gaining more and more interest, because of their interesting physical properties. Borates exhibit a high laser damage threshold and a very high transparency down into the deep UV. Several rare earth (*RE*) borates such as YBO3 and GdBO3 [1] doped with Eu3+ found application as red phosphors in plasma display panels (PDPs). LiB3O5 (LBO) [2] and *β*-BaB2O4 (BBO) [3] are known for their excellent nonlinear-optical behavior. In this context, a new compound Li3K3Y7(BO3)9 was prepared by high-temperature solid state synthesis at 900 °C in a platinum crucible from lithium carbonate, potassium carbonate, boric acid, and yttrium(III) oxide. The compound crystallizes in the orthorhombic space group *Pca*21 (no. 29) ($Z = 4$). The structure was refined from single-crystal X-ray diffraction data: $a = 20.743(8)$, $b = 6.387(4)$, $c = 17.474(4)$ Å, $V = 2315.2(2)$ Å3, $R1 = 0.0473$, and $wR2 = 0.0637$ for all data. The crystal structure of Li3K3Y7(BO3)9 consists of LiO6 octahedra which are connected *via* planar BO3 groups forming [Li3B4O21]*n* units in the *ac* plane. These units are linked by additional BO3 groups, building up a [Li12B32O96]*n* network. The network is forming tunnels of twelve-membered rings by six LiO6 octahedra and six BO3 groups, in which one isolated BO3 group is positioned. The Y atoms are eight- and nine-fold coordinated to oxygen anions, whereas the K+ ions are coordinated to ten, eleven, and twelve oxygen anions. Both cations are arranged in layers along the *a*-axis.

References:
[1] G. Chadeyron, M. El-Ghozzi, R. Mahiou, A. Arbus, J. Cousseins, *J. Solid State Chem.* **1997**, *128*, 261.
[2] S. Lin, Z. Sun, B. Wu, C. Chen, *J. Appl. Phys.* **1990**, *67*, 634.
[3] C. Chen, B. Wu, A. Jiang, G. You, *Sci. Sin. B-Chem. B. A. M.* **1985**, *28*, 235.

Figure 1

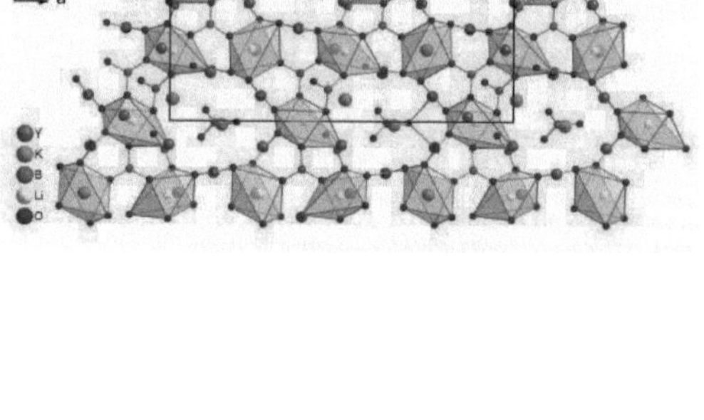

P138

Synthesis and characterization of a novel ferrite $K_2Fe_4O_7$

M. Gogolin[1,2], M. M. Murshed[1,2], T. M. Gesing[1,2]
[1]University of Bremen, Institute of Inorganic Chemistry and Crystallography, Bremen, Germany
[2]MAPEX Center for Materials and Processes, Bibliothekstraße 1, Universität Bremen, Bremen, Germany

Due to their commercial importance, globally as much as 300000 tons of ferrites are produced each year [1], in particular for use as sintered permanent magnets and data storage. In search of new members of this loosely defined family of materials, potassium-containing diferrite with an idealized chemical formula of $K_2Fe_4O_7$ has been hydrothermally synthesized. The crystal structure determined from single crystal X-ray diffraction data [2] was refined using least squares methods ($R1$ = 0.026). The compound crystallizes in the space group *P-3*, with hexagonal lattice parameter a = 518.71(2) pm, c = 691.98(3) pm and Z = 1. The structure was confirmed by X-ray powder diffraction refinements and has further been characterized using SEM/EDX, FTIR and Raman spectroscopy. $K_2Fe_4O_7$ is closely related to barium diferrite $BaFe_4O_7$ [3], described in the non-isomorphic supergroup *P63/m* with an approximately doubled c-cell parameter. Amid such differences, the arrangement of atoms is strikingly similar, with iron being equally distributed over the octahedral and the tetrahedral sites which form layers in the *ab*-plane as shown in the figure. The K^+ cations are located in the interstitial sites between the layers, analogous to the Ba^{2+} cations in $BaFe_4O_7$. The occupancy factor obtained from the single crystal refinement, in combination with OH-stretching modes observed in the IR-spectra point towards a chemical composition of $K_{1.78(8)}H_{0.22(8)}Fe_4O_7$.

Figure: Crystal structures of $BaFe_4O_7$ (left) and $K_2Fe_4O_7$ (right) showing the empty spaces in barium diferrite compared to the potassium analogue.

[1] R.C. Pullar, Prog. Mater. Sci. 57 (2012) 1191. [2] G. M. Sheldrick, Acta Crystallogr. A64 (2008) 112. [3] S. Okamoto, S. I. Okamoto, T. Ito, Acta Crystallogr. B29 (1973) 832.

Figure 1

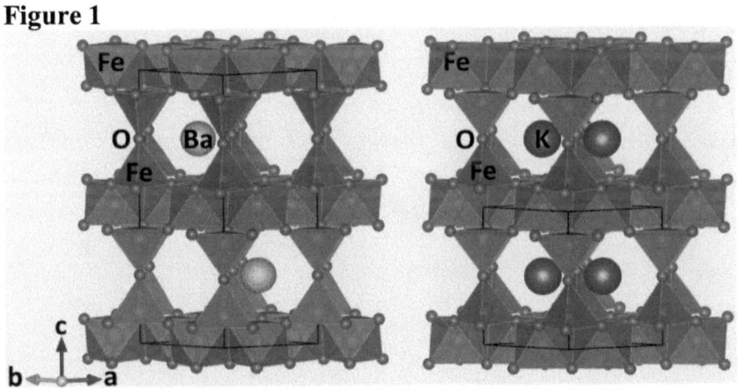

P139

Experimental electron density in bromomalonic aldehyde

R. Wang[1], J. George[1], R. Dronskowski[1], U. Englert[1]
[1]RWTH Aachen, Institue fuer Anorganische Chemie, Aachen, Germany

To understand the cohesive forces in bromomalonic aldehyde [1] beyond chemical intuition, intensity data are accurately measured at low temperature (100K) and up to high resolution. Multipole refinement [2,3] and interpretation of the experimental electron density have been done according to Bader's *AIM* theory [4].

The transition from intra- to inter-molecular geometry could follow chemical intuition: the classical hydrogen bond (scheme) links translationally equivalent molecules along the polar crystallographic c axis and is doubtlessly strong (O...O = 2.628(3) Å). In the b direction, weaker intermolecular halogen bonds (O...Br = 3.149(2) Å) and non-classical C–H...O bonds (H...O = 2.65 Å) connect the hydrogen-bonded strands to layers in the bc plane. Stacking of these layers finally leads to the three-dimensional structure (figure right).

Acknowledgements

Financial support from the DFG is gratefully acknowledged.

References:

[1] V. L. Deringer, F. Pan, J. George, P. Müller, R. Dronskowski, U. Englert, *CrystEngComm* **2014**, 16, 135-138.
[2] N. K. Hansen, P. Coppens, *Acta Crystallogr. Sect. A* **1978**, 34, 909-921.
[3] A. Volkov; P. Macchi; L. J. Farrugia; C. Gatti; P. Mallinson; T. Richter; T. Koritsanszky, XD2016 - A Computer Program Package for Multipole Refinement, Topological Analysis of Charge Densities and Evaluation of Intermolecular Energies from Experimental and Theoretical Structure Factors, 2016.
[4] R. F. W. Bader, 'Atoms in Molecules - a Quantum Theory', Clarendon Press, 1990.

Figure 1

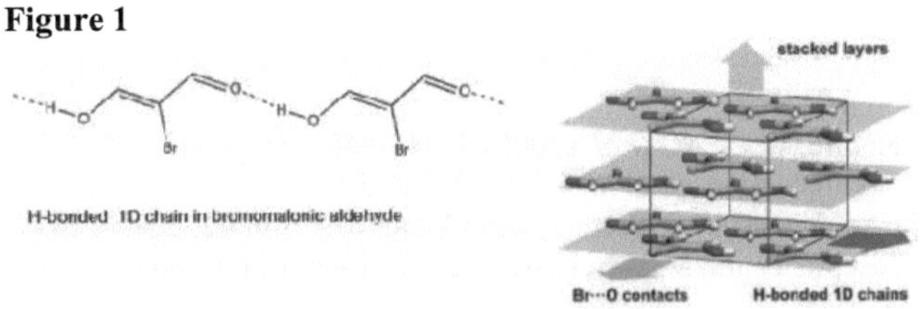

P140

Crystal growth of Mn-doped melilites

P. Teichmann[1], M. Burianek[1], R. Fischer[1]
[1]Universität Bremen, Bremen, Germany

The large family of tetragonal, noncentrosymmetric melilite-type compounds represents an interesting class of materials due to its magneto-electric and multiferroic properties. The chemical variability of the melilite family of crystal structures offers a wide range of possibilities for the design of new crystalline materials. The general composition is given by the formula $M_2^{[8]} T^{[4]} T''_2{}^{[4]}X_7$ with M = Ca, Na, Sr, Pb, Ba, RE; T = Be, Mg, Cu, Co, Zn, Fe^{2+}, Fe^{3+}, Mn^{2+}, Cd, Al, Ga; T" = Si, Ge, Al, Fe^{3+}, Ga, Be; and X = O, F, N, S [1, 2]. A comprehensive overview of the known melilite-type crystals is given by Kaminskii (2008) and Röthlisberger [1], [2].

Interesting magnetic, optical, and dielectric properties have been reported for Mn-containing melilites [3-6]. However, these results are mainly based on polycrystalline materials. In order to systematically study the magnetic structure as well as the physical and chemical properties, large single crystals of Mn-doped melilites are needed.

We have successfully grown single crystals of the host mineral åkermanite ($Ca_2MgSi_2O_7$) using the Czochralski method. In first experiments Mn-doped single crystals were grown from melts containing 0.2-1.5 mol % MnO.

Crystals were grown under reducing conditions with forming gas (98%N_2/2%H_2). From this difficulties arose by an increased solubility of forming gas in the melt thus complicating the growth process. Finally single crystals were grown with very low growth rates (0.5 mm/h) to avoid the incorporation of gas bubbles into the crystal.

Mn-incorporation was confirmed by Microprobe analyses. In subsequent experiments optical and magnetic properties will be determined.

References:
[1] Kaminskii, A.A., *et al.* (2008) **5**, 845-868.
[2] Röthlisberger, F., *et al.* (1990) **2**, 585-594.
[3] Hao, Y. and Y.H. Wang (2007) Mater Res Bull **42**, 2219-2223.
[4] Joseph, T. and M.T. Sebastian (2010) J Am Ceram Soc **93**, 147-154.
[5] Koo, H.J. (2012) Solid State Commun **152**, 1116-1118.
[6] Koo, H.-J., *et al.* (2012) **324**, 3716-3718.

P141

Structural and Magnetic Properties of Eu_2TeO_6

M. E. Bohem[1,2], B. Blaschkowski[1], T. Schleid[1]
[1]University of Stuttgart, Institute for Inorganic Chemistry IAC, AK Schleid, Germany
[2]Fraunhofer Institut für Produktionstechnik und Automatisierung, Abt. Beschichtungssystem- und Lackiertechnik, Stuttgart, Germany

Phase-pure powders and single crystals of Eu_2TeO_6 were synthesized from europium sesquioxide and tellurium trioxide via a simple one-step solid-state reaction with T_{max} = 1050 K. The resulting transparent crystals have been identified as isostructural to other members of the series *RE*$_2$TeO$_6$ (*RE* = Y, La–Nd, Sm–Tm), representing the so-called A-type structure [1–3]. The compound crystallizes in orthorhombic space group $P2_12_12_1$ with the unit-cell parameters a = 534.28(4), b = 918.91(8), c = 1008.37(9) pm for Z = 2. Its structure contains two crystallographically independent Eu^{3+}-cation positions (Figure 1), both in sevenfold oxygen coordination (d(Eu1–O) = 230–255 pm; d(Eu2–O) = 226–262 pm). The Te^{6+} site shows six bonds to oxygen, resulting in almost regular octahedral $[TeO_6]^{6-}$ units (d(Te–O) = 191–196 pm; Figure 1, C). The six independent oxygen sites are not equally coordinated, since O1, O3, O5 and O6 reside in a trigonal planar cation environment, but O2 and O4 are surrounded by cationic tetrahedra. The photoluminescence spectra have been recorded at room temperature and the non-centrosymmetric nature of the *ortho*-oxotellurate(VI) is confirmed by the presence of $^5D_0$ -> $^7F_2$ as the strongest peak for the Eu^{3+} line luminescence [4].

Measurements of the molar magnetic susceptibilities χ_{mol} for Eu_2TeO_6 in a temperature range between 5 and 300 K at a static magnet field of 500 Oe show the typical van Vleck paramagnetic behaviour for Eu^{3+} cations [5, 6]. With decreasing temperature χ_{mol} increases until about 100 K, where the nearly temperature-independent plateau is reached (Figure 2). To better understand the processes behind, further studies will be of need. The structure determination of Eu_2TeO_6 belongs to a larger framework, which deals with the detailed analysis of new host lattices for pigment applications.

References
[1] J. Llanos, R. Castillo, D. Barrionuevo, D. Espinoza, S. Conejeros, *J. Alloys Compd.* **2009**, *485*, 565–568.
[2] P. Höss, Th. Schleid, *Acta Crystallogr.* **2007**, *E63*, i133–i135.
[3] P. Höss, Th. Schleid, *Z. Kristallogr.* **2008**, *Suppl. 28*, 165–165.
[4] K. Binnemans, *Coord. Chem. Rev.* **2015**, *295*, 1–45.
[5] H. Taira, *J. Mater. Chem.* **2002**, *12*, 148–152.
[6] J. H. van Vleck, The Theory of Electric and Magnetic Susceptibilities, *Oxford University Press*, Oxford (UK), **1932**.

Figure 1. Oxygen coordination polyhedra of $(Eu1)^{3+}$ (A), $(Eu2)^{3+}$ (B) and Te^{6+} (C).

Figure 2. Plot of the magnetic susceptibility χ against the temperature for Eu_2TeO_6 powder samples.

Figure 1

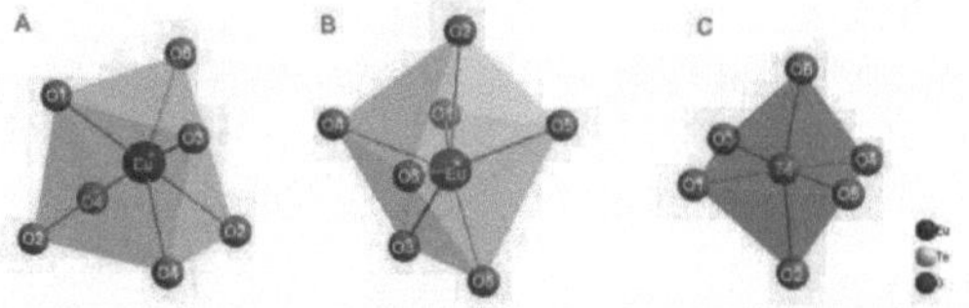

Figure 2

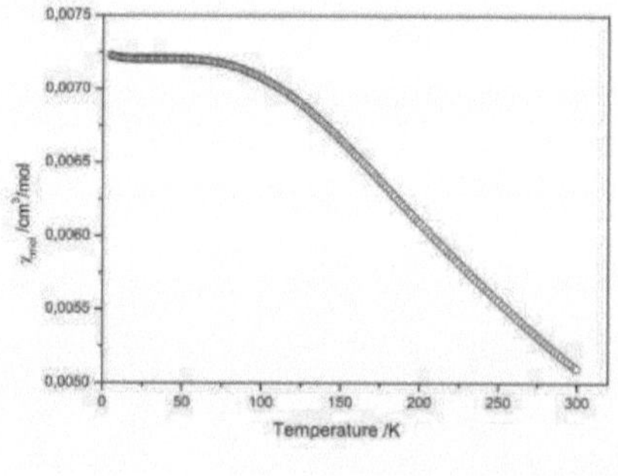

P142

Crystal and magnetic structures of multiferroic melilites: $Ba_2CoGe_2O_7$ versus $Ba_2MnGe_2O_7$

A. Sazonov[1], V. Hutanu[1], M. Meven[1], G. Roth[2]
[1]Institute of Crystallography, RWTH Aachen University and Jülich Centre for Neutron Science (JCNS) at Heinz Maier-Leibnitz Zentrum (MLZ), Garching, Germany
[2]Institute of Crystallography, RWTH Aachen University, Aachen, Germany

Recently, several members of the melilite family, including $Ba_2CoGe_2O_7$ and $Ba_2MnGe_2O_7$ have been found to exhibit static as well as dynamic magnetoelectric effects. The precise information for the atomic positions as well as about the spin order (crystal and magnetic structures) is essential to unravel the complex physics behind the magnetoelectric behavior of those systems. However, this information is still missing for many melilites.

Here, we present the results of the low-temperature (2.2 K – 10 K) single-crystal neutron diffraction studies of $Ba_2XGe_2O_7$ (X = Co, Mn). The experiments were performed on the four-circle diffractometer HEiDi at the hot-neutron source of the FRM II reactor, Heinz Maier-Leibnitz Zentrum (MLZ).

Both compounds have many similarities. Thus, the same tetragonal space group P-42_1m describes their crystal structures at room temperature and down to at least 10 K. At low temperatures, long-range antiferromagnetic order develops with $T_N \approx 6.7$ K for X = Co and $T_N \approx 4$ K for X = Mn. No evidence for a structural change upon the magnetic phase transition was observed. Nevertheless, different transition metal ions lead to different magnetic structures: While both compounds show a primary antiferromagnetic ordering in the (001) planes, the magnetic coupling between planes is ferromagnetic for X = Co, but antiferromagnetic for X = Mn. A non-collinear spin arrangement due to small canting within the (001) planes is allowed by symmetry in both cases. Differences between the magnetic structures are discussed based on the concept of Shubnikov groups (magnetic space groups). The reported structural parameters can serve as a profound experimental and theoretical basis to develop microscopic models describing the multiferroic nature and the peculiar magnetoelectric phenomena in melilites.

P143

Structure of the charge-density wave in CuV_2S_4

S. Ramakrishnan[1], A. Schönleber[1], H. A. B. Nyguen[1], F. Feulner[1], M. Anurova[1], S. van Smaalen[1], D. Chernyshov[2]
[1]University of Bayreuth, Laboratory of Crystallography, Bayreuth, Germany
[2]ESRF, SNBL, Grenoble, France

The symmetry of the spinel compound CuV_2S_4 at room temperature is de scribed by the space group Fd -3m [1]. At 90 K the diraction pattern shows a charge-density wave (CDW) transition with a modulation wave vector q = (3/4 + δ, 3/4 + δ, 0) [2]. The X-ray powder diraction shows a lowering of the symmetry of the lattice from cubic to tetragonal at temperatures below 90 K [1, 3]. The temperature dependence of the magnetization data report results of the decrease in magnetic susceptibility (χ) at 90 K followed by an upward jump at 54 K [4]. Further phase transitions at lower temperatures were proposed but not characterized. We have studied CuV_2S_4 by single-crystal X-ray direction pattern at temperatures down to 5 K. Here, a detailed analysis will be given of the CDW structure near the 90 K phase transition.

References:
[1] S. Kawaguchi, H. Ishibashi, N. Tsuji, J. Kim, K. Kato, M. Takata, and Y. Kubota, J. Phys. Soc. Jpn. **82** (2013) 064603.
[2] R. M. Fleming, F. J. DiSalvo, R. J. Cava, and J. V. Waszczak, Phys. Rev. B **24** (1981) 2850.
[3] H. Okada, K. Koyama and K. Watanabe, J. Phys. Soc. Jpn **73** (2004) 3227.
[4] I. Naik, C. S. Yadav, and A. K. Rastogi, Phys. Rev. B **75** (2007) 115122.

P144

High-Resolution Single-Crystal X-Ray Diffraction Investigations of *β*-Rhombohedral Boron

C. Eisele[1]
[1]University of Bayreuth, Laboratory of Crystallography, Bayreuth, Germany

The crystal structure of *β*-rhombohedral boron and its chemical bonding are still a matter of debate [1]. The most widely accepted structure model of *β*-rhombohedral boron is the one published by Slack *et al.* in 1988 [2]. *β*-boron crystallizes in space group *R-3m* and comprises 320 atoms per hexagonal unit cell. Its complex framework consists of 15 crystallographically independent boron atoms.
Additional interstitial sites, which are only partially occupied, and vacancies in the framework cause distortions from the ideal icosahedral geometry [2]. We have reinvestigated *β*-boron by high-resolution X-Ray diffraction at 100 K in order to establish a more detailed structure model, in particular with respect to the partial occupied sites (POS). The purpose is a more comprehensive concept of the POS in between the *β*-boron framework for understanding its striking thermodynamic stability at ambient pressure including all temperatures areas [1,3-6].

References
[1] T. Ogitsu, E. Schwegler, and G. Galli, Chem. Rev. **113**, 3425 (2013).
[2] G.A. Slack, C.I. Hejna, M.F. Garbauskas, and J.S. Kasper, J. Solid State Chem. **76**, 52 (1988).
[3] A. Masago, K. Shirai, and H. Katayama-Yoshida, Phys. Rev. B **73**, 104102 (2006).
[4] M.J. van Setten, M.A. Uijttewaal, G.A. de Wijs, and R.A. de Groot, J. Am. Chem. Soc. **129**, 2458 (2007).
[5] T. Ogitsu, F. Gygi, J. Reed, M. Udagawa, Y. Motome, E. Schwegler, and G. Galli, Phys. Rev. B **81**, 020102 (2010).
[6] T. Ogitsu and E. Schwegler, Solid State Sci. **14**, 1598 (2012).

P145

Introducing a new tripodal ligand including a hydroxamic acid moiety into coordination chemistry

D. Premuzic[1], M. Holynska[1]
[1]Philipps-Universität Marburg, Fachbereich Chemie, Marburg, Germany

The understanding of single-molecule magnets (SMMs) has been thoroughly extended in the last few years. The role of the molecular symmetry was underlined.[1,2] Furthermore, it was shown, that such bridging functional groups like oximes assist in generating a high spin ground state of SMMs.[3] Tripodal ligands may transfer three-fold symmetry on their metal complexes.

Combining these properties we use amine cores and extend it with functional groups leading to a semi-cryptand like coordination sphere (Fig. 1) with the ability to coordinate to further metals, allowing for multinuclear complexes synthesis. Herein we present the synthesis of the related ligand **1** and its first manganese metal complex (Fig. 2). The ligand, originally planned as oxime derivative, turned out to transform to a hydroxamic acid. A mononuclear anionic complex with the ligand and hexaaquamanganese(II) as counterion will be described.

References:

[1] T. Glaser, *Chem. Comm.*, **2011**, *47*, 116-130. [2] T. Glaser, V. Hoeke, K. Gieb, J. Schnack, C. Schröder, P. Müller, *Coord. Chem. Rev.*, **2015**, 289-290, 261-278. [3] M. Hołyńska, D. Premužić, I.-R. Jeon, W. Wernsdorfer, R. Clérac, S. Dehnen, *Chem. Eur. J.*, **2011**, *17*, 9605-9610. [4] G. Mezei, C. M. Zaleski, V. L. Pecoraro, *Chem. Rev.*, **2007**, *107*, 4933-5003.

Figure 1

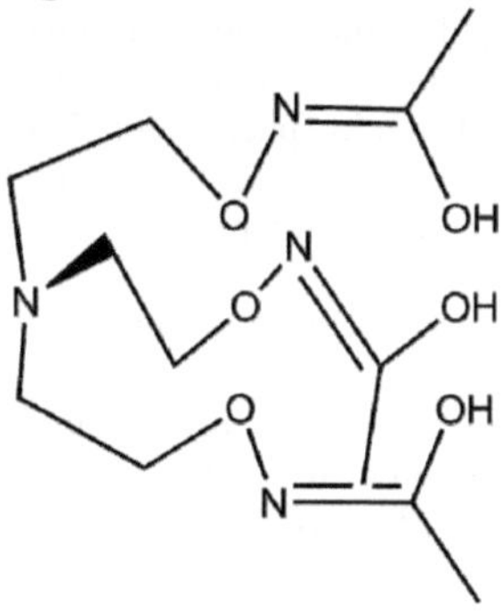

Figure

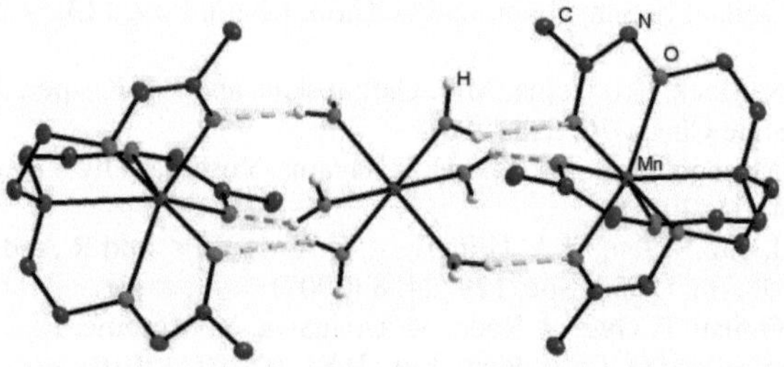

2?

P146

Structural relationships between mullite, sillimanite and "sillimullite"

S. Lenz[1], J. Birkenstock[1], L. A. Fischer[2], H. Schneider[1], R. X. Fischer[1]
[1]University of Bremen, Crystallography, Bremen, Germany
[2]Leibniz University of Hannover, Mineralogy, Hannover, Germany

The family of mullites belongs to the aluminosilicate system SiO_2-Al_2O_3 following the general composition $Al_2(Al_{2+2x}Si_{2-2x})O_{10-x}$. In this series sillimanite (x=0), 3/2-mullite (x=0.25), and 2/1-mullite (x=0.4) are described to be the main phases (e.g. Fischer and Schneider, 2005). Fischer et al. (2015) recently introduced the new mineral "sillimullite". It has a composition intermediate between sillimanite and 3/2-mullite and combines structural features of both phases. To reveal structural relations on the SiO_2-rich side of the mullite system we have investigated two crystals by single-crystal X-ray diffraction methods. Specimens were obtained from natural rock samples from the Bellerberg, Eifel, Germany and from Kladne, Czech Republic. Crystal structure analyses were carried out for both crystals. In addition to that, reciprocal sections were generated from the observed data. The crystal structure of the sample from the Eifel area was refined in the orthorhombic space group *Pnam* with a=7.4988(3) Å, b=7.6855(4) Å, c=5.7956(3) Å, V=334.02(3) Å^3 and Z=1. Structure analysis showed that it is "sillimullite". By observation of the reciprocal sections $h0l$ and $0kl$ a doubling of the c lattice parameter was evidenced by extra positions at $l \pm ½$. Thus, we could distinctly identify this crystal as "sillimullite". Referring to Fischer et al. (2015), our findings support that "sillimullite" is in fact a discrete new mineral phase. The sample from Kladne exhibits the mullite structure (*Pbam* with a=7.5439(3) Å, b=7.6958(4) Å, c=2.8879(1) Å, V=167.657(14) Å^3 and Z=1). From the lattice parameters it can be inferred that the chemical composition is near to that of stoichiometric 3/2-mullite but slightly shifted towards the SiO_2-rich side. This mineral could certainly be identified as mullite. Structure analysis revealed the existence of oxygen vacancies and triclusters. For this reason we would have expected significant diffuse X-ray scattering in the diffraction patterns. But astonishingly diffuse scattering could not be detected despite the very good counting statistics of the data. The reason for this observation is an open question at this point of our investigation.

References

[1] Fischer, R. X., and Schneider, H. *Wiley-VCH* **2005**, 1-46
Fischer, R. X., Tikhonova, V., Birkenstock, J., Fischer, L. A.,
[2] Herrmann, K., Mengel, K., and Schneider, H. *Am. Min.* **2015**, 100, 1493-1501

Phase transitions and dynamic phenomena

P147

The growth kinetics and the structure of expanded austenite in AISI316L stainless steels characterized by in-situ XRD

Z. Balogh-Michels[1], A. Faeht[2], S. Kleiner[2], J. M. Rufer[2], P. Markgraf[3], A. Dommann[1], A. Neels[1]
[1]Empa, Swiss Federal Laboratories for Materials Science and Technology, Center for X-ray Analytics, Dübendorf, Switzerland
[2]Bern University of Applied Science, Institute for Applied Laser, Photonics and Surface Technologies, Biel, Switzerland
[3]Härterei Gerster AG, Egerkingen, Switzerland

Introduction

Austenitic stainless steels are widely used in medical applications; however their low hardness leads to wearing and low lifetime [1]. Expanded austenite (EA) can be formed by low temperature N/C treatment. In EA N/C atoms occupy a large fraction of the interstitial sites [2] without the formation of nitrides/carbides, thus conserving the good corrosion resistance.

Recently significant effort is contributed to understand the properties of the EA, but it is far from being a closed topic. Key difficulties lie in the fact the multiple phenomena happens parallel to each other, in a material with sharp composition and stress gradients [3].

Objectives
We carried out experiments to study the growth of EA layer. Our key question was whether the process is diffusion limited or not? If diffusion limited, is it possible to characterize the process by a single (inter)diffusion coefficient? If other phenomena controls the growth of the EA is it possible to determine which?

Materials and methods
We used AISI316L stainless steel samples and carried out in-situ nitrocarburising in a custom built chamber attached to a Bruker D8 Advance diffractometer, with a Cu Kα source and a Vantec-1 PSD. We collected data in the 35-45° 2Θ range, including the 111 peaks of both the γ-Fe and the EA. The thickness of the EA was determined by intensity loss of the γ-Fe peak. The EA is modelled as a multilayer stack, where each 0.5 μm thick layer has a homogeneous lattice parameter. The EA XRD peak was fitted by Gaussians.

Results
The growth of the EA was modelled by linear-parabolic model. Depth profiling (Fig.1) have shown that the thickness of layer is constant in the Boltzmann-metric (η=0.5xt-1/2). This indicates a purely diffusion limited behavior. The lattice expansion however varies with the time. A scale independent master curve is possible after dividing the data with the lattice expansion of the first layer (Fig 2).

Conclusion
Our results indicate that two processes plays a key role in the growth. The thickness is determined by diffusion within the layer, while the exact composition by the uptake at the interface.

References:
[1] J. Buhigar et al., Surf. Eng. 23 (2007) 313.
[2] T. Christiansen, M.A.J. Somers, Metall. Mater. Trans. A, 37 (2006) 575.
[3] D. Manova, et al., J. Phys. D: Appl. Phys., 47 (2014) 365301.

Figure 1

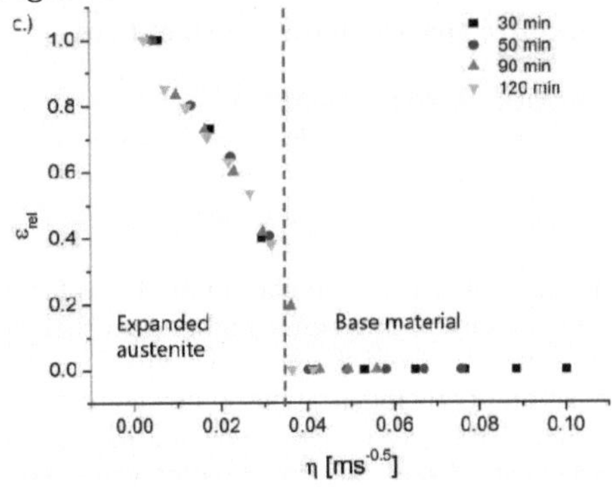

P148

A two dimensional dynamic instability of titanite, $CaTiSiO_5$

T. Malcherek[1], M. Fischer[2,3]
[1]Universität Hamburg, Mineralogisch-Petrographisches Institut, Hamburg, Germany
[2]University of Bremen, Crystallography group, Department of Geosciences, Bremen, Germany
[3]University of Bremen, MAPEX Center for Materials and Processes, Bremen, Germany

Phonon dispersion of titanite, $CaTiSiO_5$, has been calculated using the density functional perturbation theory as implemented in CASTEP. In the *C2/c* aristotype structure, imaginary frequencies are calculated for a phonon branch that is dominated by Ti-motion parallel to the Ti-O1 bond direction. The unstable part of the phonon branch is associated with wave vectors pointing normal to the octahedral chain direction, [001], of the titanite crystal structure. At the Y Brillouin zone boundary point, the unstable mode generates the distorted ground state titanite structure in space group symmetry $P2_1/c$. All calculated phonon modes are stable for this distorted structure and reasonable agreement with the experimentally observed structure parameters as well as with Raman and infrared spectra is obtained. Typical for the generalized-gradient-type exchange-correlation approximations employed in the calculation, lattice parameters are overestimated by several percent in comparison to experiment.
The phonon branch that is found to be unstable in the *C2/c* titanite structure is stable in similar calculations carried out for malayaite, $CaSnSiO_5$, emphasizing the crucial role of Ti for the occurence of the structural instability. While the Born effective charges are close to the nominal charges in malayaite, anomalously large and anisotropic Born effective charges are calculated for titanite, giving rise to giant LO-TO splitting similar to the perovskite ferroelectrics [1]. The results of the first principle calculations are discussed in the context of anisotropic thermal diffuse X-ray scattering observed at elevated temperatures and the $P2_1/c \leftrightarrow C2/c$ phase transition of titanite [2].

References:
[1] W. Zhong, R.D. King-Smith, D. Vanderbilt, Phys. Rev. Lett. 72(22), 3618, 1994.
[2] T. Malcherek, C. Paulmann, M.C. Domeneghetti, U. Bismayer, J. Appl. Cryst. 34, 108, 2001.

P149

Thermodynamic and kinetic studies on adsorption of Chromium(VI) onto the olive pomace in the region of Beni Mellal

A. MENICHI[1], A. CHAGRAOUI[1], A. MOUSSAOUI[1]
[1]University Hassan II-Casablanca, Laboratory of Physical-Chemistry of Applied Materials (LPCMA), Faculty of Science Ben M'sik, Casablanca, Morocco

Most liquid effluents from tanneries contain chromium. In its form chromium (VI), it shows toxicity to 500 times higher than that of chromium (III). Consequently, to avoid the harmful impact on the public health, the elimination of this metal to the source constitutes a major concern of the industrial world. Moreover, the interest is centered on effective processes of application and with lower costs.

The present work focuses on the study of the adsorptive removal of Chromium (VI) to the lignocellulosic olive pomace (GOLC) prepared with olive pomace of crude (GOB) in the region of Beni Mellal [2]. The important parameters influencing the adsorption power of (GOLC) the concentration of adsorbent, pH, and temperature of the medium were examined in a static reactor.

The kinetic study carried out by using the equations of pseudo-first, pseudo-second order as well a modeling of the isotherms of chromium retention in different temperatures by applying models known such as Langmuir, Freundlich, Temkin and Dubinin Radushkevich.

The results of this work showed that the prepared material may be used as an effective adsorbent for the waste water treatment because of its high adsorption properties.

P150

Femtosecond diffraction of solid flame reactions at HED at European XFE

W. Morgenroth[1], L. Bayarjargal[1], B. Winkler[1], D. Varentsov[2], K. Appel[3], U. Zastrau[3]
[1]Goethe University Frankfurt, Institute of Geosciences, Frankfurt, Germany
[2]GSI Helmholtzzentrum für Schwerionenforschung GmbH, Darmstadt, Germany
[3]European XFEL, Schenefeld, Germany

Self-propagating high-temperature synthesis (SHS) allows the synthesis of a very large variety of compounds and new materials in an energy-efficient way [1]. They have therefore been extensively studied. SHS reaction fronts propagate with 1 – 150 mm/s, reach temperatures up to 4 500 K and the heating rates vary between 1 000 and 100 000 K/s. Many SHS reactions have been well characterized with respect to the underlying equilibrium thermodynamics. In contrast, the characterization of the transient processes before, in, and after the reaction fronts have not been characterized yet due to the extreme reaction rates, temperatures and heating rates.

The European XFEL in Schenefeld will be operated at a repetition rate of 10 Hz. The number of electron bunches per pulse can be as high as 2 700. The pulse length will vary between a few and 100 fs and bunches are separated by 220 ns. The number of photons per pulse is in the order of 1012. Due to these beam characteristics, diffraction experiments with approx. 25 keV photon energy and femtosecond time resolution will become possible.

Our goal is to contribute to the infrastructure for time resolved diffraction experiments in the femtosecond region including an appropriately fast radiometric pyrometry at the High Energy Density science (HED) instrument [2]. The SHS experiments will be carried out in a small chamber located in the interaction area 2. For the temperature determination, a fast pyrometer built by the HED group of GSI will be adopted [3]. First experiments during commissioning of the instrument will include the formation of silicides [4], and will focus on structural changes before, in, and after the reaction front.

The authors gratefully acknowledge financial support by BMBF (project 05K16RFB).

References:

[1] Merzhanov, A. and Averson, A. The present state of the thermal ignition theory: An invited review. Combustion and Flame, Elsevier, 1971, 16, 89-124
[2] http://www.xfel.eu/research/instruments/hed
[3] Ni, P. et al. Pyrometric system for temperature measurements of HED matter generated by intense heavy ion beams. Journal de Physique IV (Proceedings), EDP Sciences, 2006, 133, 977-980
[4] Bernert, T. Kinetische Parameter selbstfortpflanzender Hochtemperatur-Festkörpersynthesen und mechanochemische Darstellung von Ti_5Si_3, Ta_5Si_3 und TiB_2 sowie Struktur-Eigenschafts-Beziehungen von Hexaphenyldisilan. Dissertation, Goethe-Universität Frankfurt, 2014

P151

Multiferroic Hübnerite with unusual high molybdenum doping studied by x-ray powder diffraction and Raman spectroscopy

F. Ziegler[1], L. M. Gomell[1], H. Gibhardt[1], G. Eckold[1]
[1]Georg-August-Universität Göttingen, Institut für Physikalische Chemie, Göttingen, Germany

Introduction

The physical properties of Hübnerite ($MnWO_4$) have been widely investigated over the last years since this material serves as model substance for type II multiferroics with spiral spin order. While their coupling of magnetic and electric properties seems promising for future applications, they all suffer from impractical low phase transition temperatures. Recently, it was shown for MnWO4 that these temperatures can be increased when tungsten atoms are partially substituted by molybdenum [1].

Objective

The transition temperatures increase linearly with the amount of molybdenum. The published maximum doping concentration varies in the range of $0.1 \leq x \leq 0.3$ [1,2]. The aim of the present investigation is to increase the molybdenum content in the wolframite phase using different synthesis strategies and avoiding the formation of a molybdate phase.

Methods

We prepared powder samples with a variety of different molybdenum contents in the range of $0 \leq x \leq 1$. All samples were produced by solid state reaction with a variation of the reaction temperature and characterized by x-ray powder diffraction and Raman spectroscopy.

Results

The maximum doping concentration strongly depends on the reaction path and reaction temperature. With a careful synthesis from the pure metal oxides at low temperatures it is possible to suppress the formation of the molybdate phase and reach higher doping concentrations than previously published. Indeed, it is possible to prepare a sample with more than 65% molybdenum in the wolframite phase as reflected by the x-ray pattern (see Fig. 1) as well as by characteristic stretching vibrations in the Raman spectra. In contrast, the same synthesis at higher temperatures yields both phases completely demixed. At lower temperatures, it is kinetically preferable for molybdenum to become incorporated in the wolframite phase.

Conclusion
We were able to produce kinetically induced metastable Hübnerite with unusual high molybdenum doping. Future investigations will show if this leads to even higher phase transition temperatures and stronger magnetic interactions between the manganese ions.

References:
[1] L. Meddar *et al.*, Chem. Mater., **2012**, 24, 353–360.
[2] L. L. Y. Chang, Mineral. Mag., **1968**, 36, 992-996.

Fig. 1: X-ray powder diffraction pattern for different amounts of molybdenum.

Figure 1

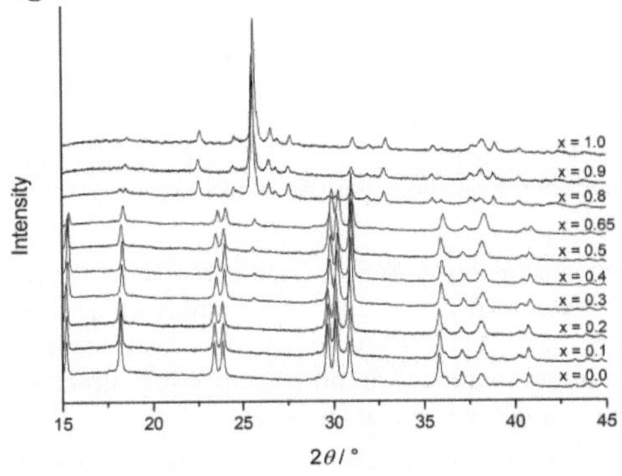

P152

Low temperature behaviour of $Sb_2Se_{3-x}Te_x$ mixed crystals

M. Herrmann[1], K. Friese[1], K. Glazyrin[2]
[1]FZ Juelich, JCNS-2, Juelich, Germany
[2]DESY, P.02, Hamburg, Germany

Antimony telluride, Sb_2Te_3, is a phase-change material and hence characterized by a fast switching between an amorphous and a crystalline state [1]. Four polymorphs of Sb_2Te_3 at ambient temperature and pressure between 0-19 GPa are known [2]. At about 5 GPa α-Sb_2Te_3 (*R-3m*) undergoes a phase transition to β-Sb_2Te_3 (*C2/m*) [2]. β-Sb_2Te_3 is stable up to 15 GPa and transforms to γ-Sb_2Te_3 (*C2/c*) [2]. δ-Sb_2Te_3 (*Im-3m*) is stable above 19 GPa [2]. The ^{121}Sb and ^{125}Te Phonon DOS of α- and δ-Sb_2Te_3 were determined in a previous nuclear inelastic scattering (NIS) experiment [3]. Similar results for β- and γ-Sb_2Te_3 are unknown. As NIS measurements are only possible at low temperature and/or high pressure, a knowledge of the *p-T* phase diagram is mandatory.The lattice parameters of α-Sb_2Te_3 at temperature between 50-290 K were measured by Bessas *et al.*, (2012) [3]. The low temperature behaviour of the lattice parameter of the $Sb_2Se_{3-x}Te_x$ has not been studied.

Powder of nonstoichiometric $Sb_2Se_{3-x}Te_x$ mixed crystals were synthesized in quartz glass ampoules at 923 K over 120 h. Stoichiometric amounts of elemental Sb, Te and Se were weighted and the ampoules were filled under a Ar pressure of 0.8 bar. The samples were quenched in water. The compositions of the samples were verified by XFA measurements. The products are isostructural to α-Sb_2Te_3.

Figure 1 shows the temperature-dependent normalized lattice parameter of

Fig. 1. Normalized lattice parameter of the nonstoichiometric $Sb_2Se_{3-x}Te_x$ samples.

nonstoichiometric $Sb_2Se_{3-x}Te_x$ samples. For all samples pronounced anomalies were determined at about 50, 125, 175 and 250 K. While the lattice parameters *a* change significantly with increasing Se content, the relative lattice parameter *c* is less influenced by changes in composition. Our LTXRPD measurements have confirmed that α-Sb_2Te_3 is stable up to at least 20 K.

In order to understand the lattice dynamics of β- and γ-Sb_2Te_3 and the $Sb_2Se_{3-x}Te_x$ solid solution in more detail, we are planning high pressure experiments between 0-20 GPa to perform *in situ* NIS and NFS measurements. We are currently working on the design of a suitable DAC.

References:
[1] Lencer, D. *et al.*. Nature, 7:972-977, (2008)
[2] Ma, Y. *et al.*, J. Phys.: Conden. Matter, 24:475403, (2012)
[3] Bessas, D. *et al.*, Phys. Rev. B, 86:224301, (2012)

Figure 1

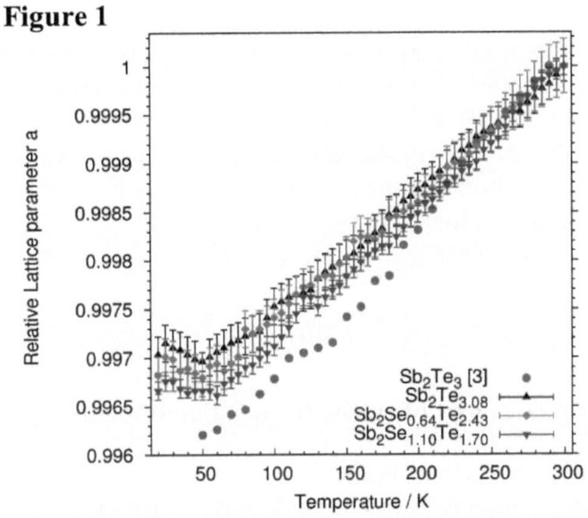

Figure 2

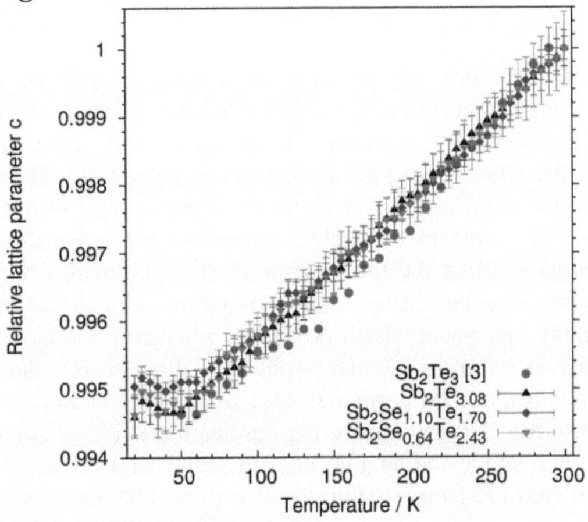

P153

Structure variations within certain rare earth-disilicides

M. Nentwich[1], M. Zschornak[1,2], M. Sonntag[1], R. Gumeniuk[1], S. Gemming[2], T. Leisegang[1,3], D. C. Meyer[1]
[1]TU Bergakademie Freiberg, Experimental Physics, Freiberg, Germany
[2]Helmholtz-Zentrum Dresden-Rossendorf, Nichtgleichgewichts-Thermodynamik, Dresden, Germany
[3]Samara National Research University, Samara Center for Theoretical Materials Science, Samara, Russian Federation

The dimorphism of the RSi_2 and R_2TSi_3 compounds is a well known phenomenon (R is an alkaline earth metal, rare earth metal or actinoide, T is a transition metal). They crystallize in structures, which derive from hexagonal AlB_2 or tetragonal $ThSi_2$ prototypes. Despite their local similarities, these two prototypes do not have a common root in the Bärnighausen diagram. These diagrams summarize the symmetry relations between the high symmetrical basic structures and their lower symmetric variations.

We performed an extensive literature research based on more than 400 structure reports of the RSi_2 and R_2TSi_3 compounds. To gain an overview of the various structure reports within these compounds we summarized composition, lattice parameters a and c, ratios c/a, formula units per unit cell, and structure types in an extensive table. To evaluate the probability of a successful synthesis, we performed DFT calculations on carefully chosen compounds that have not yet been reported. Finally, we investigated peculiarities of symmetry distribution among the RSi_2 and R_2TSi_3 compounds and several correlations related to structural parameters.

We found that the thermal treatment has a massive effect to the formation of superstructures. Furthermore, there are two different kinds of hexagonal R_2TSi_3 compounds being ionic or metallic, depending on the R element. Additionally, the main influence to the variation of the Si-T bonds is the electronic interplay between R element and Si sublattice rather than the R radii.

P154

Pseudomorphic replacement of calcite by acicular fluorite crystals

E. Trindade Pedrosa[1], C. Putnis[2,1], A. Putnis[1,3]
[1]University of Münster, Institut für Mineralogie, Münster, Germany
[2]Curtin University, Department of Chemistry, Perth, Australia
[3]Curtin University, The Institute for Geoscience Research (TIGeR), Perth, Australia

A pseudomorphic mineral replacement is a special case of a mineral phase transformation during which a mineral phase is fully or partially replaced by another maintaining the external morphology of the first. This typically occurs as a reequilibration reaction between a solid and an aqueous fluid phase, and is best explained by an interface-coupled dissolution-precipitation mechanism. Understanding the mechanism of this type of reaction is crucial in earth sciences because these reactions can affect or even control many large-scale earth processes whenever fluids are present, such as during metamorphism, diagenesis, and hydrothermal alteration. The particular case of the replacement of calcite by fluorite has also an engineering application since calcium carbonate has been suggested as a successful seed material for the sequestration of fluoride from contaminated waters. The aim of the present work is to investigate aspects of the replacement of calcium carbonate by fluorite to better understand the mechanism of this reaction.

Small (~9 mg) calcite single crystals were reacted with two 0.25 M fluoride-containing solutions (NaF and NH_4F) for different reaction times (4 to 32 hours), at 200°C. After reaction all samples were washed with milli-pure water and dried overnight at 40°C. To analyse the extent of reaction, part of the samples were mounted in epoxy and polished until half of their size for SEM analysis. For a more detailed examination of the microstructure of the product phase, other samples were simply cleaved in several direction for direct visualization at the SEM. XRD was also used to analyse the final product phase(s).

After reaction, all samples preserved their size and external morphology (a pseudomorphic replacement) and XRD confirmed the presence of a single product phase (fluorite). SEM images showed that the product phase consisted of needle-like shaped fluorite crystals that conferred porosity to the reacted rim. At the reaction front the needles are oriented perpendicular to the reaction interface, and as the reaction interface moves forward into the crystal, the needles heal or recrystallize forming sheets of fluorite parallel to the reaction front. The replacement reaction was faster when using NH_4F than when using NaF.

P155

High pressure behaviour of Pb_2SnO_4 using Raman spectroscopy

D. Spahr[1], L. Bayarjargal[1], W. Morgenroth[1], N. Schrodt[1], R. Luchitskaia[1], B. Winkler[1]
[1]Goethe Universität Frankfurt, Kristallographie, Frankfurt, Germany

Studies of REE doped stannates, M_2SnO_4 with M = Ca, Sr, Ba, have been studied extensively, as they are employed as long afterglow phosphors (e.g. [1]). In contrast, for lead stannate (Pb_2SnO_4, [2]), a pigment employed in oil paintings [3], structure-property relations have not been investigated yet.

We synthesized powder samples by solid state reaction [2] and single crystals (up to 150 µm) hydrothermally [4]. We doped Pb_2SnO_4 with Eu^{3+}. The sample quality was checked by SEM, powder and single crystal diffraction. Incorporation of the Eu was demonstrated by fluorescence spectroscopy. Pressure-dependent Raman spectra (fig. 1) were obtained using diamond anvil cells up to 17 GPa. The experimental studies were complemented by density functional theory based calculations. High pressure X-ray diffraction was carried out at beamline P02.2 (PETRA III).

The ambient pressure Raman spectra are in good agreement with earlier studies [5]. The DFT calculations reproduce the spectra accurately. Pressure increase causes significant changes of the Raman spectra at $\approx$ 2 GPa and 5 GPa, which are indicative of structural phase transitions. Increasing pressure also leads to a drastic colour change (fig. 2). At ambient pressure, the crystals are colourless. They are yellow at about 3 GPa, and on further increase become orange, red and opaque. This is consistent with the calculated closure of the band gap. All pressure induced changes are fully reversible on pressure release. The analysis of the synchrotron diffraction data, which will allow us to benchmark the structural changes predicted by the DFT calculations, is currently on-going. The Eu^{3+} doped sample shows strong fluorescence. Fluorescence life time measurements are currently on-going in order to allow a comparison with the other stannates [1].

Funding by the BMBF (05K13RF1), DFG (RA2585/1-1) and Adolf Messer Stiftung is gratefully acknowledged. DS is grateful for support from E. Haussühl, M. Stękiel and D. Zimmer.

Fig. 1: Raman-spectra between 0 and 17(1) GPa
Fig. 2: Colour change in DAC between 0 and 17(1) GPa

References:
[1] Stanulis et al. *Opt. Mater.*, 36:1146-1152, 2014.
[2] Gavarri et al. *J. Solid State Chem.*, 36:81-90, 1981.
[3] Agresti et al. *Wiley Periodicals, Inc.*, 41:226-231, 2016.
[4] Wu et al. *Mater. Res. Bull.*, 34:1135-1142, 1999.
[5] Pelosi et al. *e-Preserv. Sci.*, 7:108-115, 2010.

Figure 1

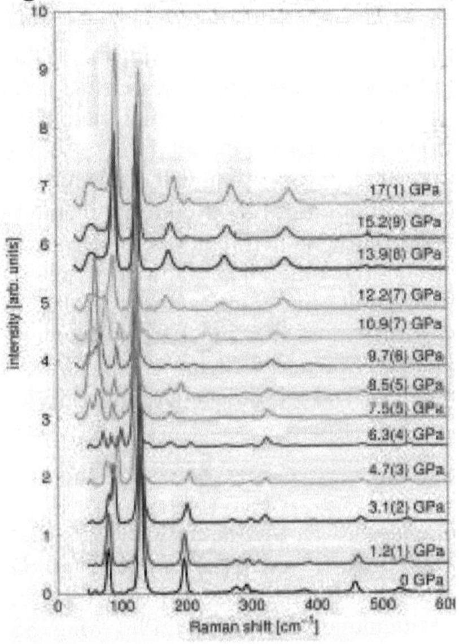

Figure 2

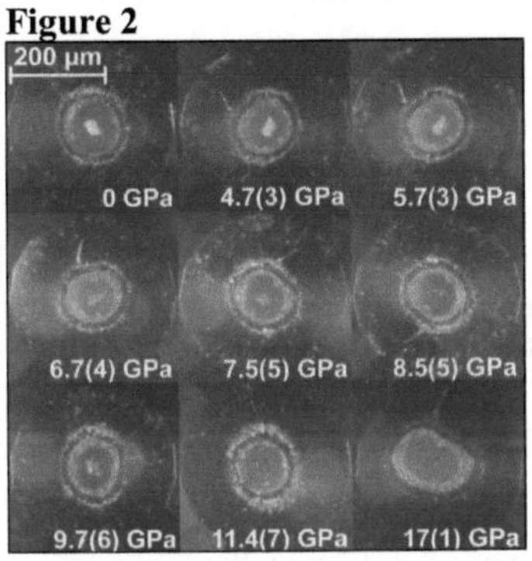

P156

Tricritical phase transition in Norsethite

M. Ende[1], R. Miletich[1], H. Effenberger[1]
[1]Universität Wien, Institut für Mineralogie und Kristallographie, Wien, Austria

Introduction

Although the natural occurrence of norsethite, $BaMg(CO_3)_2$, a double carbonate phase with cations of very different sizes, is restricted to only a few localities worldwide, it appears that this phase may be a more common authigenic component of selected natural low-temperature environments [e.g. 1]. Furthermore, norsethite early revealed its close structural relationship to $CaCO_3$ [2].

Objectives

However, the space-group symmetry of norsethite remained unclear a long time and recently a phase transition was suspected around 360 K and investigated with Raman, FTIR spectroscopy and powder X-ray diffraction [3].

Materials & Methods

Powder and single crystal samples of norsethite were used on a STOE-StadiVari X-ray diffractometer system with a Dectris Pilatus 300K pixel detector for both, the single-crystal and powder X-ray diffraction studies in a temperature region between 100 K and 500 K (± 1 K). The measurements were performed using a 50 W molybdenum air-cooled Incoatec IµS microfocus (110 µm FWHM) source with high-brilliance 2D-focussing Quazar multilayer mirrors. In powder diffraction mode, 190 mm sample-detector distance was used instead of the 60 mm standard distance.

Results & Conclusion

High sensitive single-crystal X-ray diffraction measurements reveal the superstructure reflections of norsethite, for the first time at non-ambient temperatures, which confirm the suspected space group change from R-3c to R-3m around 365 K. Thermodynamic analyses reveal the phase transition near 365 K. It can be shown from symmetry arguments that the spontaneous strain (Fig. 1) should be proportional to the square of the order parameter Q [e.g. 4]. Thus, order parameter analyses using power law fits onto the spontaneous strains (Fig. 1) reveal critical exponents near ideal tricritical values.

Figure 1: Spontaneous strain derived from subtraction of thermal EoS from lattice parameters of norsethite.

References:
[1] Schmidt, B.C., Gehlken, P.-L., Böttcher, M.E. (2013). Eur. J. Mineral., 25, 137-144.
[2] Mrose, M.E. Chao, E.T.C., Fahey, J.J. and Milton, C. (1961). Am. Mineral., 46, 420-429.
[3] Effenberger, H., Pippinger, T., Libowitzky, E., Lengauer, C.L., Miletich, R. (2014). Mineral. Mag., 78(7), 1589-1611.
[4] Putnis, A. (1998) Introduction to Mineral Sciences. Cambridge University Press, Cambridge.

Figure 1

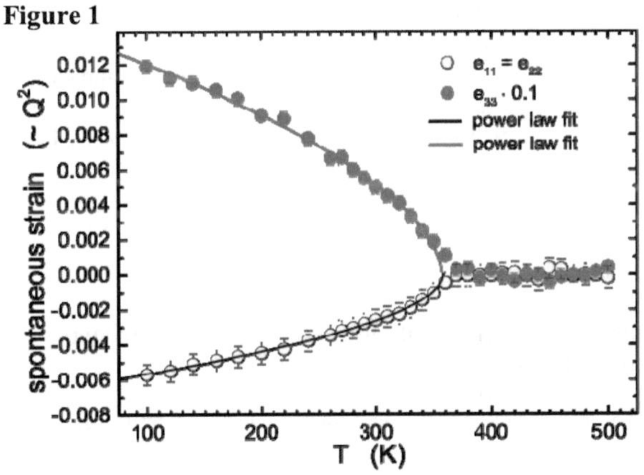

P157

Zinc oxide phase transitions under non-ambient conditions

D. Zagorac[1,2], J. C. Schoen[3], J. Zagorac[1,2], M. Rosic[1,2], M. Cebela[1,2], J. Lukovic[1,2], B. Matovic[1,2]
[1]"Vinča" Institute of Nuclear Sciences, University of Belgrade, Department of Materials Science, Belgrade, Serbia
[2]Center for the synthesis, processing and characterization of materials for use in extreme conditions "CEXTREME LAB", Laboratory for Theoretical Investigation of Materials (L-TIM),, Belgrade, Serbia
[3]Max Planck Institute for Solid State Research, Stuttgart, Germany

In order to investigate phase transitions under non-ambient conditions of zinc oxide, we have performed several sets of calculations using different approaches. First, we have performed crystal structure prediction for ZnO using simulated annealing (SA), with an empirical potential and local optimization on *ab initio* level. [1,2,3] We have found the experimentally observed structures wurtzite, sphalerite, and rock salt in agreement with previous ZnO research. In addition, full analysis of the energy landscape was performed and many new interesting non-equilibrium modifications were found in different regions of the energy landscape, at elevated pressures and/or temperatures.

Very important aspect beyond the identification of equilibrium ZnO structure (global minimum) is the analysis of the energy barriers and energy landscape surrounding the global minimum, and of the possible transition paths connecting the major locally ergodic regions. [4,5] Therefore, we have performed threshold algorithm (TA) calculations for various pressures and temperatures, and for different numbers of formula units of ZnO. Furthermore, we have performed calculations using the prescribed path algorithm (PP), where connections between experimental structures on the energy landscape, and in particular transition states, were investigated in detail.

References:
[1] D. Zagorac, J. C. Schön, K. Doll, and M. Jansen, *Acta Phys. Pol. A*, **120**, 215 (2011).
[2] D. Zagorac, J. C. Schön, J. Zagorac, and M. Jansen: *Phys. Rev. B* **89**, 075201 (2014).
[3] D. Zagorac, J. C. Schön, J. Zagorac, and M. Jansen, *RSC Advances*, **5**, 25929 (2015).
[4] D. Zagorac, J. C. Schön, J. Zagorac, I. Pentin, and M. Jansen, *Processing and Application of Ceramics,* **7**, 111 (2013).
[5] D. Zagorac, J. C. Schön, and M. Jansen, *J. Phys. Chem. C*, **116**, 16726 (2012).

P158

Room temperature aging and tempering of highly nitrogen-supersaturated ferrite

M. Wetzel[1], A. Leineweber[1]
[1]Institute of Materials Science, TU Bergakademie Freiberg, Freiberg, Germany

Mechanical alloying by high-energy ball milling represents an effective method to form microstructures that are far from thermodynamic equilibrium. As a result the alloying components may show increased solid solubility or even form metastable phases. It has been shown that ball milling of α-Fe in N_2 or NH_3 atmosphere [1,2] as well as mixtures of α-Fe and iron nitrides [3] leads to the formation of a highly supersaturated ferrite-like phase. Information on the actual crystal structure of the solid solution, however, are conflicting.

The objective of this work is a thorough examination of diffraction phenomena related with the nitrogen supersaturation of ferrite and the changes in the line profiles caused by room temperature aging and tempering of the samples.

Powder mixtures of α-Fe and γ′-Fe_4N containing up to 10 at.% N have been mechanically alloyed in a planetary ball mill. The resulting samples were investigated by means of X-ray diffraction and changes in line profiles and positions were examined in order to characterize their aging and tempering behavior.

The milled powder mixtures show powder diffraction patterns which appear to consist of N-supersaturated α-Fe-N solid solution and minor amounts of γ-Fe-N. Pronounced line broadening and asymmetries are encountered, both increasing with increasing nitrogen content. More interestingly, the resulting ferritic solid solution shows peculiar anisotropic line shifts, which cannot be explained by a mere anisotropy in lattice strain and do not seem to be compatible with a tetragonal distortion of the ferrite unit cell occasionally proposed in the literature [2]. Room temperature aging experiments have, furthermore, shown that the anisotropy in the line shifts reduce with aging time. The lattice expansion vanishes entirely upon tempering at temperatures above 200°C as soon as enhanced γ′-nitride formation sets in. Although several approaches to explain the diffraction phenomena are checked, a comprehensive model for the crystal or defect structure of the supersaturated ferrite remains to be found.

References:
[1] Koyano, T., Lee, C., Fukunaga & T., Mizutani, U. (1992). Mat. Sci. For. 88-90, 809-816. [2] Rawers, J., Govier, D. & Cook, D. (1995). J. Mater. Synth. Process. 3, 263-283. [3] Aufrecht, J., Leineweber, A., Foct, J. & Mittemeijer, E.J. (2008). Phil. Mag. 88, 1835-1855.

P159

Lab-based x-ray diffractometer for the investigation of low temperature magnetic field-induced phase transitions

T. Faske[1], W. Donner[1]
[1]TU Darmstadt, Materials Science, Structure Research, Darmstadt, Germany

Investigations of crystal structures of polycrystalline specimens under magnetic field are most often done at synchrotron or neutron facilities. High magnetic fields are rarely reached with equipment available for use with laboratory-based x-ray diffractometers. The presented setup assembled at TU Darmstadt not only enables the user to perform x-ray diffraction experiments under magnetic fields of up to 5.5 T but also temperature-dependent experiments ranging from room temperature down to 15 K.

The powder x-ray diffractometer is set up in a modified Debye-Scherrer transmission geometry using Mo $K_{\alpha 1}$ radiation, samples glued on aluminium foil and an on-site readable high resolution image plate detector spanning up to 90° 2θ, originally developed for HASYLAB/DESY[1]. The specimen is mounted on the cold finger of a rotating closed-cycle He-cryostat which itself is placed into the superconducting magnet.

The setup has been successfully tested on samples with phase transitions at known temperatures and the accuracy of both magnetic field strength and temperature have been confirmed and are shown by the example of $LaFe_{11.4}Si_{1.6}$. The intermetallic $LaFe_{11.4}Si_{1.6}$ exhibits a giant spontaneous magnetostriction of about 1% at its first order metamagnetic phase transition at T_c = 195 K as can be seen in Fig. 1. The ferromagnetic low-temperature phase is stabilized by an external magnetic field causing T_c to increase by approximately 4.3 K/T applied magnetic field, see Fig. 2.

Fig. 1: Spontaneous magnetostriction of $LaFe_{11.4}Si_{1.6}$ at H = 0 T and magnetic field induced magnetostriction at T = 192 K.

Fig. 2: Temperature-dependent magnetization of $LaFe_{11.4}Si_{1.6}$ at different external magnetic fields showing the shift of the paramagnetic to ferromagnetic transition to higher temperatures by approximately 4.3 K/T.

References:

[1] M. Knapp, V. Joco, C. Baehtz, H. H. Brecht, A. Berghaeuser, H. Ehrenberg, H. von Seggern, H. Fuess, Nucl. Instrum. Meth. A 521 (2004) 565-570.

Figure 1

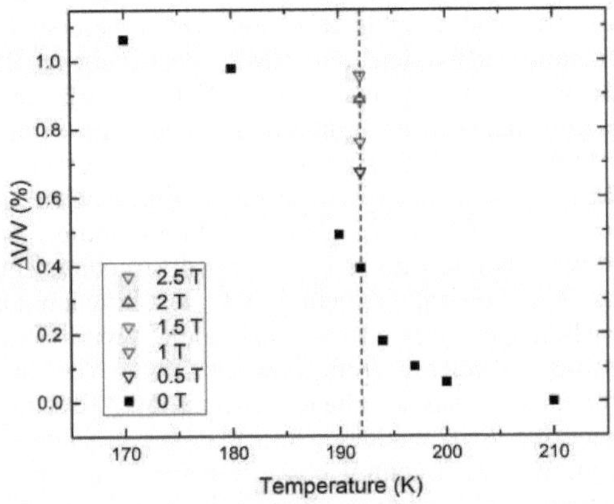

Figure 2

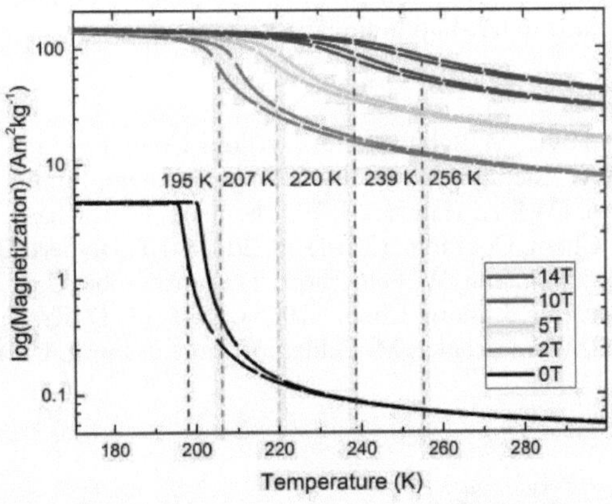

Spectroscopy as supporting tool in crystallography

P160
Substitution and doping in iron pnictides

M. Merz[1], P. Schweiss[1], P. Nagel[1], M. J. Huang[1], A. Plog[1], R. Eder[1], T. Wolf[1], H. von Löhneysen[1], S. Schuppler[1]
[1]Karlsruher Institut für Technologie, Institut für Festkörperphysik, Eggenstein-Leopoldshafen, Germany

The composition-dependent electronic structure of iron pnictides, in particular the question if and how charge carriers are introduced to the system upon substitution – by Sr or alkali metals (AM) for Ba; by transition metals (TM) for Fe; and/or by P for As – continues to provide surprises. Our systematic study of spatial structure and electronic states by x-ray diffraction and x-ray absorption, performed on a large number of compositions in the (Ba,Sr,AM)(Fe,TM $)_2$(As,P$)_2$ family of compounds, shows, for instance, that valences are often different from expectations, that doping can be "effective" or not, depending on substituent, and that many doping effects are "site decoupled", meaning that they occur either on the Fe site or on the As site but not on both simultaneously. Furthermore, the energy-level schemes derived from electronic and spatial structure differ from the present "standard" assumptions. In all, it appears that indirect, structural effects of substitution may in many cases be more important for magnetism and superconductivity in iron pnictides than direct, charge-carrier doping effects.

P161
The First Zn(II) Complex of 4-Methoxypicolinic Acid: Synthesis, Crystal Structure and IR spectrum

î TAMER[1], D. AVCI[1], Y. ATALAY[1]
[1]Sakarya University, Physics, Sakarya, Turkey

The crystal structure of the newly synthesized Zn(II) complex of 4-methoxypicolinic acid has been characterized by X-Ray diffraction technique. Figure 1 demonstrates that the Zn(II) ion surrounded by two pyridine nitrogens, two carboxylate oxygens and two aqua oxygens, resulting in a distorted octahedron. C7-O2 and C7-O3 bond lengths have been observed as 1.257 (3) and 1.239 (3) Å, respectively. So, C7-O2 bond has a single bond character while C7-O3 bond has a double bond character. Additionally, the optimized structure at the ground state energy has been obtained by density functional theory (DFT) method. A substantial agreement has been found between the experimental and theoretical results. In order to support the X-Ray crystal structure and determine the functional group vibrations, experimental and theoretical infrared spectra have been presented in this study. The wavenumber interval between the symmetrical and asymmetrical COO- stretching vibrations demonstrated that the carboxylate groups coordinate to Zn(II) ion as a monodentate ligands.

Figure 1. The ORTEP III structure of Zn(II) complex of 4-methıxypicolinic acid

Figure 1

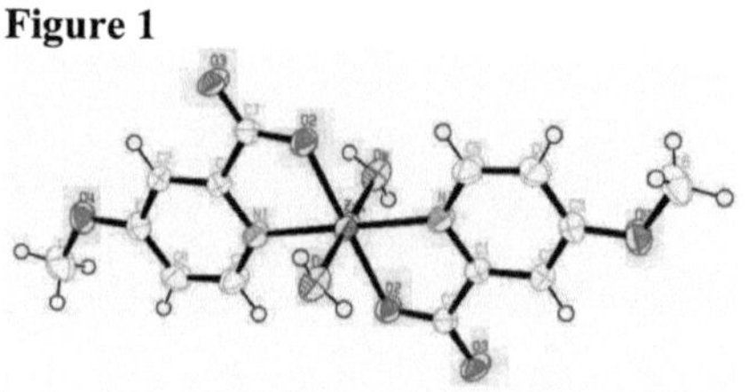

P162
X-ray Diffraction in Combination with DFT, Vibrational and Nuclear Magnetic Resonance Spectroscopy for the Investigation of Structures and Phase Transitions of Molecular and Complex Aluminium Hydrides

T. Bernert[1], M. B. Ley[1], M. Fischer[2], B. Zibrowius[1], M. Felderhoff[1], C. Weidenthaler[1]
[1]Max-Planck-Institut für Kohlenforschung, Mülheim an der Ruhr, Germany
[2]University of Bremen, Crystallography Group, Bremen, Germany

Substances bearing Al-H bonds cover a wide range of chemical compounds such as complex aluminum hydrides (alanates). Alanates are investigated for solid-state hydrogen storage.[1] Also molecular hydrides, especially group 13 Lewis-acid base compounds, are interesting for hydrogen storage, since reversible hydrogen release has been shown for other Lewis-acid base compounds.[2,3] We focus on aminoalanes, compounds containing AlH_2-N groups. The crystal structures of molecular aminoalanes mostly contain 4- and 6-membered rings formed by Al_2N_2 or Al_3N_3 units (Fig. 1). In here we present results of combined X-ray diffraction, DFT, Raman and solid-state NMR spectroscopy studies, to reveal structural properties of alanates. $CsAlH_4$ crystallizes in two polymorphic structures which were studied by X-ray and neutron diffraction[4] and NMR spectroscopy.[5] $CsAlH_4$ and $NaAlH_4$ can react to polycrystalline $NaCs(AlH_4)_2$. Indexing shows ambiguity between an orthorhombic and a hexagonal cell but results from ^{27}Al-NMR spectroscopy are consistent with a hexagonal cell. The structure was determined from powder XRD in combination with DFT calculations. Furthermore, several aminoalanes were synthesized and their crystal structures could be determined. The crystal structure of tert-butylaminoalane, $[N(C(CH_3)_3)H\text{-}AlH_2]_2$, was determined along with structural changes during H_2 release from powder XRD. The resulting tetrameric iminoalane has a cubane structure of Al_4N_4 moieties (Fig. 2). Since the Al-H stretching mode is very sensitive to its chemical environment, structural phase transitions in dependence of temperature and pressure were measured by Raman spectroscopy.

Figure 1: Molecular structures of amino adducts of group 13 elements (M) with R = H or alkyl groups, (a) dimers and (b) trimers. Figure 2: Reaction of two molecules dimeric tert-butylaminoalane to tert-butyliminoalane.

References:
[1] B. Bogdanović, M. Schwickardi. J. Alloys Compd., 253:1–9, 1997. (2) D. W. Stephan and G. Erker. Angew. Chem. Int. Ed. (2010) 49, 46. (3) J. A. Hatnean, J. W. Thomson, P. A. Chase, D. W.Stephan. Chem. Commun. (2014) 50, 301. (4) T. Bernert, D. Krech, W. Kockelmann, W. Felderhoff, T. Frankcombe, C. Weidenthaler. Eur. J. Inorg. Chem. (2015), 5545. (5) D. Krech, B. Zibrowius, C. Weidenthaler, M. Felderhoff. Eur. J. Inorg. Chem. (2014), 5683.

Figure 1

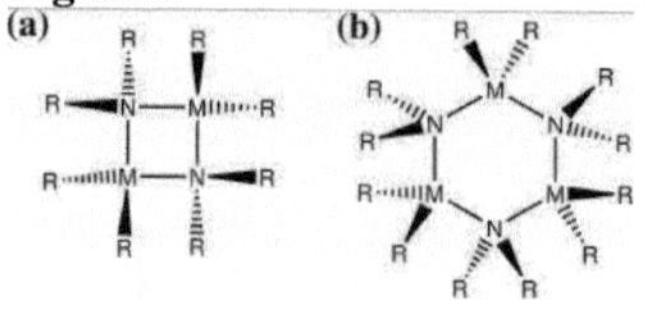

Figure 2

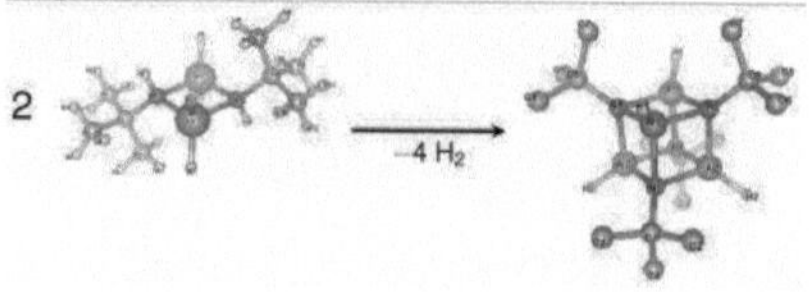

Structural biology meets chemistry

P163
Structure Analysis of Lignin-based Polyurethanes using WAXS
S. Witzleben[1], K. Walbrück[1], S. Klein[1], J. Rumpf[1], M. Schulze[1]
[1]Hochschule Bonn-Rhein-Sieg, Angewandte Naturwissenschaften, Rheinbach, Germany

Next to cellulose, lignin is the most abundant biopolymer and responsible for the strength and structure of plants cell walls. Today, it is available in large amounts as a side product of the paper and pulp industry. Lignin is a crosslinked macromolecule, arising from three phenylpropane derivatives: *p*-hydroxyphenyl, guaiacyl and syringyl. The corresponding precursors are the three aromatic alcohols *p*-coumaryl, coniferyl and sinapyl alcohol. [1]

One of the still remaining challenges is the precise analysis of the detailed chemical structure of lignin due to its dependency on various parameters, e.g. the plant source and pulping process [2].

Due to the high number of hydroxyl groups, lignin is a promising renewable source to replace diols in polyurethane (PU) synthesis. Lignin-based polyurethane coatings were synthesized according to Griffini *et al.* [3].

The lignin-based PU coatings and commercial polyurethanes were investigated using wide angel X-ray scattering (WAXS). The contribution presents WAXS studies of different commercial and lignin-based PUs. Most, lignin-based PU display only a broad diffraction signal, which is typical for various PU like thermoplastic polyurethane [4]. In contrast, three sets of lignin-based PU coatings where lignin was precipitated at pH 2 show a remarkable higher degree of crystallinity. According to the PDF database, the lignin-based PU coating indicate a polyaniline analogue structure.

Structure solutions of the synthesized PU were performed by using different Lignin precursor components.

References:
[1] S. Laurichesse and L. Avérous, "Chemical modification of lignins: Towards biobased polymers" *Prog. Polym. Sci.* 39(7): 1266–1290, 2014.
[2] B. Hansen, P. Kusch, M. Schulze, and B. Kamm, "Qualitative and quantita-tive analysis of lignin produced from beech wood by different conditions of the Organosolv process," *J. Polym. Environ.* 24(2):85–97, 2016.
[3] G. Griffini, V. Passoni, R. Suriano, M. Levi, and S. Turri, "Polyurethane Coatings Based on Chemically Unmodified Fractionated Lignin," ACS Sustain. Chem. Eng.3(6): 1145–1154, 2015.
[4] Witzleben, S., Walbrueck, K., Klein, S. E., Schulze, M. "Investigation of Temperature Dependency of Morphological Properties of Thermoplastic Polyurethane using WAXS and SAXS Monitoring", J. Chem. Chem. Eng. 9, 494-499, 2015.

P164
Current Status of Microfocus X-ray Sources for Chemical and Biological Crystallography
J. Graf[1], T. Stuerzer[2], S. Freisz[2], H. Ott[2], A. Kleine[1], J. Wiesmann[1], C. Michaelsen[1]
[1]Incoatec GmbH, Geesthacht, Germany
[2]Bruker AXS GmbH, Karlsruhe, Germany

Modern microfocus X-ray sources define the state-of-the-art for most applications in X-ray diffraction. These sources are usually combined with multilayer X-ray mirrors which are excellent X-ray optical devices for beam shaping and preserving the brightness of the source.

Microfocus rotating anode generators and liquid metal jet systems deliver flux densities in the range of 10^{11} photons/s/mm^2. However, these sources are expensive and need regular and sometimes time-consuming maintenance. Low power microfocus sealed tube sources, such as the Incoatec Microfocus source IμS, represent an interesting low-maintenance alternative to rotating anode generators. Power loads of several kW/mm^2 in anode spot sizes of < 50 µm deliver a small and bright beam. Flux densities of up to 10^{10} photons/s/mm^2 can be achieved in a focused beam suitable for most protein crystals and poorly diffracting small molecule samples. The latest generation of the IμS, the IμS 3.0, is the first microfocus X-ray source that is optimized for X-ray diffraction resulting in a gain in intensity of about 30% compared to its predecessor.

In this presentation, we will be reviewing the current performance levels of different low power microfocus X-ray sources. Further, we will be discussing the main features of the newest generation of the IμS. We will be presenting selected results to demonstrate the impact of these modern microfocus X-ray sources on the data quality for applications in chemical and biological crystallography.

P165
Calcium Carbonate Biomineralisation
O. Salim[1], C. Daniel[1]
[1]Université of Bejaia, Bejaia, Bejaia, Algeria

Introduction
A deeper understanding of the mineralisation process and the mimicking of complex structures produced by nature in laboratory may have a significant impact on many fields. Many studies showed a correlation between the calcium carbonate growth modifications and the structure of the additive organic molecules as polyacrylic acid, EDTA, etc...[1-2]. The mechanisms of how these organic molecules control CaCO3 crystal growth of the three polymorphs are still matter of conjectures [3].

Objectives
In this work, we study the crystallization of CaCO3 particles in aqueous solutions in the presence of PAA. The PAAs effects on crystals mean sizes and on the polymorphs volume contents at temperatures varying from 25° to 80° C are examined using X-ray and Rietveld method.

Materials and Methods
Two solutions of CaCl2 and K2CO3 are mixed at different temperatures: 20°, 50°, and 80° C (with the presence or not of PAA). The solid precipitates were collected, rinsed and dried under vacuum. The samples were examined using SEM and X-ray diffraction (XRD). Rietveld method and Popa anisotropic model

are used to study the mechanism of action of polyacrylic acid on the growth of different polymorphs of calcium carbonate.

Results
The Rietveld refinements on X-rays diffraction diagrams using anisotropic model for crystallites sizes shows a flattening effect on the crystallites in the case of both the vaterite and aragonite. This flattening is in the direction of the **c**-axis. This is due to the PAA-Ca+2 complexes which block the growth along the c-axis of the crystallites. The refinements show that no effect is observed on the calcite crystallite (fig. 1).

Conclusion
The mechanism of action of the polyacrylic acid was clearly demonstrated when refining with anisotropic sizes model which shows a flattening effect on the crystallites in the case of vaterite and aragonite but not in the case of calcite.

Fig 1: SEM image of vaterite particle (a) vaterite crystallite without PAA (b) vaterite crystallite with PAA.

References:
[1] N. Hosoda, A. Sugawara, T. Kato, Macromolecules 36 (2003) 6449.
[2] S. Zhang, K.E. Gonsalves, Langmuir 14 (1998) 676.
[3] Ouhenia, S., Chateigner, D., Belkhir, M.A., Guilmeau, E., Krauss, C., 2008, J. Crystal Growth 310 (11) (2008), 2832.

Figure 1

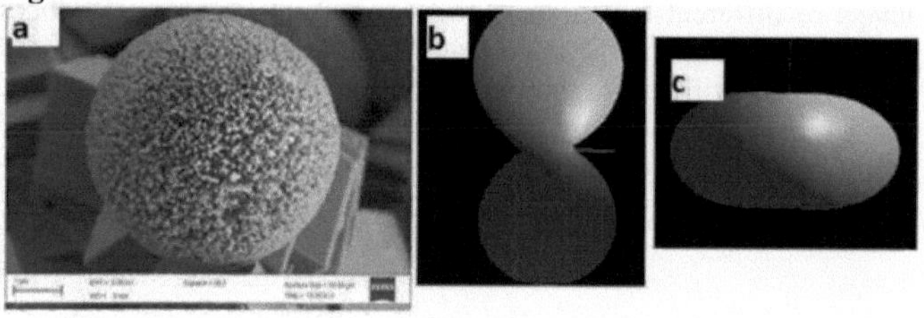

P166
A New Type of RNA Double Helix Based on U·G and C·A⁺ Wobble-Base Pairing

U. Heinemann[1], A. Garg[1]
[1]Max Delbrück Center for Molecular Medicine, Berlin, Germany

DNA and RNA double helices tolerate base pairs that deviate from the canonical Watson-Crick pairing scheme. Usually, mismatched bases occur in pairs or in isolation within a standard Watson-Crick environment, see (1,2). U·G wobble base pairs are stabilized by two hydrogen bonds and occur frequently in RNA (2). The formation of DNA or RNA double helices based exclusively on non-standard base pairs has not been demonstrated so far by high-resolution structural analysis.

We have crystallized the self-complementary oligoribonucleotide r(UGUUCUCUACGAAGAACA) and determined its three-dimensional structure by molecular replacement at 1.38-Å resolution (R / Rf = 0.132 / 0.149). Structural transitions of the RNA induced by pH change were analyzed by UV and circular dichroism (CD) thermal melting experiments, as well as differential scanning calorimetry (DSC).

Unexpectedly, the RNA did not crystallize as a stem-loop structure, but formed an 18-bp double helix in the A conformation instead. In this duplex, the 6 terminal base pairs at either end show standard Watson-Crick pairing, whereas the central six nucleotides are wobble-base paired throughout. In addition to two U·G wobble pairs, this regions includes four unusual C·A^+ pairs which are stabilized by two hydrogen bonds each and whose formation was facilitated by the acidic pH of the crystallization medium. Biophysical experiments support this interpretation by providing evidence for a pH-induced structural change in r(UGUUCUCUACGAAGAACA).

In conclusion, our study reveals a new double helical conformation of RNA at the center of an 18-bp RNA duplex where half a helical turn (6 bp) is exclusively composed of wobble-base pairs.

References:
[1] Privé, G.G., Heinemann, U., Chandrasegaran, S., Kan, L.-S., Kopka, M.L. & Dickerson, R.E. (1987) Helix geometry, hydration, and G·A mismatch in a B-DNA decamer. *Science* **238**, 498-504.
[2] Mueller, U., Schübel, H., Sprinzl, M. & Heinemann, U. (1999) Crystal structure of acceptor stem of tRNAAla from *Escherichia coli* shows unique G·U wobble base pair at 1.16 Å resolution. *RNA* **5**, 670-677.

Structural chemistry under non-ambient conditions

P168
Crystalline high-pressure phases in the Bi-Co system

L. Wiehl[1], S. Bhat[1], I. Radulov[1], K. Skokov[1], M. Dürrschnabel[1], L. Molina-Luna[1], S. Sicolo[1], L. Diop[1], D. Karpenkov[1], N. Nishiyama[2], H. J. Kleebe[1], K. Albe[1], R. Riedel[1], O. Gutfleisch[1]
[1]TU Darmstadt, FB Material- und Geowissenschaften, Darmstadt, Germany
[2]DESY, Hamburg, Germany

$RECo_5$ (RE = rare-earth) compounds are known as materials with high magnetocrystalline anisotropy [1]. They crystallize in the $CaCu_5$ structure type, space group *P 6/m m m*. In the course of searching for RE-free magnetic materials, which could replace the classical NdFeB magnets, MCo_5 type materials were considered as promising candidates. The existence of crystalline MCo_5 and MFe_5 (*M* = Bi, Ca, Zr) phases and their magnetic properties were explored. Here we report on the Bi-Co system. Samples with starting compositions (Co:Bi = 5:1 & 1:1) were subjected to high pressure (15.6 GPa) and temperature (900°C) in a multianvil press at DESY, Hamburg. The recovered products were characterized using X-ray diffraction with synchrotron radiation (ALS, Berkeley), SEM/EDX, and magnetic measurements. They proved to be a mixture of several crystalline phases, with the most prominent phase not being the expected $BiCo_5$, but Bi_3Co [2]. The search for new phases was complemented by DFT calculations.

Financial support by the German federal state of Hessen through its excellence program LOEWE RESPONSE is gratefully acknowledged.

References:
[1] K. Strnat et al., J. Appl. Phys. 38 (1967) 1001
[2] S. Tencé et al., J. Phys.: Condens. Matter 26 (2014) 395701

P169
Temperature depended Crystal Structure Prediction (CSP) of Organic Molecules

D. Hofmann[1,2]
[1]CRS4, Pula, Italy
[2]FlexCryst, Uttenreuth, Germany

Since several years it is possible to predict crystal structures from the structural formula. An overview of the actual methods can be found in the report on last blind test realized by the CCDC in Cambridge[1]. The report on some results of our specific method, which is based on data mining, will be presented by L. Kuleshova. An interesting point in the existing CSP methods is that for the moment no one method takes care explicitly about the temperature, despite experimentalists describe for an endless number of polymorphs of organic crystals under non-ambient conditions.

In the presented work we extend our former approach[2] for temperature dependent CSP to a prediction with a continuous parameter for the temperature. In our approach we use a force field. To obtain a temperature dependent force field we exploit two effects: the heat capacity and the thermal expansion. Both effects are commonly considered as almost linear and connects temperature with structural effects. To introduce the effect of heat capacity we added the term to the force field. To obtain a linear expansion for 3D crystal with temperature, we introduced a scaling factor to the 1D potentials . For the parametrization of the force field we use our standard procedure[3] of data mining on experimental structures, which will be shortly outlined during the presentation.

In the results we will show two examples: In the first example the force field is applied to experimentally known crystal structures of Aspirin. The task of the method is to recognize correctly at which temperature the crystal have been measured. In the second example the force field is applied to experimentally known crystal structures of Paracetamol. For Paracetamol a low- and a high temperature polymorph are known. Task of the method is to 'predict' correctly the transition temperature.

References:
[1] Reilly, Anthony M., et al. Acta Cryst. B: 72.4 (2016): 439-459.
[2] Hofmann, D. W. M., and L. N. Kuleshova. Crystal Growth & Design 14.8 (2014): 3929-3934.
[3] Hofmann, D. W. M., and L. N. Kuleshova. International Tables for Crystallography (submitted)

Figure 1: The two polymorphs of Paracetamol under investigation.
Figure 2: The energy of known experimental crystal structures of paracctamol as function of the temperature.

Figure 1

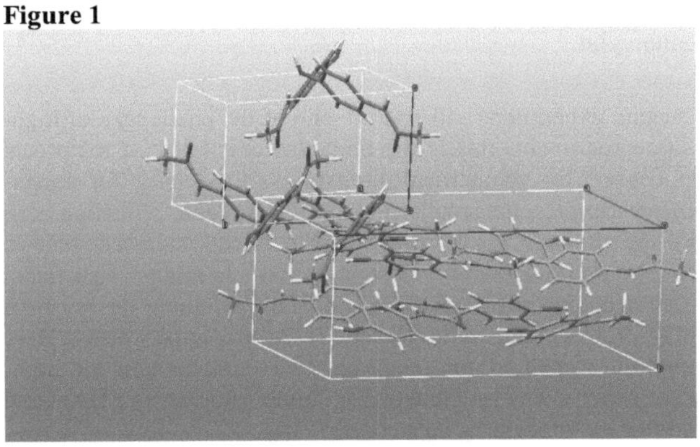

Figure 2

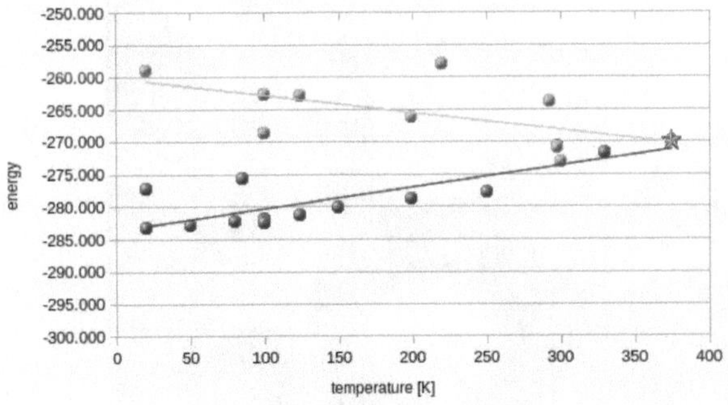

P170

Getting the most out of your high pressure experiments

M. Adam[1], H. Ott[1], M. Ruf[2], T. Stuerzer[1], J. Graf[3]
[1]Bruker AXS GmbH, Karlsruhe, Germany
[2]Bruker AXS Inc., Madison WI, United States
[3]incoatec GmbH, Geesthacht, Germany

Traditionally, high-pressure crystallography has been associated with the study of rocks and minerals from within the earth"s crust. More recently, pharmaceutical development has benefitted from organic solid-state chemistry and its perspectives on structure and function in biologically-relevant molecules, with polymorphism being of critical importance. Of growing interest is the effect of pressure on drugs, since many solid drugs are exposed to mechanical manipulation during manufacturing.

The increased interest in high-pressure crystallography for the home lab has prompted the development of high-performance software solutions that address its particular challenges, elevating high-pressure crystallography into a powerful method that can be routinely used on modern diffractometers, such as the D8 QUEST or the D8 VENTURE.

We will report on an experiment, where advanced processing methods for Diamond Anvil Cell (DAC) data were put to the test, to produce results comparable in quality to standard single-crystal data, using the less common monoclinic polymorph of the sulfonium ylid.

The second experiment will address one of the challenges of high-pressure experiments-this is the limited accessibility of reciprocal space caused by geometrical limitations. Especially for lower-symmetry samples, the smaller number of available reflections can reduce the structure quality dramatically. Recent enhancements in hardware and software design have brought significant improvements in data acquisition and data processing quality both for high-pressure and multiple-domain-sample experiments. Here we explore a method for increasing the completeness of high-pressure experiments by mounting multiple samples in a DAC and measuring and processing data concurrently.

Figure 1

P171

Closer look into close packing: pentacoordinated silicon in high-pressure polymorph of danburite

A. Pakhomova[1], E. Bykova[1], M. Bykov[2], K. Glazyrin[1], H. P. Liermann[1], L. Gorelova[3], L. Dubrovinsky[2]
[1]DESY, Hamburg, Germany
[2]Bayerisches Geoinstitut, Bayreuth, Germany
[3]Saint-Petersburg State University, Saint-Petersburg, Russian Federation

Silicates, as the most common minerals of Earth"s crust and mantle, have been under special attention since the incipience of structural crystallography. The crystal chemistry of the rock-forming silicates with four- and six-coordinated silicon is well-established, whereas very few is known about five-coordinated silicon. Here we report on high-pressure single-crystal X-ray diffraction study of danburite, $CaB_2(SiO_4)_2$, that is likely to be the first compound showing the step-wise change of silicon coordination from tetrahedral to octahedral through coordination of trigonal bipyramid.

At ambient conditions danburite possesses an orthorhombic symmetry *Pnam*. Tetrahedral framework of danburite is built on Si_2O_7 and B_2O_7 dimers that join to four- and eight-membered rings running along *c* axis. The 8-membered rings are occupied by Ca atoms. By compression to ~23 GPa the danburite framework undergoes severe distortion: 8-membered rings become strongly elliptically distorted and T-O-T angles decreases.

Above ~23 GPa the crystal structure undergoes phase transition that is indicated by abrupt change in unit-cell parameters. The phase transition to danburite-II is displasive and induced by shift of Si along *a* axis. The new polyhedra of pentacoordinated silicon possess trigonal bipyramid geometry. The SiO_5 trigonal bipyramids share common vertices and form chains running along *a* axis.

Further compression results in another phase transition above 30 GPa. The structure of danburite-III was solved and refined in *P*-1 symmetry. The *Pnam* → *P*-1 transition is reconstructive in character. Across the transition the chain composed of vertex-sharing SiO_5 trigonal bipyramids transforms into the chain of edge sharing SiO_6 octahedra.

The details of structural evolution of danburite as well as nature of rare pentacoordinated silicon will be discussed.

P172

Characterization of a new tin borate Sn2B3O6(OH) synthesized under hydrothermal conditions

S. Schönegger[1], K. Wurst[1], G. Heymann[1,2], A. Schaur[1,2,3], A. Saxer[1,4], D. Johrendt[1,4], H. Huppertz[1,4,5]
[1]Inorganic chemistry, Innsbruck, Austria
[2]Leopold-Franzens-University, Inorganic Chemistry, Innsbruck, Austria
[3]Leopold-Franzens-University, Konstruktion und Materialwissenschaften, Innsbruck, Austria
[4]Department Chemie, Ludwig-Maximilians-University, Inorganic Chemistry, München, Germany
[5],, Austria

So far, several tin borates have been reported, including SnO-B2O3 [1], SnB2O4 and Sn2B3O6 [2] which were synthesized under high-temperature conditions. Moreover, one tin borate, β-SnB4O7 [3], was synthesized under high-pressure conditions. Furthermore, two interesting crystalline phases of mixed tin borates have been studied such as CaSn1-xTix(BO3)2 [4] and Ni5SnB2O10 [5]. Up to now, no tin borates have been synthesized under hydrothermal conditions in the Sn-B-O system. Our goal was to synthesize a new tin borate under mild hydrothermal conditions. Based on this idea, we obtained a new compound with the composition Sn2B3O6(OH).

The compound Sn2B3O6(OH) was synthesized in a stainless-steel autoclave with a Teflon inlet. A mixture of SnO, H3BO3, H2O, with pH value adjusted to 14.0 by the addition of a few drops of a 2 M aqueous KOH solution was heated at 513 K and kept for 2 days. Afterwards, the reaction mixture was cooled at a rate of three degrees per hour down to 323 K. The new compound crystallizes in the centrosymmetric space group *P*21/*n* (no. 14) with the lattice parameters a = 443.5(2), b = 1102.5(4), c = 1396.4(5) pm, and β = 95.31(1)°. Structurally, Sn2B3O6(OH) features infinite chains of B3O8 groups connected along the *a* axis. Furthermore, adjacent anionic borate chains are connected by O–H···O hydrogen bond interactions into a three-dimensional structure. The Fundamental Building Block (FBB) itself consists of two corner-sharing BO4 tetrahedra and one BO3 triangle group described as [6]. The Sn2+ cations are arranged in form of layers between the anionic borate chains.

This poster will present the new tin borate Sn2B3O6(OH) synthesized under hydrothermal conditions including its characterization by single-crystal X-ray diffraction and vibrational spectroscopy.

References:
[1] A. Hayashi, M. Nakai, M. Tatsumisago, T. Minami, Y. Himei, Y. Miura, M. Katada, *J. Non-Cryst. Solids* **2002**, *306*, 227-237.
[2] C. Gejke, J. Swenson, R. G. Delaplané, L. Börjesson, *Phys. Rev. B* **2002**, *65*, 212201.
[3] J. S. Knyrim, F. M. Schappacher, R. Pöttgen, J. Schmedt auf der Günne, D. Johrendt, H. Huppertz, *Chem. Mater.* **2007**, *19*, 254-262.
[4] T. Kawano, H. Yamane, *J. Alloys Compd.* **2010**, *490*, 443-447.
[5] K. Bluhm, H. Müller-Buschbaum, *Monatsh. Chem.* **1989**, *120*, 85-89.
[6] P. C. Burns, F. C. Grice, F. C. Hawthrone, *Can. Mineral* **1995**, *33*, 1131-1151.

Figure 1

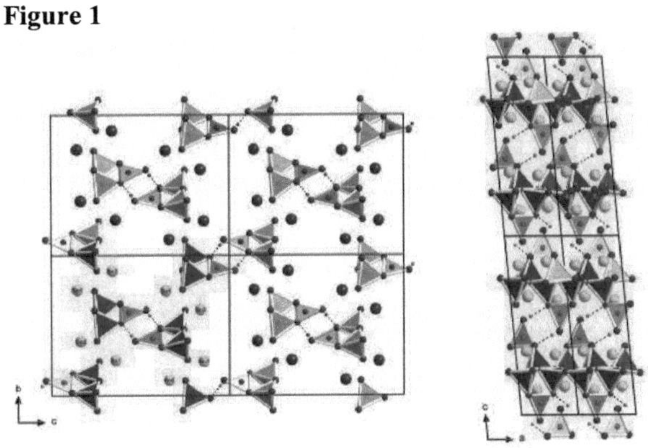

P173

Electron density investigation of the M-Nitroacetanilide compound via Accurate X-ray Diffraction Data

N. BOUKABCHA[1], Y. MEGROUSS[1], S. YAHIAOUI[1], A. CHOUAIH[1], F. HAMZAOUI[1]
[1]University of mostaganem, Process Engineering, Belaasel, Algeria

In this work, we present electron density analysis and electrostatic properties of M-Nitroacetanilide compound. However, the determination of electrostatic properties of organic nonlinear optical materials requires knowledge of the distribution of the electron density with high precision from high quality X-ray data. The electron density distribution of the title compound is determined by single-crystal X-ray refinements using aspherical structure factors and multipolar model of Hansen & Coppens implemented in the MoPro program. From the deformation density map, accumulations of density are visible in bonding as well as in the lone-pair regions of oxygen atoms (O4, O5 and O6). The electron density study allows determination of electrostatic parameters. The following crystal properties are determined: net atomic charges, dipole moment, electrostatic potential, gradient of electron density and critical points. These parameters are basic for understanding the nature of intra- and intermolecular charge transfer. Hereafter, we give figures of the deformation density, the electrostatic potential and the gradient of the electron density.

Keywords: M-nitroacetanilide, electron density, dipole moment, MoPro program.

Figure 1

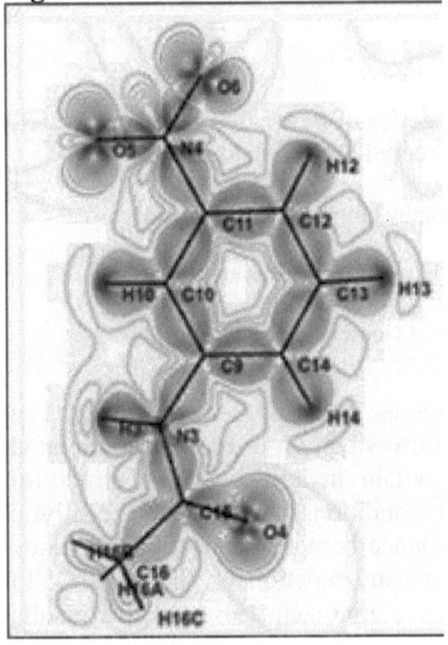

Figure 2

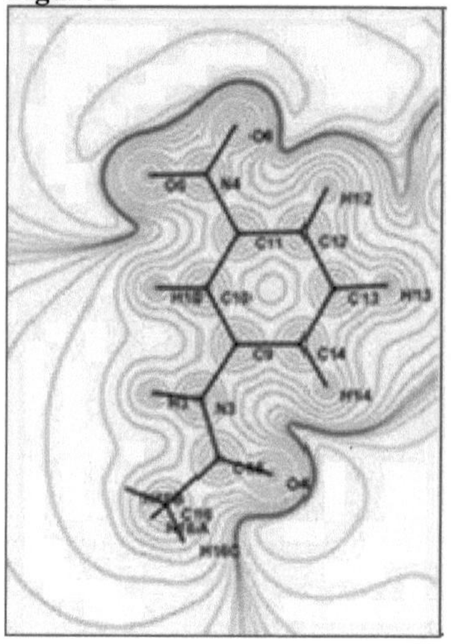

P174

Solid Solvents: Crystal Structures of Acetic Anhydride, 2-Methylfuran, Furfural and Furfurylamine

N. Nöthling[1], R. Goddard [1], C. W. Lehmann[1], R. Seidel[1]
[1]Max-Planck-Institut für Kohlenforschung, Chemische Kristallographie, Mülheim an der Ruhr, Germany

Acetic anhydride, 2-methylfuran, furfural and furfurylamine are, respectively, an industrial feedstock and acetylating reagent, a possible alternative fuel, an important renewable non-petroleum based chemical and a useful starting material. Despite the importance of these liquids, the crystal structures of their frozen solids are unknown. Crystal structures provide an opportunity to investigate the interactions between molecules in the solid and hence throw some light on the interactions that the molecules can undergo. Here we describe the crystal structures of solid acetic anhydride (*T*m = -73.1 °C) [1], 2-methylfuran (*T*m = -89 °C), furfural (*T*m = -37 °C) and furfurylamine (*T*m = -70 °C). Crystals were grown from the respective liquid by *in situ* crystallization [2] in an open glass capillary by zone melting in a cold N2 gas stream.

References:
[1] R. W. Seidel, R. Goddard, N. Nöthling, C. W. Lehmann, *Acta Cryst.* **2016**, C72, 753-757.
[2] R. Boese, *Z. Kristallogr.* **2014**, 229, 595-601.

P175

Accurate structures and energetics of neutral-framework zeotypes from dispersion-corrected DFT calculations

M. Fischer[1,2], R. J. Angel[3]
[1]University of Bremen, Crystallography group, Department of Geosciences, Bremen, Germany
[2]University of Bremen, MAPEX Center for Materials and Processes, Bremen, Germany
[3]University of Padova, Department of Geosciences, Padova, Italy

Computational chemistry methods have proven to be a very valuable tool for the theoretical investigation of zeolites and related materials (zeotypes), e.g. to obtain insights into the structural chemistry under non-ambient conditions or into their catalytic properties.[1] Due to its good balance between achievable accuracy and computational cost, density-functional theory (DFT) constitutes a particularly attractive approach. However, the results are strongly dependent on the choice of the exchange-correlation functional and the inclusion (or omission) of a dispersion correction. In previous work, we have shown that a combination of the widely used PBE functional[2] with a pairwise dispersion correction delivers a very good prediction of the structural parameters of neutral-framework zeotypes (all-silica zeolites and aluminophosphate zeotypes).[3] In the present contribution, we expand on these findings by extending our work in the following directions:

(1) We focus on two dispersion-corrected variants of the PBEsol functional (PBEsol-D2 and PBEsol-TS).[4–6] These functionals are found to give even better agreement with experimental structure data than the best-performing approach of our previous work.

(2) We assess the temperature dependence of lattice parameters and bond lengths for those zeotypes where sufficient experimental data are available for such an analysis. In most instances, the agreement between DFT and experiment improves if the experimental data are corrected for the effects of thermal motion, and if low-temperature rather than room-temperature structure data are used as reference.

(3) We calculate the relative stabilities of a total of 16 all-silica zeolites with respect to α-quartz, and compare the DFT results to experimental enthalpies of transition. While PBEsol-TS tends to overestimate the energy differences, PBEsol-D2 gives excellent agreement with experiment.

References:
[1] Van Speybroeck, V.; Hemelsoet, K.; Joos, L.; Waroquier, M.; Bell, R. G.; Catlow, C. R. A. *Chem. Soc. Rev.* **2015**, *44*, 7044–7111.
[2] Perdew, J. P.; Burke, K.; Ernzerhof, M. *Phys. Rev. Lett.* **1996**, *77*, 3865–3868.
[3] Fischer, M.; Evers, F. O.; Formalik, F.; Olejniczak, A. *Theor. Chem. Acc.* **2016**, *135*, 257.
[4] Perdew, J.; Ruzsinszky, A.; Csonka, G.; Vydrov, O.; Scuseria, G.; Constantin, L.; Zhou, X.; Burke, K. *Phys. Rev. Lett.* **2008**, *100*, 136406.
[5] Csonka, G. I.; Ruzsinszky, A.; Perdew, J. P.; Grimme, S. *J. Chem. Theory Comput.* **2008**, *4*, 888–891.
[6] Al-Saidi, W. A.; Voora, V. K.; Jordan, K. D. *J. Chem. Theory Comput.* **2012**, *8*, 1503–1513.

P177

Thermal expansion of Co olivine compared with Mn and Fe olivine: An assessment of equations-of-state

P. Schmid-Beurmann[1], H. Kroll[1], A. Sell[1], J. Büscher[1], A. Kirfel[2]
[1]Institut für Mineralogie, Universität Münster, Münster, Germany
[2]Steinmann Institut, Universität Bonn, Bonn, Germany

The thermal expansion of Co olivine has been studied by powder diffraction at ambient pressure and temperatures between 25°C and 1000°C. The anisotropy of the thermal expansion was analysed using various *isothermal* equations-of-state (EoS) coupled with the *thermal* Mie-Grüneisen EoS.

The axial expansivities α are controlled by the axial compressibilities β_T and thermal pressures $(\partial P/\partial T)_V$ according to the thermodynamic identity $\alpha = \beta_T\ (\partial P/\partial T)_V$. For olivines, the structural features suggest a sequence of compressibilities $\beta_{T,b} > \beta_{T,c} > \beta_{T,a}$ (SG *Pbnm*). Usually, such ranking is also found for the axial expansivities. In this case, however, only Mn and Mg olivine meet the expectation contrary to Fe olivine and to a lesser degree to Co olivine. The reason for the unusual behaviour is found in the specific anisotropies of the axial thermal pressures that could be derived from the coupled isothermal-thermal EoSs.

Helffrich (1999) pointed out that planetary scientists need a reliable means to extrapolate thermal expansion data to the high temperatures encountered in planetary interiors from the lower temperatures where thermal expansion is usually measured. With regard to their extrapolation behaviour we compared five EoSs originally designed as isothermal high-pressure EoSs (Birch-Murnaghan etc.) with two formulations that have explicitly been devised to describe thermal expansion at ambient pressure (Wallace, Suzuki). The first five EoSs were found to deliver similar results upon extrapolating thermal expansivity up to 3000 K, whereas the other two EoSs clearly deviate.

A suitable means for judging the high-temperature performance of an EoS is to compare isothermal bulk moduli predicted by the respective EoS, $K_T(T,P=0) =$

$-V(dP/dV)_T$, with those derived from measured elastic constants. K_T values obtained from the Wallace and Suzuki EoSs severely deviate from measured bulk moduli. By contrast, good agreement is seen with the first five EoSs. Thus, we have the paradox that those EoSs, which were explicitly designed to describe thermal expansion fail on extrapolation, whereas the high-pressure EoSs nicely perform with respect to K_T.

References:
Helffrich G (1999) Phys Earth Planet Int 116, p133-136

P178

Mechanochemical Synthesis of Cubic Li_2TiO_3

D. Becker[1], R. Haberkorn[1], G. Kickelbick[1]
[1]Universitaet des Saarlandes, Anorganische Festkoerperchemie, Saarbruecken, Germany

In addition to classical solid state reactions at elevated temperatures, solid-solid reactions also can be induced by mechanical energy, for instance by high energy ball milling (*hebm*) in a planetary ball mill [1]. The high local energy inputs may initiate processes similarly to high temperatures and pressures and thus promote the formation of new phases, like high temperature or high pressure polymorphs. If two different metal oxides are milled together they may form a ternary compound. Such a new phase typically has a very small grain size and high surface area. The mechanochemical synthesis of nanocrystalline cubic Li_2TiO_3 was achieved via different routes from binary and ternary oxides. This structure is related to the NaCl type. The experiments were conducted in a high energy planetary ball mill with varying milling parameters and different grinding tools. Lithium hydroxide and titanium dioxide (rutile or anatase) were used as starting materials in the milling process. Milling of $Li_4Ti_5O_{12}$ with a spinel structure or Li_2TiO_3 with a monoclinic structure lead to the formation of cubic Li_2TiO_3, too, but also causes formation of byproducts. The syntheses of the spinel and monoclinic phase were carried out by conventional solid state reactions. Appropriate amounts of lithium carbonate and anatase were homogenized by wet milling and heated in a platinum crucible [2][3].

Microstructure and crystallographic structure were characterized by powder X-ray diffraction via Rietveld analysis. Cubic Li_2TiO_3, produced by milling processes, tends to show a lithium deficiency and can be described more properly by the formula $Li_{2-4x}Ti_{1+x}O_3$ [4]. Due to the metastable state of cubic Li_2TiO_3 at ambient conditions, a phase transformation from the cubic to the monoclinic phase is observed after heating, as reported for similar products from hydrothermal processes [5].

References:
[1] A. F. Fuentes, L. Takacs, *J. Mater. Sci. 48*, **2013**, 598-611.
[2] G. Lang, *Z. Anorg. Allg. Chem. 276*, **1954**, 77-94.
[3] I. A. Leonidov, O. N. Leonidova, L. A. Perelyaeva, R. F. Samigullina, S. A. Kovyazina, M. V. Patrakeev, *Physics of the Solid State 45*, **2003**, 2183-2188.
[4] G. Vitins, G. Kizane, A. Lusis, J.Tiliks, *J. Solid State Electrochem. 6*, **2002**, 311-319.
[5] A. Laumann, K. T. Fehr, M. Wachsmann, M. Holzapfel, B. B. Iversen, *Solid State Ionics 181*, **2010**, 1525-1529.

P179

Lattice Thermal Expansion of $Bi_2Fe_4O_9$, $Bi_2Fe_2Mn_2O_{10}$ and $Bi_2Mn_4O_{10}$

M. M. Murshed[1,2], C. M. N. Kumar[3], A. Huq[4], T. M. Gesing[1,2]
[1]University of Bremen, Institute of Inorganic Chemistry and Crystallography, Bremen, Germany
[2]MAPEX Center for Materials and Processes, Universität Bremen, Leobener Straße /NW2, D-28359, Bremen, Germany
[3]Jülich Centre for Neutron Science (JCNS), Outstation at SNS, Oak Ridge National Laboratory, Oak Ridge, United States
[4]Chemical and Engineering Materials Division, Oak Ridge National Laboratory, Oak Ridge, United States

Both $Bi_2Fe_4O_9$ and $Bi_2Mn_4O_{10}$ were reported as multiferroics although their centrosymmetric crystal structures are not typical for ferroelectric behavior as the dipoles of each constituent polyhedra orient in different directions. We report the temperature-dependent nuclear and magnetic structures of $Bi_2Fe_4O_9$, $Bi_2Mn_4O_{10}$ and $Bi_2Fe_2Mn_2O_{10}$. The structural and vibrational features were studied using TOF neutron powder diffraction between 10 K and 300 K and Raman spectroscopy. Temperature-dependent structural data provide possible ferroelectric displacements and magnetic features (moment, structure, couplings) as well as their thermodynamic behaviors. Whether the ferroelectric property is related to the stereochemical activity of the $6s^2$ lone electron pairs of Bi^{3+} the Liebau density vector and the Wang-Liebau eccentricity parameter [1] of the $Bi^{3+}O_6$ octahedra were calculated and compared. Beside considerable conflicts on the structure of $Bi_2Fe_2Mn_2O_{10}$ we determined its nuclear and magnetic structures. Rietveld analyses showed that the cell parameters lie close the respective mean value of the end-members. The EDX results showed a Mn/Fe ratio of 1:1, and both Raman and FTIR spectra show complementary compound-bands produced from the two end-members. The lattice thermal expansion was simulated using Grüneisen first-order approximation for the zero pressure equation of state, where the internal energy was calculated by the DEA model [2-4]. In this model the Grüneisen parameter accounts for the isothermal anharmonicity while a low perturbed quantum anharmonicity term measures the isochoric anharmonicity. The characteristic frequencies extracted from the lattice thermal expansion agree well with those calculated from the averaged mass-weighted values of the isotropic atomic displacement parameters. The temperature-dependent magnetic moment was scaled to fit the magneto-volume regions of the respective cell volumes.

Reference:
[1] Curti et al., Z. Kristallogr. 228 (2013) 629.
[2] Murshed et al., Mater. Res. Bull. 48 (2013) 3284.
[3] Murshed et al., Mater. Res. Bull. 59 (2014) 170.
[4] Murshed et al., J. Solid State Chem. 229 (2015) 87.

P180

The crystal structure of sodium hexachlororhodate

M. Etter[1]
[1]Deutsches Elektronen Synchrotron, PETRAIII, Hamburg, Germany

The crystal structure of sodium hexachlororhodate (Na_3RhCl_6) was solved using laboratory powder X-ray diffraction. It was found that the crystal structure of Na_3RhCl_6 crystallizes in space group *P*-31*c* isostructural to the crystal structures of Na_3CrCl_6, Na_3InCl_6 and Na_3MoCl_6 (see figures 1 and 2). Structural details as well as temperature dependent investigations will be presented.

Fig. 1: Crystal structure of Na_3RhCl_6 presented along the (110) direction.

Fig. 2: Crystal structure of Na_3RhCl_6 presented along the *c*-axis.

Figure 1

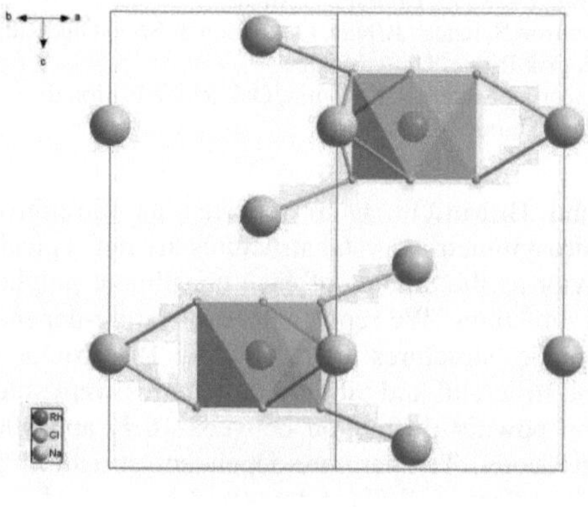

Figure 2

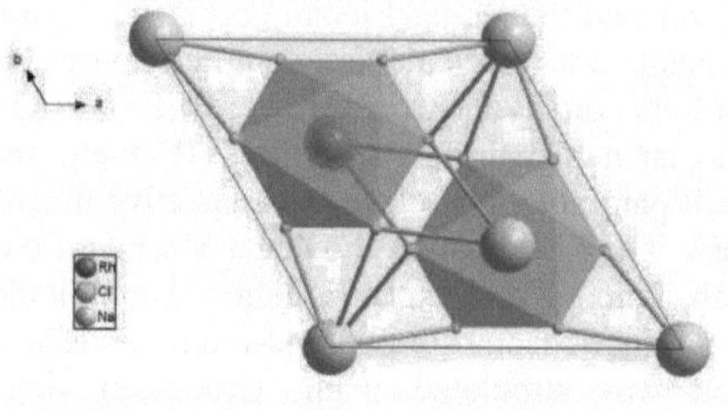

P181

Novel High-pressure Polymorph of NiAs-Type FeN

W. P. Clark[1], S. Steinberg[2], R. Dronskowski[2], C. McCammon[3], I. Kupenko[4], M. Bykov[3], L. Dubrovinsky[3], L. G. Akselrud[5], U. Schwarz[5], R. Niewa[1]
[1]Institute of Inorganic Chemistry, Stuttgart, Germany
[2]Institute of Inorganic Chemistry, Aachen, Germany
[3]Bayerisches Geoinstitut, Bayreuth, Germany
[4]ESRF, Grenoble, France
[5]Max Planck Institute for Chemical Physics of Solids, Dresden, Germany

The combination of laser-heated diamond anvil cells and synchrotron Mössbauer source spectroscopy was used to investigate phase formations of iron and iron nitride with a nitrogen pressure medium above 1000 K under pressures up to 45 GPa at Nuclear Resonance beamline ID18, ESRF. Samples heated up to 1300 K, with a pressure below 10 GPa, were observed to react with the pressure medium and result in an iron nitride with a composition close to ζ-Fe_2N/ε-$Fe_3N_{1.4}$.[1] At pressures larger than 10 GPa, a new phase was observed after annealing, which showed magnetic hyperfine splitting. This splitting appeared to be reversible, upon annealing at pressures approximately below 10 GPa, and remained stable up to 45 GPa.

In situ-diffraction measurements at varying pressures allowed characterization of this new phase, allowing the unit cell-pressure dependence to be determined, as well as give an indication to the presence of a homogeneity range. From structural refinements, the phase was determined to belong to the space group *P*63/*mmc*, with unit cell dimensions at 13.3 GPa of $a = 2.737(3)$ Å, $c = 4.933(5)$ Å. At lower pressures the unit cell was refined to $a = 2.775(7)$ Å, $c = 4.982(1)$ Å and $a = 2.800(2)$ Å, $c = 5.015(5)$ Å at 4.44 GPa and 0.1 MPa, respectively. This phase adopts the NiAs-type structure, resulting in the idealized chemical composition of FeN. This FeN polymorph remains stable at room temperature and lower pressures, including atmospheric pressure. This is the first time a NiAs-type 3*d* transition-metal nitride has been synthesized and characterized.

Electronic-structure calculations, from first principles, were conducted to compare the stability of this new NiAs-type FeN with the known ZnS-type polymorph. According to calculations, there is a distinct preference for FeN to adopt the ZnS-type rather than the NiAs-type structure at standard conditions. However, a pressure-induced phase transition to NiAs-type FeN has been calculated. Magnetism and chemical bonding, as addressed by the calculations, will be discussed.

[1] V. Ksenofontov, S. Reiman, M. Waldeck, R. Niewa, R. Kniep, P. Gütlich, *Z. Anorg. Allg. Chem.* **2003**, *629*, 1787–1794.

Figure 1

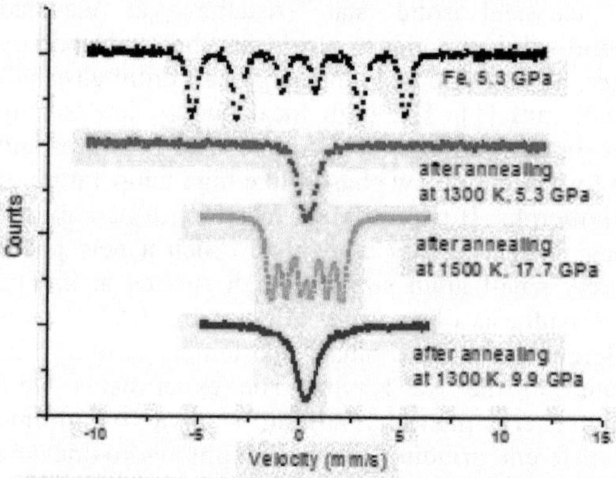

Fig. 1: Mössbauer spectra of iron, with a N_2 pressure medium, at various pressures before and after heat treatment. Top: Initial iron at 5.3 GPa prior to laser heating. Top-middle: The spectra after laser heating at 5.3 GPa, with a N_2 pressure medium, corresponds to the spectra for ζ-Fe_2N/ε-Fe_3N_{1+x}. Bottom-middle: New phase after laser heating at 17.7 GPa with N_2 pressure medium. Bottom: Possible re-formation of ζ-Fe_2N/ε-Fe_3N_{1+x} after laser heating at nitrogen pressures below 10 GPa.

P182

High-Pressure Synthesis of the Rare-Earth Metal(III) Hydroxide *Ortho*-Oxidotungstates(VI) *RE*(OH)[WO4] (*RE* = Dy, Ho)

K. V. Dorn[1], T. Schleid[1], I. Hartenbach[1]
[1]University of Stuttgart, Institute for Inorganic Chemistry, Stuttgart, Germany

Rare-earth metal(III) chloride oxidotungstates(VI) with the formula $RE_3Cl_3[WO_6]$ (*RE* = La–Nd, Sm–Tb) provide an interesting field of research for luminescent applications due to their isolated trigonal prismatic $[WO_6]^{6-}$ anions and the fact that they can host luminescent cations [1, 2]. Since synthesis experiments under ambient conditions lead to no results for *RE* = Dy–Lu for the $RE_3Cl_3[WO_6]$ series, attempts using a piston-cylinder apparatus [3] at 700–800 °C and a pressure of 15–20 kbar for four days under moist atmosphere yielded unexpected single crystals of *RE*(OH)[WO4] (*RE* = Dy, Ho). The crystal structure of the first member of this series, Nd(OH)[WO4], was already published in 1969 by *Klevtsova* and *Borisov* [4], but no more further results have been announced in the meantime.

Both compounds Dy(OH)[WO4] and Ho(OH)[WO4] crystallize monoclinically in space group *P*21/*c* (*Z* = 4) (Figure 1). The RE^{3+} cations are surrounded by six oxide and two hydroxide anions forming distorted square antiprisms $[RE(OH)_2O_6]^{11-}$, whereas the W^{6+} cations are located in the centers of oxygen tetrahedra forming isolated $[WO_4]^{2-}$ anions (Figure 2, *top*). In the crystal structure two $[RE(OH)_2O_6]^{11-}$ polyhedra are edge-connected to form planar $[(OH)_2RE_2]^{4+}$ rhombuses (Figure 2, *mid*) completed by their surrounding oxygen environment. These bipolyhedra are further interconnected via common edges to layers with the composition 2D-$\{[RE(OH)_{2/2}O_{4/2}O_{2/1}]^{6-}\}$ running orthogonal to [010] (Figure 2, *bottom*).

Due to the fact that the title compounds are isotypic with rare-earth metal(III) fluoride oxidomolybdates(VI) *RE*F[MoO4] (*RE* = Y, Sm–Lu) [5, 6], there should be also the chance to obtain the so far unknown fluoride oxidotungstates with the empirical formula *RE*F[WO4].

References:
[1] K. V. Dorn, Th. Schleid, I. Hartenbach, *Z. Anorg. Allg. Chem.* **2016**, *642*, 1045–1045.
[2] K. V. Dorn, Th. Schleid, B. Blaschkowski, K. Förg, H. A. Höppe, I. Hartenbach, *Eur. J. Inorg. Chem.* **2017**, submitted for publication.
[3] H.-J. Massonne, W. Schreyer, *Neues Jb. Miner. Abh.* **1986**, *153*, 177–215.
[4] R. F. Klevtsova, S. V. Borisov, *Kristallografiya* **1969**, *14*, 904–907.
[5] Th. Schleid, S. Strobel, P. K. Dorhout, P. Nockemann, K. Binnemans, I. Hartenbach, *Inorg. Chem.* **2009**, *47*, 3728–3735.
[6] I. Hartenbach, S. Strobel, P. K. Dorhout, Th. Schleid, *J. Solid State Chem.* **2008**, *181*, 2828–2836.

Figure 1

Lattice constants of the rare-earth metal(III) hydroxide *ortho*-oxidotungstates(VI) *RE*(OH)[WO4] (*RE* = Dy and Ho).

Compound	*a* [pm]	*b* [pm]	*c* [pm]	*β* [°]
Dy(OH)[WO4]	524.96(2)	1241.15(6)	675.25(3)	114.680(2)
Ho(OH)[WO4]	523.49(2)	1237.36(6)	672.87(3)	114.740(2)

Figure 2

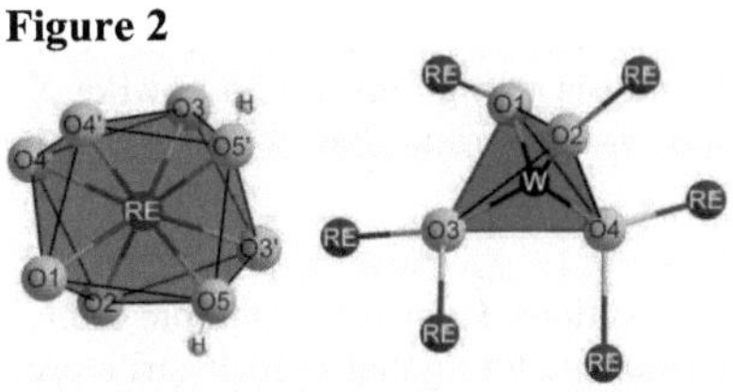

Coordination sphere of the RE^{3+} cations (*left*) and the tetrahedral $[WO_4]^{2-}$ anions (*right*) in the crystal structure of the *RE*(OH)[WO4] representatives (*RE* = Dy and Ho).

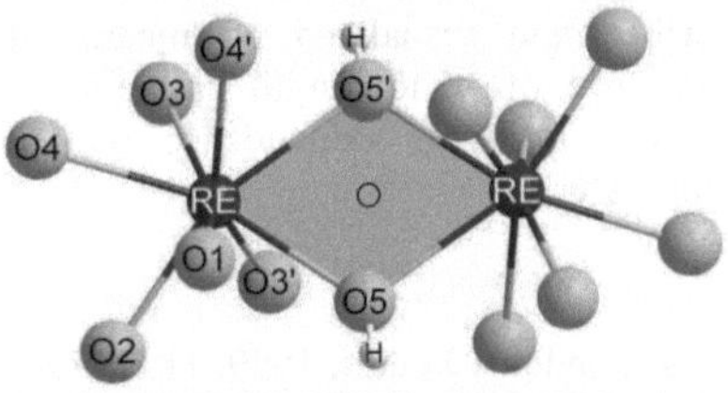

Two edge-connected $[RE(OH)_2O_6]^{11-}$ polyhedra forming planar $[(OH)_2RE_2]^{4+}$ rhombuses completed by their surrounding oxygen environment (o = inversion center).

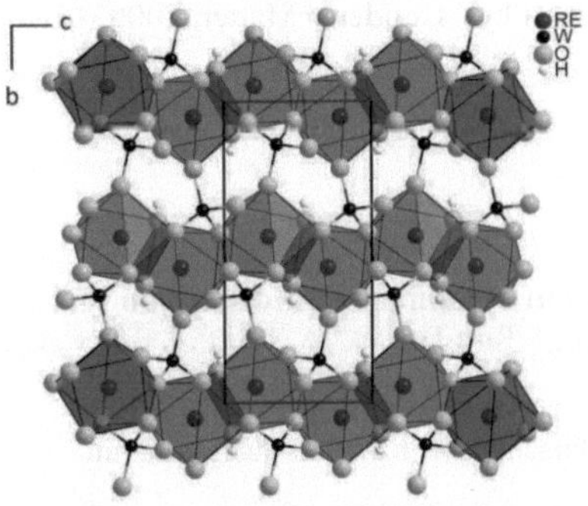

View at the layers orthogonal to [010] formed by edge-connected $[RE(OH)_2O_6]^{11-}$ polyhedra in the crystal structure of the *RE*(OH)[WO4] representatives (*RE* = Dy and Ho).

P183

In-situ HT-PXRD studies of Ca(Fe,Mn,Ti)O3-δ perovskites: Phase transitions and interplay between octahedral tilting and distortion

A. M. Schaller[1], S. Stöber[1], H. Pöllmann[1]
[1]Martin-Luther-University Halle-Wittenberg, Institute for Geosciences and Geography, Department Mineralogy and Geochemistry, Halle, Germany

Compounds with the perovskite structure (ABX_3) and its derivatives exhibit numerous physical, magnetic and electronic properties and offer a wide range of technological applications due to their chemical and structural variability. Most of them do not adapt the ideal cubic structure at standard ambient temperature and pressure (SATP) but a distorted one of lower symmetry; e.g. $CaTiO_3$ is orthorhombic (*Pbnm*) under SATP but undergoes structural transitions via orthorhombic *Cmcm* to tetragonal *I4/mcm* to the cubic *Pm-3m* structure at elevated temperatures [1]. These transitions are often central for technological applications and are studied extensively for this reason.

The objective of this study is to characterize the structural development and the complex interplay between octahedral tilting and distortion in Ca(Fe,Mn,Ti)$O_{3-\delta}$ perovskites as a function of temperature and composition; a special focus lies on the orthorhombic *Pbnm* to *Cmcm* transition, which has been discussed controversially in the past (e.g. for $CaTiO_3$ [1,2,3] and $SrZrO_3$ [4,5]).

The crystal structures of $CaFe_xMn_{(1-x)/2}Ti_{(1-x)/2}O_{3-\delta}$ ($0.08 \leq x \leq 0.31$) perovskites, prepared by a Pechini-type reaction route, were therefore refined by Rietveld analysis of high resolution powder X-ray diffraction data collected *in-situ* from 296 to 1273 K in capillary transmission geometry.

The samples undergo phase transitions from orthorhombic *Pbmn* (at SATP) to *Cmcm* to an intermediate tetragonal *I4/mcm* structure, stable only in a narrow temperature range, to the cubic *Pm-3m* structure upon heating. The transition temperatures vary as a function of the chemical composition / oxygen stoichiometries. The structural changes are described and visualized by means of octahedral tilting and the variance of bond length and angles, polyhedral volumes and volume ratios in dependence of the chemical composition and temperature.

References:
[1] Kennedy, B.J. et al.; J. Phys. Condens. Matter, 1999, 11, 1479-1488. [2] Ali, R.; Yashima, M; J. Solid State Chem., 2005, 178, 2867–2872. [3] Yashima, M.; Ali, R.; Solid State Ion., 2009, 180, 120-126. [4] Kennedy, B.J. et al.; Phys. Rev. B, 1999, 59, 4023-4027. [5] Howard, C.J. et al.; J. Phys. Condens. Matter, 2000, 12, L677-L683.

P184

Synthesis and characterisation of monazite solid solution series

A. Hirsch[1], P. Kegler[2], I. Alencar[3], J. Ruiz-Fuertes[3], A. Thust[3], J. Bauer[3], C. Schausten[4], L. Peters[1], G. Roth[1]
[1]Institut für Kristallographie, RWTH Aachen, Aachen, Germany
[2]Institut für Energie- und Klimaforschung (IEK6), Forschungszentrum Jülich, Jülich, Germany
[3]Institut für Geowissenschaften, Goethe Universität Frankfurt am Main, Frankfurt am Main, Germany
[4]Institut für Gesteinshüttenkunde, RWTH Aachen, Aachen, Germany

Monazite (*Ln*PO_4: *Ln* from La to Gd; $P2_1/n$) type ceramics are considered as potential ceramic storage materials for high-level nuclear waste. They are known for their high chemical durability, high resistance to radiation damages and long term stability. Natural analogues incorporate up to 30 wt% of ThO_2 and UO_2 without amorphisation [1]. In this contribution, synthetic monazites are investigated, using La, Pr, Nd, Sm, Gd, and Eu as surrogates for the radioactive elements Np, Am, and Cm. Samples of solid solutions ($Ln^1_{1-x}Ln^2_xPO_4$: Ln^1 and Ln^2 from La to Gd) were synthesised as powders, single crystals, and ceramics and characterised by room- and high-temperature and high-pressure XRD (X-ray diffraction), EPMA (Electron probe micro analysis), spectroscopy (Raman, Ultrasound, Fluorescence), calorimetry (high-temperature solution and low-temperature micro calorimetry), as well as dilatometry.

The various solid solutions give an insight into the structural, vibrational, thermochemical and physical behaviour of monazite. Lattice parameters show a linear decrease with increasing ionic *Ln* radii. Raman spectroscopic analyses of the local environment of the phosphorous atom in the PO_4 tetrahedron yield increasing frequencies with increasing ionic *Ln* radii [2]. Calculated coefficients of thermal expansion from *in-situ* powder XRD are compared to dilatometry measurements on ceramics of the same composition. High-pressure XRD on a $LaPO_4$ single crystal revealed the phase transition to post-monazite ($P2_12_12_1$) [3]. Additionally, a Schottky contribution to the heat capacity is found due to the thermal excitation of *f* electrons of Pr [4]. All data suggest that monazite forms almost ideal solid solutions.

The authors thank the BMBF (02NUK021D, 02NUK021E, 02NUK021F) for financial funding.

References:
[1] Ewing, R.; Weber, W. & Clinard, F., 1995. Radiation effects in nuclear waste forms for high-level radioactive waste. *Progress in nuclear energy, Elsevier,* **29**, 63-127
[2] Hirsch et al., 2017. Structural, vibrational, and thermochemical properties of the monazite-type solid solution $La_{1-x}Pr_xPO_4$. *Journal of Solid State Chemistry, Elsevier,* **245**, 82-88
[3] Ruiz-Fuertes et al., 2016. High-pressure phase of $LaPO_4$ studied by x-ray diffraction and second harmonic generation. *Physical Review B, APS,* **94**, 134109
[4] Bauer et al., 2016. Schottky contribution to the heat capacity of monazite type (La,Pr)PO_4 from low-temperature calorimetry and fluorescence measurements. *Chemical Physical Letters, Elsevier,* **654**, 97-102

P185

High-pressure X-ray diffraction studies on cuprous sulfides

D. Zimmer[1], W. Morgenroth[1], J. Ruiz-Fuertes[2], L. Bayarjargal[1], J. Zhang[3], C. Mühle[4], D. Santamaría-Pérez[5], E. Haussühl[1], C. Jin[3], B. Winkler[1]
[1]Goethe University, Crystallography, Frankfurt am Main, Germany
[2]Universitat de València, Departament de Física Aplicada, València, Spain
[3]Chinese Academy of Sciences, Institute of Physics, Beijing, China
[4]Max-Planck Institute for Solid State Research, Quantum Materials, Stuttgart, Germany
[5]Universitat de València, Instituto Universitario de Ciencia de los Materiales, València, Spain

Cuprous sulfides ($Cu_{2-x}S$; $0 \leq x \leq 0.25$) have been the focus of numerous studies for several decades due to their outstanding physical properties and their geoscientific relevance as one of the most important copper ore on Earth. However, because of the high mobility of copper atoms and vacancies within the sulfur sublattice, their crystal structures and phase relations are complex. Our knowledge of the crystal structures of the copper sulfide compound family as a function of pressure is limited to two recent high pressure studies [1,2] on chalcocite, Cu_2S, reporting two pressure-induced phase transitions up to 25 GPa.

The aim of this study is to extend our knowledge of the influence of pressure on the crystal structures of copper sulfides by determining the crystal structure of anilite [3], Cu_7S_4, djurleite [4], $Cu_{31}S_{16}$, digenite [5], Cu_9S_5, and tetragonal copper sulfide [6], Cu_2S, as a function of pressure by means of single crystal and/or powder X-ray diffraction (XRD).

Synthetic samples of anilite, digenite and tetragonal copper sulfide were characterized by in-house XRD. The evolution of crystal structures as a function of pressure was investigated by *in situ* XRD in diamond anvil cells using synchrotron radiation at the PETRA III beamline P02.2 [7].

Powder diffraction indicates that anilite undergoes five pressure-induced phase transitions up to 74(2) GPa. Tetragonal copper sulfide transforms into the known high-pressure phases of chalcocite [2] with pressure. Digenite transforms into djurleite at ~ 1 GPa, which transforms into a high-pressure polymorph of chalcocite (HP1) [2] upon further compression.

The authors gratefully acknowledge financial support by the DFG (Wi1232 & BA4020/2-1) and the BMBF (05K13RF1).

Figure 1: Powder XRD patterns of anilite on increasing pressure.

Figure 2: Le Bail fits [8] of powder XRD patterns of tetragonal copper sulfide as a function of pressure.

References:
[1] S. Wang *et al.* (2002). *Mater. Chem. Phys.* **75**, 32
[2] D. Santamaría-Pérez *et al.* (2014). *J. Alloys Compd.* **610**, 645
[3] K. Koto, N. Morimoto (1970). *Acta Cryst.* **B26**, 915
[4] H. T. Evans Jr. (1979). *Z. Kristallogr.* **150**, 299
[5] G. Donnay *et al.* (1958). *Am. Mineral.* **43**, 228
[6] A. Jánosi (1964). *Acta Cryst.* **17**, 311
[7] H.-P. Liermann *et al.* (2015). *J. Synchrotron Radiat.* **22**, 908
[8] A. Le Bail *et al.* (1988). *Mat. Res. Bull.* **23**, 447

Figure 1

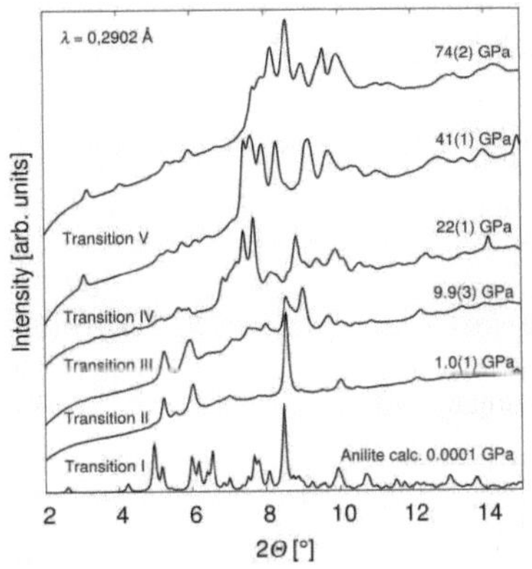

Figure 2

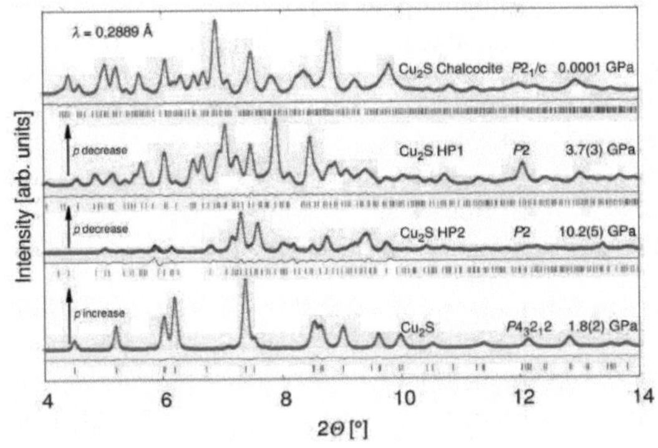

P186

Synthesis, compressibility and stability of $W_7Re_{13}B$

C. Neun[1], B. Petermüller[2], L. Bayarjargal[1], W. Morgenroth[1], H. Huppertz[2], B. Winkler[1]
[1]Goethe University Frankfurt, Dpt. of Crystallography, Frankfurt, Germany
[2]University of Innsbruck, Institute of General, Inorganic and Theoretical Chemistry, Innsbruck, Austria

Transition metal borides are a group of potentially high-performance materials, of which the binary compounds have attracted considerable interest [1]. Ternary systems have only been rarely investigated and only a few are known up to now. Among those is W7Re13B, a phase which has been first described in 1968 [2] and which has been characterized with respect to its low-temperature properties, such as the heat capacity and the transition temperature into the superconducting state [3].

In contrast, the compressibility and the high (p,T)-behavior of $W_7Re_{13}B$ are still unknown. The aim of this study was to expose $W_7Re_{13}B$ at high pressures and temperatures in order to establish its bulk modulus and high (p,T)-stability. We studied phase pure samples with synchrotron X-ray diffraction at the "extreme conditions beam line" P02.2@PETRA III at high pressures using diamond anvil cells. Conventional powder diffraction was carried out on samples quenched from high temperatures after heating them up to 2073 K.

Heat treatment led to two new polymorphs. Samples quenched from temperatures above 873 K exhibited split main reflections of the cubic parent phase. Synchrotron diffraction patterns have already been collected for this phase and the data evaluation is currently in progress. A second quenchable modification was observed after annealing at 2073 K (Figure 1). This new modification is isostructural to $MnRe_3B_2$ with space group *Fddd.*

The bulk modulus of cubic $W_7Re_{13}B$ was obtained by fitting a 3rd order Birch-Murnaghan equation of state to the unit cell volume and is $B0$ = 323(3) GPa with B'' = 5.5(2). We did not observe a pressure-induced structural phase transition up to pressures of 30 GPa (Figure 2). The new *Fddd* modification is (meta-)stable at ambient temperature up to at least 41 GPa with $B0$ = 327(4) GPa and B'' of 7.3(4).

We gratefully acknowledge funding for this project by the DFG and the FWF in the framework of an ERA-chemistry project (WI 1232-401-1 and I 1636-N19), and by the BMBF in project 05K13RF1.

Figure 1: Powder diffraction patterns after heating cubic $W_7Re_{13}B$ at different temperatures.

Figure 2: Pressure dependence of the normalized unit cell volume of the cubic and the new $W_7Re_{13}B$ polymorph.

References:
[1] Chung *et al., Science* ***316*** *(2007)*
[2] Kuz"ma *et al., Powder Metal Met. Ceram.* **7** (1968)
[3] Kawashima *et al., J. Phys. Soc. Jpn.* ***74*** (2005)

Figure 1

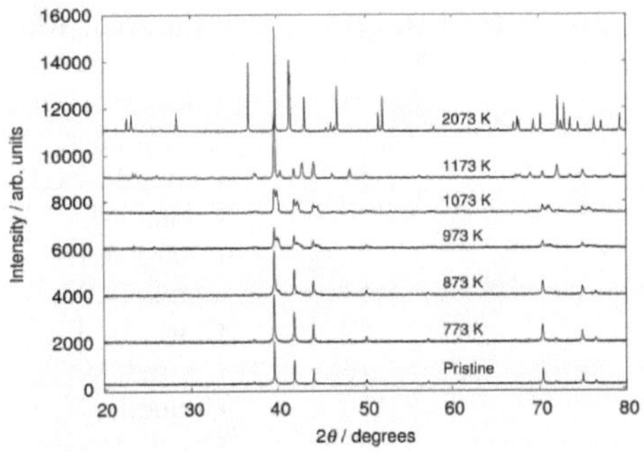

Figure 2

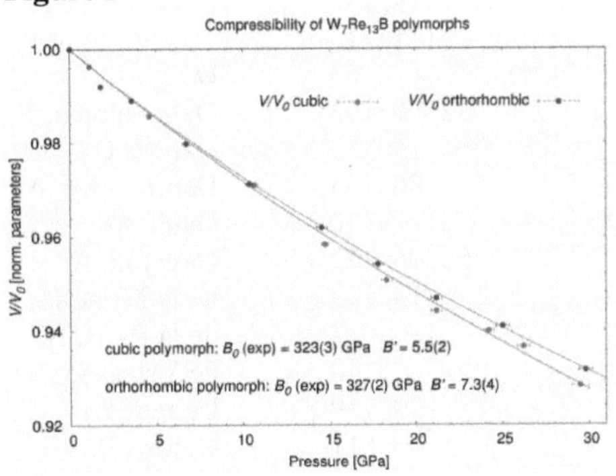

Index of Authors

Supplement Issue No. 37
Zeitschrift für Kristallographie
ISBN 978-3-11-054604-0

Bei Fragen zur Produktsicherheit wenden Sie sich bitte an:
If you have any questions regarding product safety,
please contact:

Walter de Gruyter GmbH
Genthiner Straße 13
10785 Berlin
productsafety@degruyterbrill.com